Modern NMR Approaches To The Structure Elucidation of Natural Products

Volume 1: Instrumentation and Software

Modern NMR Approaches to the Structure Elucidation of Natural Products
Volume 1: Instrumentation and Software

Edited by

Antony J. Williams
ChemConnector Inc., USA
Email: tony27587@gmail.com

Gary E. Martin
Merck Research Laboratories, USA
Email: gary.martin2@merck.com

and

David Rovnyak
Bucknell University, USA
Email: drovnyak@bucknell.edu

ROYAL SOCIETY OF CHEMISTRY

THE QUEEN'S AWARDS FOR ENTERPRISE: INTERNATIONAL TRADE 2013

Print ISBN: 978-1-84973-383-0
PDF eISBN: 978-1-84973-518-6

A catalogue record for this book is available from the British Library

Published by The Royal Society of Chemistry,
Thomas Graham House, Science Park, Milton Road,
Cambridge CB4 0WF, UK

Registered Charity Number 207890

Visit our website at www.rsc.org/books

Printed in the United Kingdom by CPI Group (UK) Ltd, Croydon, CR0 4YY, UK

AJW dedicates this volume to his mother Eirlys, his sister Rae and his sons Taylor and Tyler.

DR is grateful to Jennifer, Henry and Holly for their support.

GEM dedicates this volume to his wife Linda and his sons Joshua and Casey.

Contents

Modern NMR Approaches To The Structure Elucidation of Natural Products,
Volume 1: Instrumentation and Software
Edited by Antony J. Williams, Gary E. Martin and David Rovnyak

Published by the Royal Society of Chemistry, www.rsc.org

Part 1
Hardware

CHAPTER 1

New Directions in Natural Products NMR: What Can We Learn by Examining How the Discipline Has Evolved?

GARY E. MARTIN,*[a] ANTONY J. WILLIAMS[b] AND DAVID ROVNYAK[c]

[a] Merck Research Laboratories, Process & Analytical Chemistry, NMR Structure Elucidation, Rahway, NJ 07065, USA; [b] ChemConnector Inc., Wake Forest, NC 27587, USA, Email: tony27587@gmail.com; [c] Department of Chemistry, Bucknell University, Lewisburg, PA 17837, USA, Email: drovnyak@bucknell.edu
*Email: gary.martin2@merck.com

In 1992, in a laboratory that is set in the Amazonian Rainforest, Sean Connery (in the guise of researcher Robert Campbell and the movie *Medicine Man*[1]) made an injection into a mass spectrometer and identified a new natural product. Almost 25 years later, even with the miniaturization of analytical instrumentation and the incredible achievements in both sensitivity and mass resolution, it is *not* possible to elucidate a natural product structure in an automated manner using mass spectrometry. The only way that a complex natural product structure can be identified using this technique is by ensuring significant fragmentation and validating masses against a database. The dream of being able to use spectroscopy to elucidate automatically a molecular structure that is embodied in the movie *Medicine*

Modern NMR Approaches To The Structure Elucidation of Natural Products, Volume 1: Instrumentation and Software
Edited by Antony J. Williams, Gary E. Martin and David Rovnyak

Published by the Royal Society of Chemistry, www.rsc.org

Man is, however, much closer to reality in a laboratory setting using NMR spectroscopy.

Ultimately, the capability of modern NMR as a technique is the sum total of the assembly of a group of technologies paired with scientific acumen. NMR spectroscopists drive hardware and software in synergy to perform both data generation and analysis. With regard to natural product structure elucidation, the ultimate goal of NMR spectroscopists is to extract knowledge *via* the manipulation of magnetization and to determine, to the greatest extent possible, molecular-level detail regarding the spatial distribution, atom-to-atom connectivities, and orientations of atoms and bonds. What is achievable today using analytical spectroscopy tools applied to structure elucidation challenges is, relative to just a few years ago, truly breathtaking.

From this point forward, it is extremely difficult to predict how the future of natural products NMR will evolve in the decades to come. Twenty years ago, could we have imagined that acquiring long-range ^{1}H–^{15}N heteronuclear multiple-bond correlation (HMBC) spectra would become a routine and integral part of alkaloid structure elucidation studies?[2] Even with one of the authors having a direct hand in inaugurating those experiments, he would not have dared to predict the now routine usage of them.[3] Indeed, 5 years ago, would we have predicted the now burgeoning development of pure shift NMR methods?[4] No, we probably would not have had the foresight to predict where we are now with these methods and the impact that they are already having on our ability to probe the structures of increasingly complex natural products, often available in only very limited quantities. Could we have anticipated the significantly increased "reach" afforded an investigator for ^{1}H–^{13}C or ^{1}H–^{15}N heteronuclear shift correlation studies when using the newly reported LR-HSQMBC[5] experiment developed as a complement to the venerable HMBC experiment? Again, probably not. Most recently, would we have even dared to imagine that it would be possible to perform ^{13}C–^{15}N correlations at natural abundance using ^{1}H detection in the just reported HCNMBC experiment?[6] Despite one of the authors' own considerable experience acquiring 1,1- and 1,*n*-ADEQUATE experiments on sub-milligram samples, he would not have had the nerve to make such an audacious forecast.[7] Nevertheless, there is an example of just such a spectrum performed on a 4 mg sample of strychnine using a 1.7 mm MicroCryoProbe (Bruker) in the chapter on alkaloids (see Volume 2, Chapter 10)! None of the three techniques just noted existed when we started to assemble these volumes just four short years ago.

Similar events can be pointed to in terms of hardware developments. In the 1980s, NMR studies were routinely conducted in 5 mm NMR tubes. That changed in 1992 with the introduction of 3 mm NMR probes in one of the authors' laboratories,[8] and changed several additional times with the introduction of 1.7, 1.0, and microcoil NMR probes.[9] Smaller diameter tube formats foreshadowed the even more profound change embodied in the development and now widespread availability of helium-cooled cryogenic

NMR probes, and more recently the liquid nitrogen-cooled Prodigy probes offered by Bruker BioSpin.[10] Then we saw the diameter of cryogenic probes "shrink" to 3 mm first and then to 1.7 mm.[11] With the shrinking coil diameter of cryogenic NMR probes, sample requirements have correspondingly plummeted. Using a 1.7 mm MicroCryoProbe, one of the authors has demonstrated the acquisition of pure shift HSQC spectra of a sample of ~3 μg of a "heavy" drug metabolite (MW 661) generated using incubation with a recombinant enzyme in 14 h.[12] Natural products, of course, can also be interrogated at the same level. Readers interested in very low-level natural product structure investigations are directed to several relatively recent reviews by Molinski and co-workers.[13] Individually, all of these changes have been significant. In concert, the impact of cryogenic NMR probe technology has had a profound effect on what is possible in terms of natural product structure elucidation.

By way of providing a real-world example of what is feasible today regarding the structure elucidation of a complex natural product, we next consider what modern NMR experiments can provide in terms of connectivity and correlation data. When applied in tandem and in conjunction with accurate mass measurements, these data underpin the elucidation process, whether it be manual analysis or, as discussed later, performed by a computer. Beyond fundamental 1D proton and carbon reference spectra, there is a plethora of 2D NMR experiments available to an investigator sitting at the console of a modern high-field NMR spectrometer. Where to begin?

Strategies will vary from one investigator or laboratory to the next. One of the authors (G.M.) prefers to run a proton spectrum immediately followed by a multiplicity-edited HSQC spectrum.[14] Within the past year, the HSQC experiment choice has become more powerful with the availability of pure shift variants of the experiment, which collapse all but anisochronous geminal methylene resonances to singlets, thereby improving both resolution and sensitivity.[4]

If we employ strychnine as a model compound, the information content of strychnine can be described as illustrated in Figure 1.1.

To illustrate the homonuclear decoupling in the pure shift HSQC spectrum, a segment of the aliphatic region encompassing the H12, H23a/b, H16 and H8 resonances is shown in Figure 1.3. An expansion of the contour plot is shown in Figure 1.3a. The H12, H16 and H8 correlations are collapsed to singlets while the 23-methylene resonances are collapsed from doubled doublets to a pair of doublets. The vicinal coupling of both H23a and H23b to the H22 vinyl proton (^{1}H–^{12}C) is collapsed since the likelihood of a ^{13}C resonance being adjacent to the detected ^{1}H–^{13}C resonant pair is 1 in 10 000. In contrast, for the methylene protons, both are on the same ^{13}C and hence are unaffected by the BIRD-based decoupling applied during acquisition, leaving them as a pair of doublets. Figure 1.3b shows the high-resolution proton spectrum (A) and the phased traces extracted at the ^{13}C shifts of C12 (B) and C23 (C).

Following the acquisition of some form of an HSQC spectrum, typical structure elucidation strategies will probably next acquire COSY data, with

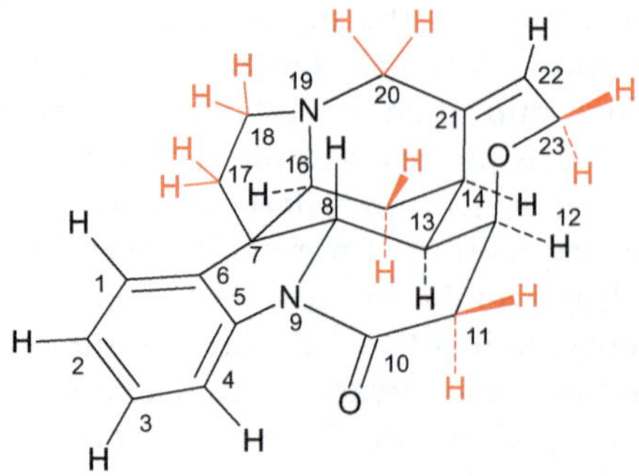

Figure 1.1 The structure of strychnine is shown with resonance multiplicity represented by black for CH/CH_3 resonances (there are no methyls in the structure) and red for methylene resonances. In a multiplicity-edited HSQC spectrum, the phase can be manipulated such that methine and methyl resonances will have positive phase whereas methylenes will be visualized with the opposite (negative) phase and a contour plot can be readily prepared in which the color coding of the correlations reflects the color coding in this figure (see Figure 1.2).

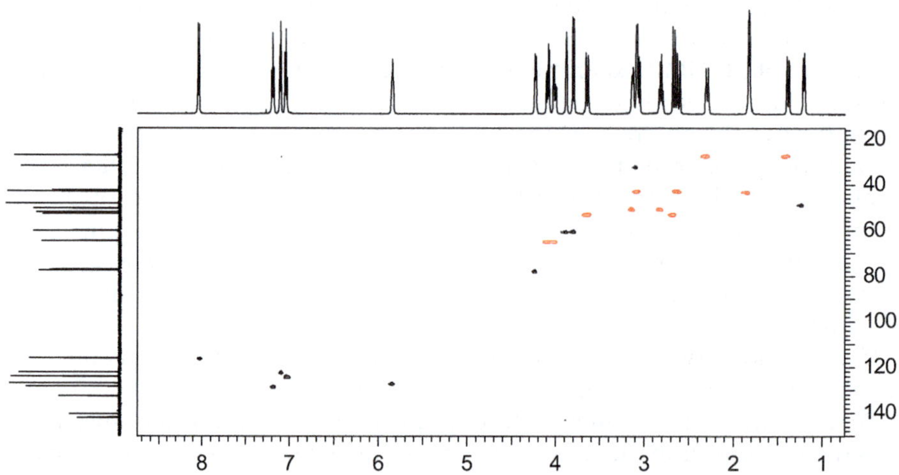

Figure 1.2 Multiplicity-edited pure shift HSQC spectrum of strychnine. The data are multiplicity edited with CH/CH_3 resonances having positive phase and plotted in black and the CH_2 resonances inverted and plotted in red. The data were acquired using "chunked" acquisition with BIRD pulses followed by hard 180° pulses interspersed during the acquisition to accomplish the homonuclear decoupling of all but the geminal methylene protons, which are unaffected by the BIRD-based decoupling since both protons are attached to the same ^{13}C resonance.

which most readers will likely be familiar. Homonuclear correlation data can be used to subgroup the proton resonances into discrete spin systems. For strychnine, the various spin systems in the structure of the molecule are

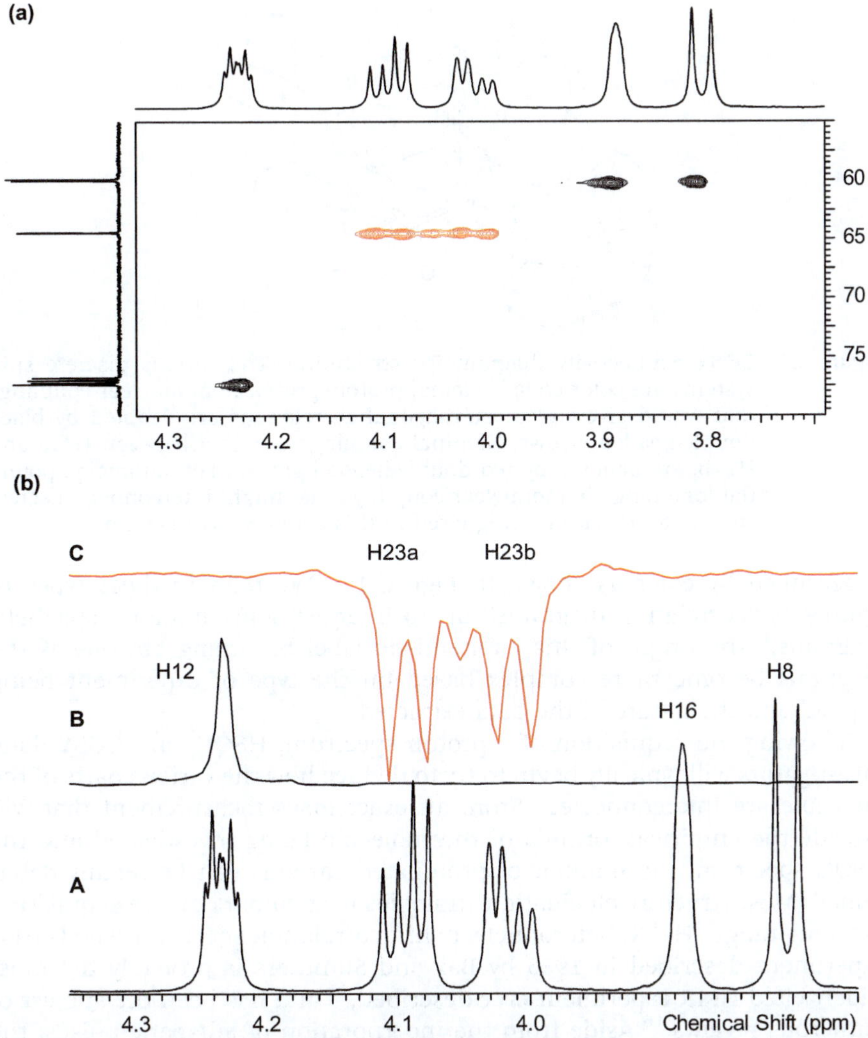

Figure 1.3 (a) Segment of the multiplicity-edited pure shift HSQC spectrum of strychnine showing the correlations for the H12, H23a/b, H16, and H8 resonances. Methine resonances are plotted in black while methylene resonances have negative phase and are plotted in red. (b) Proton reference spectrum (A) with slices extracted from the 2D plot at the ^{13}C chemical shifts of C12 and C23 (B and C). All of the vicinal couplings of the H12 resonance (B) are collapsed by the BIRD pulse/hard 180° pulse sequence element applied during the "chunked" data acquisition. In contrast, for the H23 methylene protons (C), the vicinal coupling to the H22 vinyl proton is collapsed while the geminal coupling is unaffected by the BIRD pulse/hard 180° pulse sequence element applied during acquisition. Hence the 23 methylene resonances are observed as a pair of doublets rather than as a pair of fully decoupled singlets.

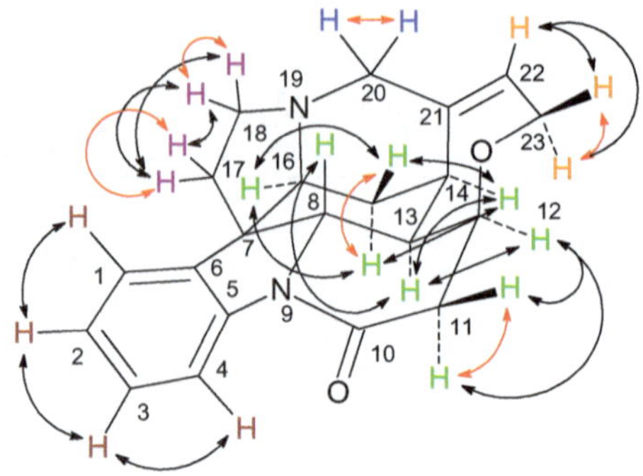

Figure 1.4 COSY connectivity diagram for strychnine. The various discrete spin systems are color coded. Vicinal proton–proton homonuclear couplings that would give rise to off-diagonal correlations are denoted by black double-headed arrows. Geminal couplings, *e.g.* that between H11a and H11b, are denoted by red double-headed arrows. For simplicity, potential long-range homonuclear couplings that might interconnect discrete spin systems have been ignored in this connectivity diagram.

subgrouped by color as shown in Figure 1.4. We refer to these types of figures as "correlation diagrams" or, to interject some humor, "spaghetti diagrams," the origin of this euphemistic label becoming obvious as the diagrams become more complex based on the type of experiment being applied and the nature of the data extracted.

Following the acquisition of a proton spectrum, HSQC, and COSY data, investigators will typically begin to try to deduce how the various parts of the molecule are interconnected. From an exact mass measurement that will provide the empirical formula of the molecule being investigated and the HSQC spectrum, the number of protonated carbons can be readily determined. Most structure elucidation strategies next embark on the acquisition of a long-range ^{1}H–^{13}C heteronuclear shift correlation spectrum. The HMBC experiment described in 1986 by Bax and Summers is probably the most widely cited NMR experiment ever described,[15] and has been the subject of numerous reviews.[16] Aside from the incorporation of adiabatic pulses, the experiment has changed relatively little since its inception and there is currently not a real-time pure shift version of the experiment available, although the pseudo-3D tilt-HMBC was reported in 2013 by Sakhaii *et al.*[17]

Prior to embarking on the acquisition of HMBC data, a decision must be made on the optimization of the long-range delay. Typically, 10 Hz has probably been the most commonly used optimization, although the data in the example that follows were acquired with an 8 Hz optimization, with the low-pass *J*-filter that that is utilized to reject unwanted $^{1}J_{CH}$ correlations optimized for 145 Hz. Despite the inclusion of a low-pass *J*-filter in the pulse sequence, signal-to-noise ratios with modern cryogenic NMR probes are so high that as one goes down towards the noise floor of the spectrum, the ^{13}C satellites can frequently be observed symmetrically displaced about the proton and carbon chemical shift coordinates, *i.e.* the location of the direct correlation response in the HSQC spectrum. The data shown were acquired with 256 increments of the evolution time used to digitize the second frequency domain. Unlike the HSQC experiment, which affords only one-bond correlations governed by the ${}^{1}J_{CH}$ coupling constant, there is no such filtering of long-range correlations in an HMBC spectrum and ${}^{n}J_{CH}$ correlations where $n = 2$–4 are routinely observed. In contrast, a recently reported modification of the 1,*n*-ADEQUATE experiment, inverted ${}^{1}J_{CC}$ 1,*n*-ADEQUATE does allow the differentiation of ${}^{1}J_{CC}$ correlations form ${}^{n}J_{CC}$ correlations where $n \geq 2$.[18] Most commonly, ${}^{2}J_{CH}$ and ${}^{3}J_{CH}$ correlations will be observed in HMBC spectra with longer range correlations, *e.g.* ${}^{4}J_{CH}$ and ${}^{5}J_{CH}$ correlations, being observed less commonly. As we shall see with strychnine, however, the rigid skeletal framework of the molecule greatly facilitates the observation of longer range couplings. All of the correlations extracted from an 8 Hz optimized HMBC spectrum of strychnine are shown superimposed on the structure in Figure 1.5, and it should now be obvious where the euphemistic label "spaghetti diagram" mentioned earlier came from.

When the data contained in the 8 Hz HMBC spectrum of strychnine are interpreted, correlation diagrams can be constructed that are limited to the number of bonds involved in the correlation. First, consider Figure 1.6, which highlights only the ${}^{2}J_{CH}$ correlations.

Going from the manageable number of ${}^{2}J_{CH}$ correlations shown on the structure in Figure 1.6 to Figure 1.7, which shows all of the ${}^{3}J_{CH}$ correlations,

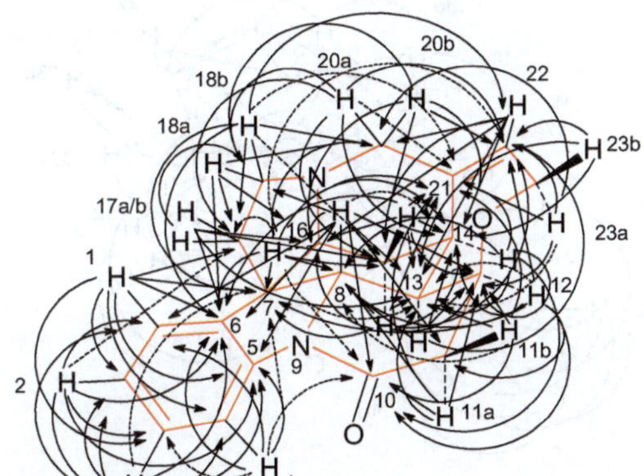

Figure 1.5 The connectivity diagram shows all of the observed correlations from an 8 Hz optimized HMBC spectrum superimposed on the structure of strychnine. There is, to a novice, a bewildering wealth of information in such a spectrum that more-or-less resembles a tangled bowl of spaghetti. To quote Woodward *et al.*'s paper describing the first synthesis of strychnine, "The tangled skein of atoms which constitutes its molecule provided a fascinating structural problem that was pursued intensively during the century just past, and was solved finally only within the last decade."[19] The same can be said for the wealth of information embodied in an HMBC spectrum! Weak correlations are designated by dashed arrows.

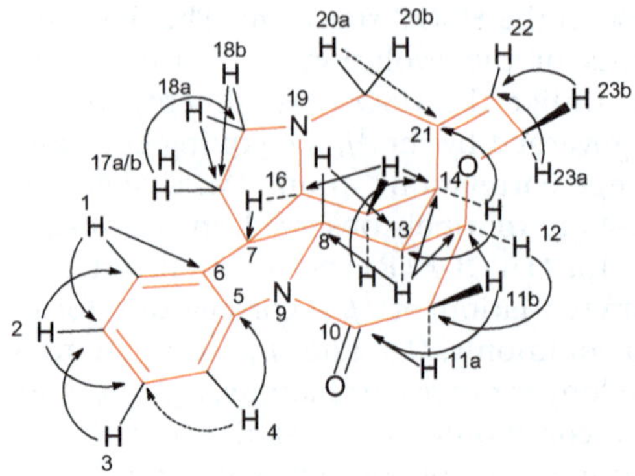

Figure 1.6 When only the two-bond ($^2J_{CH}$) HMBC correlations are superimposed on the structure, a much simpler array of data is available. Unfortunately, there is no simple way to go from the tangled web of correlations that nearly obscure the structural framework of the molecule in Figure 1.5 to this array of data short of interpreting the spectra. Clearly, the vast array of data contained in a complex HMBC spectrum makes a compelling argument for the utilization of computer-assisted structure elucidation (CASE) methods when dealing with complex structure elucidation problems.[20] Dashed arrows denote weak correlations.

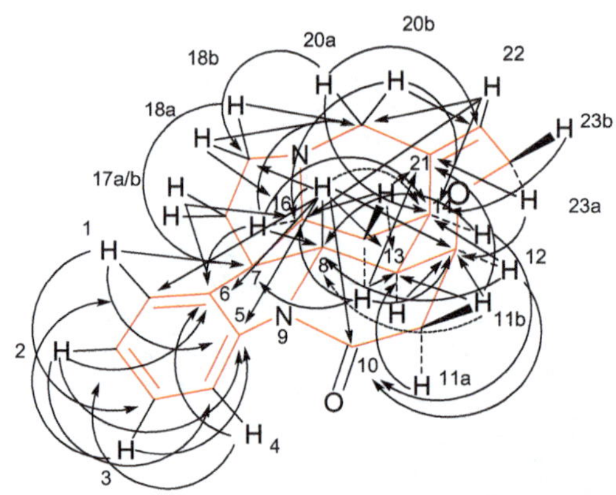

Figure 1.7 Connectivity diagram showing the $^3J_{CH}$ correlations observed in an 8 Hz optimized HMBC spectrum of strychnine. The number of correlations is nearly double the number of correlations in Figure 1.6. Dashed arrows denote weak correlations.

the level of complexity in the interpretation of HMBC data becomes more apparent. There are significantly more $^3J_{CH}$ than $^2J_{CH}$ correlations in the typical HMBC spectrum. Usefully, however, $^3J_{CH}$ correlations span heteroatoms incorporated into the skeletal framework and also make it possible to begin linking structural moieties together that would not generally be possible from HSQC and COSY data alone.

The rigid skeletal framework of strychnine facilitates a significant number of $^4J_{CH}$ correlations. Indeed, the number of $^4J_{CH}$ correlations observed is essentially identical with the number of $^2J_{CH}$ correlations (Figure 1.8). As molecules become more complex, they also in many cases become more proton deficient. When the ratio of hydrogens to heavy atoms (C, N, O, S) falls below 2, a postulate known as the "Crews Rule" was suggested, which states that such a hydrogen to heavy atom ratio may render the structure of a molecule difficult and in some cases impossible to deduce.[21] However the Crews rule was based on what were "standard" data sets acquired for structure elucidation at the time that it was suggested. Now, there are a number of experiments available with longer "reach" that probably mandate adjusting the Crews Rule ratio downwards.[22–24]

It is interesting that for strychnine, there are also a significant number of $^5J_{CH}$ correlations, as shown in Figure 1.9. In part, the large number of $^4J_{CH}$ and $^5J_{CH}$ correlations can be attributed to the acquisition of these data using a cryogenic NMR probe, for this example a 600 MHz Bruker TXI 1.7 mm gradient triple resonance MicroCryoProbe.[11b]

As noted in the caption of Figure 1.8, in 2014 Williamson and co-workers[24] developed the LR-HSQMBC experiment. That experiment is a refocused single quantum-based long-range experiment. The refocusing facilitates one-band decoupling during acquisition as in the D-HMBC[25] experiment and refocusing also prevents small heteronuclear couplings from being antiphase at the end of the pulse sequence as in the HMBC experiment. The antiphase character of weak long-range correlations can lead to their cancellation when HMBC data are magnitude calculated for presentation and interpretation. In contrast, LR-HSQMBC spectra are phase sensitive.

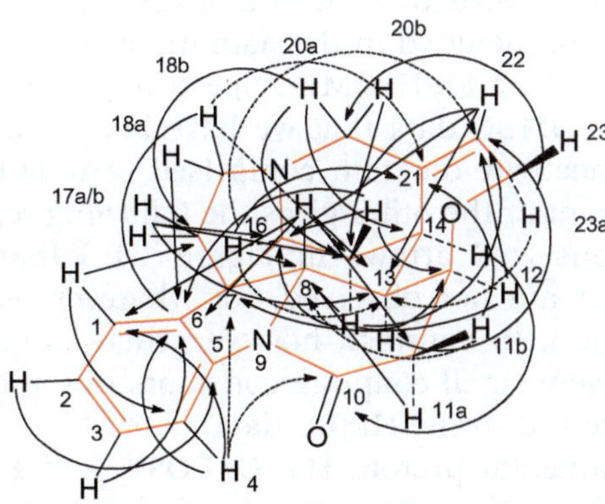

Figure 1.8 Connectivity diagram showing the $^4J_{CH}$ correlations observed in an 8 Hz optimized HMBC spectrum of strychnine. In a comparison made by Williamson and co-workers[24] when the LR-HSQMBC experiment was described in 2014, it was noted that the number of $^4J_{CH}$ correlations in the 8 Hz optimized HMBC spectrum of strychnine is essentially the same as the number of $^2J_{CH}$ correlations. Dashed arrows denote weak correlations.

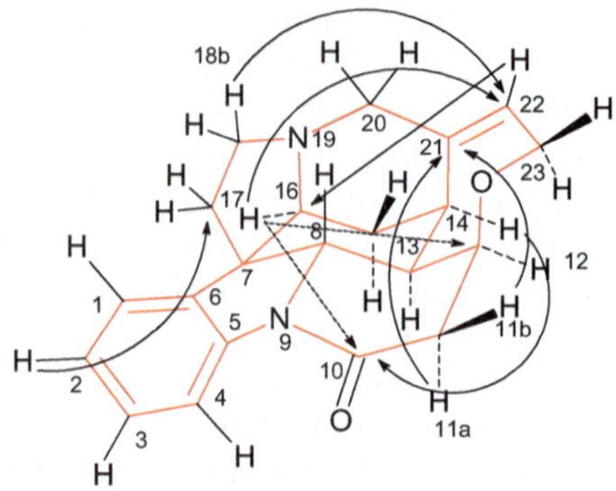

Figure 1.9 Connectivity diagram showing the $^5J_{CH}$ correlations observed in an 8 Hz optimized HMBC spectrum of strychnine. Dashed arrows denote weak correlations.

Generally, LR-HSQMBC should be considered a "second-tier" long-range heteronuclear shift correlation experiment. HMBC data should be utilized to "prune" the number of correlations in the LR-HSQMBC spectrum to just those that were not observable in an HMBC experiment, which will typically be the longer range, smaller correlations. In the initial investigation of the LR-HSMQBC experiment, cervinomycin A_2,[22] which is quite proton deficient, was used as a model compound. From DFT calculations carried out in conjunction with the NMR investigation, heteronuclear coupling constants <0.5 Hz were readily visualized by the LR-HSQMBC experiment when it was optimized at 2 Hz. The availability of such correlation data for very small coupling constants was later shown to have a significant impact on calculation times for the Structure Elucidator CASE program when the 2 Hz LR-HSQMBC data were included in the data input file.[26]

When a 2 Hz optimized LR-HSQMBC spectrum of strychnine was compared with the 8 Hz HMBC data that we have been using as an example, constructing a connectivity diagram comprised of only those correlations seen in one experiment or the other gives the following result. Clearly, there are some correlations (red arrows in Figure 1.10) that are observed in the HMBC data that are not observed, for whatever reason, in the 2 Hz LR-HSQMBC spectrum. In contrast, however, there are significantly more correlations due to very small coupling constants in the LR-HSQMBC spectrum than were observed in the HMBC data.

Beyond the fundamental proton, HSQC, COSY, and HMBC spectra, the paths that can be chosen in a structure elucidation study can be highly divergent depending on the amount of sample available, the nature of the problem in hand, and the experience of those doing the research. For example, in the event that there are resonance overlap problems that make the utilization of COSY data difficult, proton–proton homonuclear connectivity networks can be sorted by ^{13}C chemical shifts using one of the variants of the HSQC–TOCSY experiment that are available. Again using strychnine as an

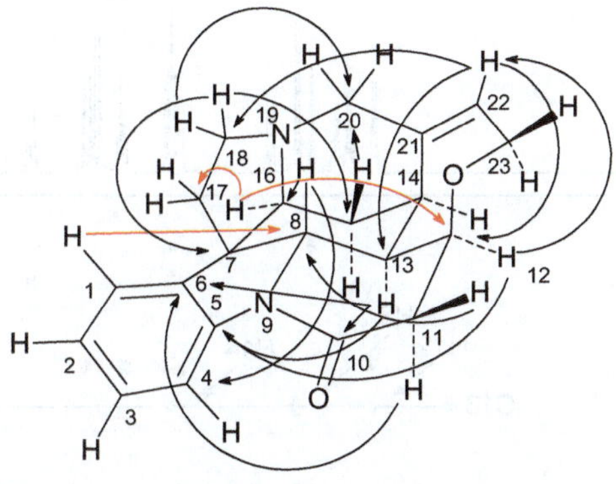

Figure 1.10 Connectivity diagram showing a comparison of the correlations observed in an 8 Hz optimized HMBC spectrum of strychnine not observed in a 2 Hz optimized LR-HSQMBC spectrum (red arrows) *versus* the correlations observed in the latter spectrum that were not observed in the HMBC data (black arrows). As is readily apparent, there were 18 correlations observed in the LR-HSQMBC spectrum that were not observed in the HMBC data, as opposed to only three correlations that were observed in the HMBC spectrum that were not visualized in the LR-HSQMBC spectrum.

example, several IDR (inverted direct response)-HSQC–TOCSY spectra were recorded.[27] The pulse sequence fundamentally establishes direct proton–carbon coherences *via* $^{1}J_{CH}$, after which magnetization is propagated *via* the homonuclear proton couplings. At the end of the sequence, a π pulse "sandwich" is applied to invert the direct correlations. Hence the spectrum is "edited." Direct responses are inverted and plotted in red while TOCSY correlations between the directly attached proton and its neighbors (vicinal and further removed as a function of the mixing time) have positive intensity and are plotted in black in the figures that follow. As shown in the contour plot of the 12 ms IDR-HSQC–TOCSY spectrum presented in Figure 1.11, correlations can be traced horizontally, *e.g.* those from the H13–C13 heteronuclide pair to the three vicinal proton neighbors, H12, H8, and H14. As will be noted from the relative intensities of the correlations along the horizontal (F_1) axis defined by the ^{13}C shift of C13, the correlation corresponding to the scalar (J) coupling between H13 and H8 is considerably more intense than the correlations between H13–H12 and H13–H14. By examining the structure of strychnine, it will be noted that there is a transdiaxial relationship between H13–H8, whereas the relationships of H13–H12 and H13–H14 are both *gauche*. This geometry is consistent with a much larger coupling between H13–H8 ($J = 10.6$ Hz), which, in turn, translates to a more efficient transfer of magnetization between H13 and H8 during the 12 ms TOCSY interval, leading to the more intense response for that correlation. At longer mixing times, the correlations between H13 and both H12 and H14 will become more intense.

Other experimental choices abound and illustrations of a number of the experiments that can be applied to strychnine can be found in the chapter

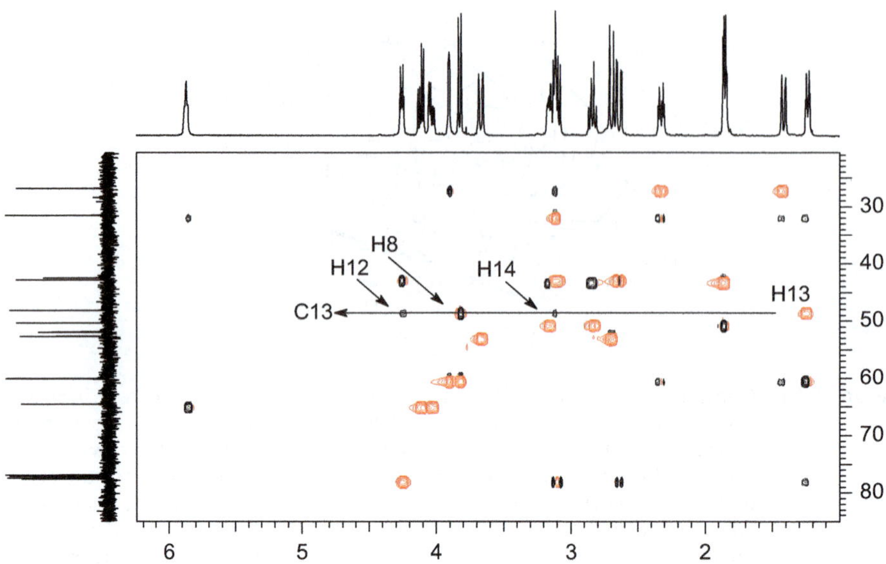

Figure 1.11 A 12 ms IDR-HSQC–TOCSY spectrum of strychnine in chloroform. Direct responses are inverted and plotted in red whereas protons with scalar couplings to H13 have positive intensity and are plotted in black. The three protons vicinally coupled to H13, H12, H8, and H14 are observed in second frequency domain, F_1, at the ^{13}C shift of the C13 resonance. Longer mixing times will propagate magnetization further from the directly correlated proton as shown in Figure 1.12.

dealing with alkaloids (Volume 2, Chapter 10). To illustrate just how far it is possible to take structure characterization experiments, perhaps one of the least sensitive 2D NMR experiments that is likely to be applied individually to a natural product structure elucidation problem is the INADEQUATE ^{13}C–^{13}C double quantum correlation experiment first described in 1980 by Bax *et al.*[29] This experiment exploits the $^{1}J_{CC}$ homonuclear coupling at ^{13}C natural abundance. Statistically, the sample pool is 1 : 10 000 of the ensemble of molecules contained in the NMR tube. In other words, the experiment is extremely insensitive. Nevertheless, using a 25 mg sample of strychnine dissolved in 600 μL of deuterochloroform in a 5 mm NMR tube and a 500 MHz NMR spectrometer equipped with a cryogenic 5 mm gradient inverse NMR probe, an INADEQUATE spectrum of strychnine was recorded over a long weekend in 74 h and is shown in Figure 1.13. The spectrum was intentionally highly folded in the second frequency domain to mitigate F_1 digitization requirements since there are no resonances contained in the region from approximately 80–120 ppm in the ^{13}C NMR spectrum of strychnine. Correlations in the second frequency domain are observed at the algebraic sum of the offsets of the coupled resonances relative to the transmitter frequency. Hence, as shown by the diagonal segments superimposed on the spectrum, the correlation axis runs through the spectrum in

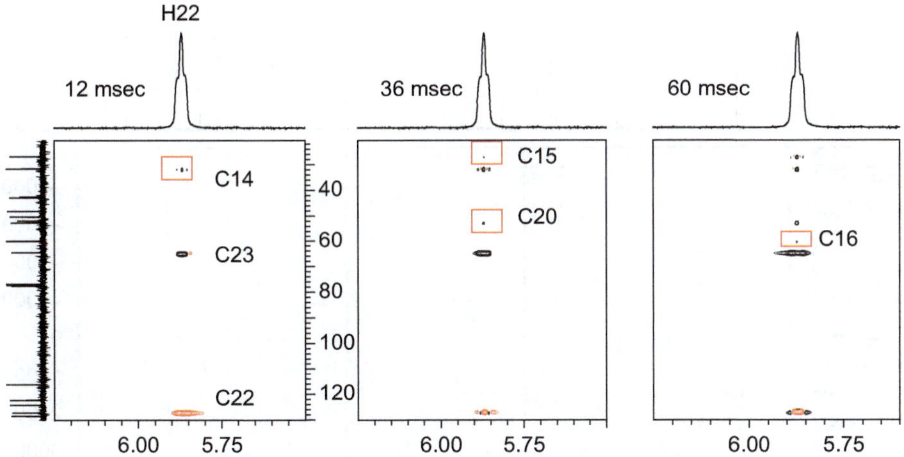

Figure 1.12 In addition to being able to interpret data horizontally along a carbon chemical shift in F_1, HSQC–TOCSY can also be interpreted vertically at the proton chemical shift.[28] As shown in the left panel, which presents data for the H22 vinyl proton of strychnine in the 12 ms IDR-HSQC–TOCSY spectrum shown in Figure 1.11, responses are observed at the F_1 shift of C23, and weakly at the chemical shift of C14 (red boxed correlation). Resorting to longer mixing times, *e.g.* the 36 ms spectrum shown in the middle panel, magnetization is propagated further. The correlations at the C23 and C14 chemical shifts have become more intense, and responses are beginning to be observed at the ^{13}C shifts of C15 and C20. Finally, in the 60 ms spectrum shown in the right panel, a correlation is also beginning to be observed at the ^{13}C chemical shift of C16.

a manner that will be foreign to many readers. Correlations in the F_2 dimension are symmetric about the folded diagonal. Dashed vertical lines shown in Figure 1.13 show the continuation of the diagonal, to provide a visual aid to the continuity of the multiply folded ^{13}C–^{13}C INADEQUATE spectrum. See also Figure 1.14.

The spectral examples shown afford a glimpse of the type of detailed structural information that can be extracted today from a series of 2D NMR experiments available on a modern NMR spectrometer to assist in the structure elucidation process. As noted on the back cover of this volume, a structure elucidation problem that took well over a century to resolve using synthetic approaches can now be resolved fairly simply with small amounts of material, a combination of analytical approaches (mass spectrometry and NMR spectroscopy), and the manipulation of magnetization to probe both homo- (^{1}H–^{1}H and ^{13}C–^{13}C) and heteronuclear (^{1}H–^{13}C, ^{1}H–^{15}N, and very recently ^{13}C–^{15}N) direct and long-range correlations in the molecule. Indeed, even ^{13}C–^{15}N heteronuclear correlations can now be probed at the natural abundance of both nuclides.[6] What will be possible in the future when we have even more tools in our armory and still greater sensitivity?

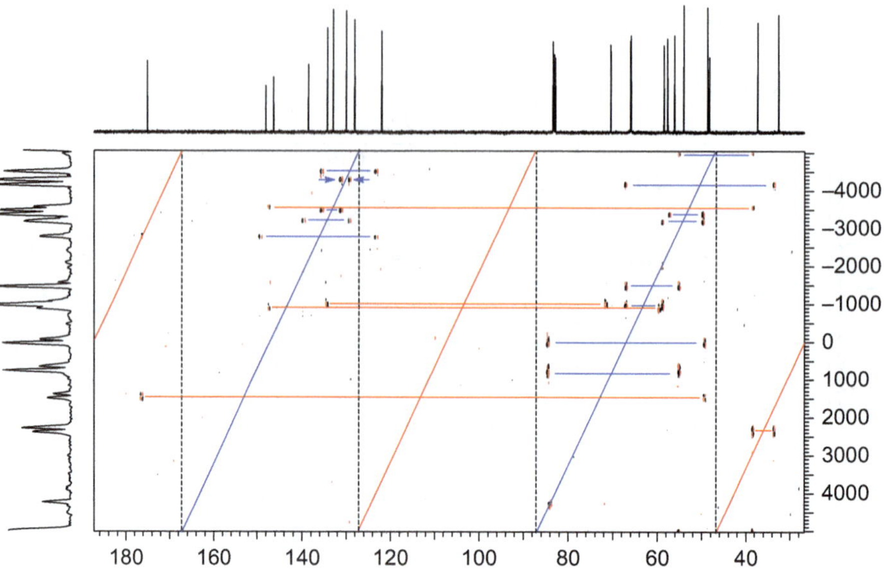

Figure 1.13 ^{13}C–^{13}C double quantum INADEQUATE spectrum of a 25 mg sample of strychnine acquired in 74 h over a long weekend at a ^{13}C observation frequency of 125 MHz using a 5 mm cryogenic NMR probe. Carbon–carbon correlations are symmetrically located about the diagonal of the experiment. The diagonal is folded four times in F_1 to afford better digital resolution across the ~40 kHz F_1 spectral range normally encompassed by this spectrum. Segments of the diagonal are alternately color coded red and blue. Correlations between pairs of resonances are designated by horizontal red or blue lines color coded as a function of the segment of the diagonal about which the resonances are symmetrically disposed. In the case of, for example, the C10 carbonyl correlation to the C11 methylene, the correlation is symmetric about the center red diagonal segment but the individual responses are outside the blue diagonal segments on either side, which might be confusing for a novice user.

In terms of sensitivity, although hyperpolarization has, in general, been applied primarily in the medical imaging field, the immense opportunities that general hyperpolarization transfer approaches offer to provide enhanced sensitivity have been of interest to the NMR community. Work is already under way to develop approaches using para-hydrogen as the polarization carrier[31] and to afford the resulting improvements in sensitivity to general NMR spectroscopy.

Data processing has also undergone some significant changes. Obviously, the processing power of the computers used for NMR data processing has increased monumentally.[32] In the 1980s, following the advent of 2D NMR, processing a simple COSY spectrum on a dedicated instrument computer could consume 10–15 min or more for only a 512 × 512 point spectrum. Data had to be processed stepwise with a transposition manually initiated by the

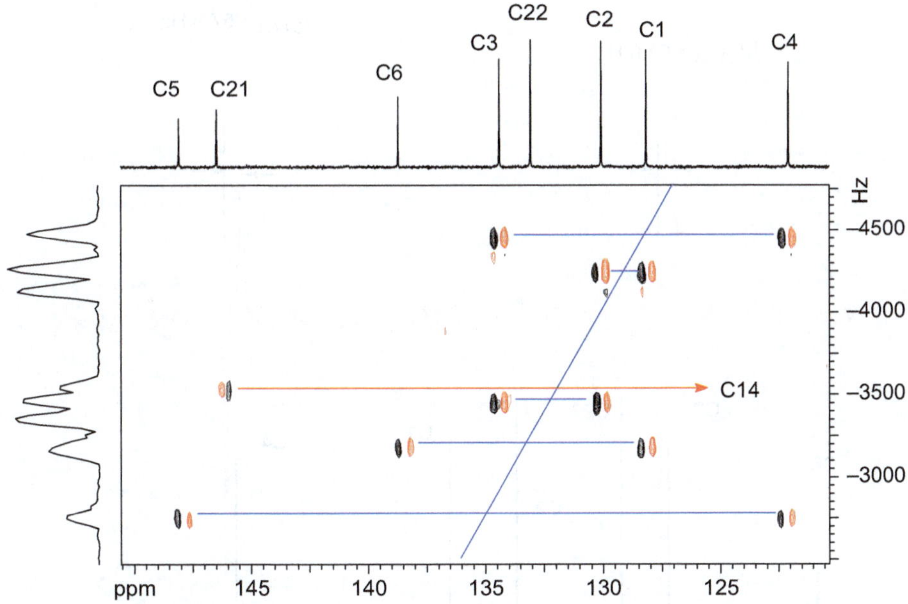

Figure 1.14 Expansion of the aromatic/vinyl correlations in the upper left corner of the spectrum shown in Figure 1.13. The scale in F_1 is arbitrary whereas the chemical shift scale in the F_2 dimension reflects the actual ^{13}C shifts of the aromatic carbons of strychnine. Note that the individual ^{13}C–^{13}C doublets are antiphase. As shown in Figure 1.15, the splitting of the correlations directly reflects the $^1J_{CC}$ coupling constant that matches what can be measured using a *J*-modulated ADEQUATE experiment.[30]

user prior to the second Fourier transformation. Now, that same processing is done in seconds, with the entire data matrix loaded into memory, with the intervening steps transparent (unrealized?) to the casual user. How many other facets of modern NMR are unrealized by workers more newly arrived in the field? During a recent conversation with a post-doctoral fellow about an illustration for a graphical abstract, one of the present authors found it necessary to explain what a white-washed stack plot (Figure 1.16) was. Although commonly encountered during the infancy of 2D NMR, they have been totally supplanted by topographic contour plots for the presentation of data, with presentations now relegated to book covers and graphical abstract illustrations and such.

It is also easy to forget that linear prediction was once viewed as computationally intensive, and is now a trivial operation for improving the appearance of many 2D-NMR spectra, taking a few seconds or less on standard desktop computers.[31] More recent advances in non-uniform sampling (NUS) methods, the benefits of which for resolution and sensitivity are now becoming widely exploited for improving small molecule NMR, are following a similar path. One of the present authors (D.R.) remembers well setting up

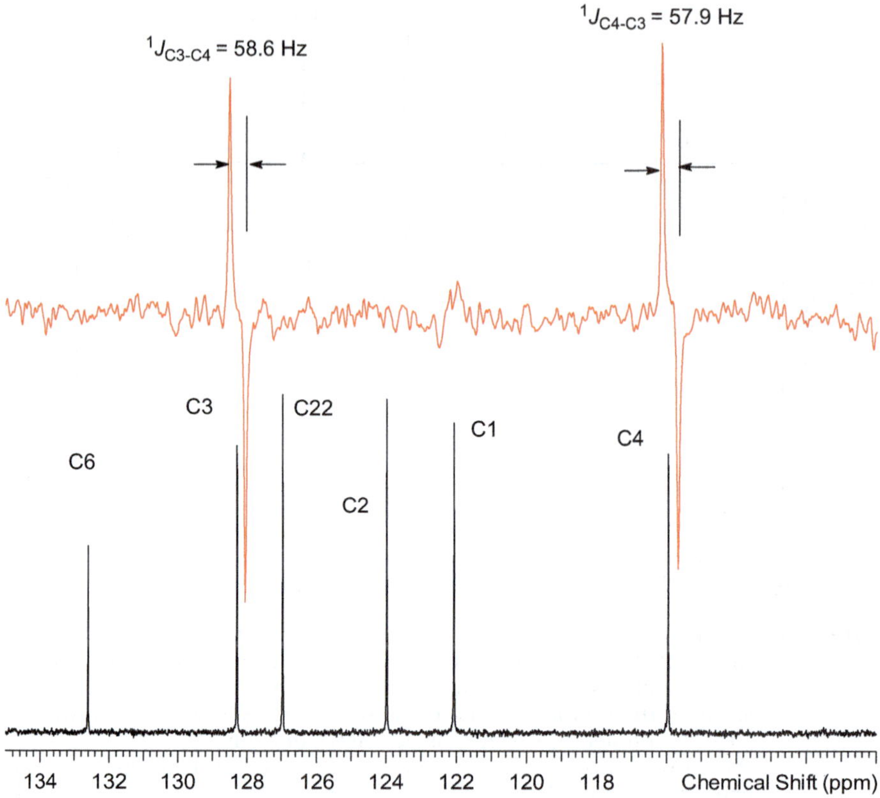

Figure 1.15 The bottom panel shows a segment of the aromatic region of the ^{13}C reference spectrum of strychnine. Plotted in red above the reference spectrum is the F_1 slice extracted from the ^{13}C–^{13}C INADEQUATE spectrum (see Figure 1.14) for the C3–C4 correlation. As can be seen, the correlations are antiphase doublets symmetrically disposed about the ^{13}C resonance frequencies. The splitting of the antiphase doublets corresponds to the $^{1}J_{CC}$ coupling, which in this case is approximately 58 Hz.

overnight jobs on Unix workstations to generate a frequency spectrum of NUS data.[36] Today, his students bemoan the 5–15 s that these same routines now consume on modern computers. Indeed, these advances have allowed major software vendors to make NUS acquisition and processing nearly as transparent to the user as linear prediction now is. As computation and algorithms continue to advance, alternative sampling and processing will continue to take on greater importance in making structure solving itself more transparent to the user. As a good example, covariance processing techniques[37] have begun to have an impact in the form of indirect covariance,[38] unsymmetrical indirect covariance,[39] and generalized indirect covariance methods.[40] Indeed, we have begun to see the combination of covariance processing with modified experimental methods to obtain some

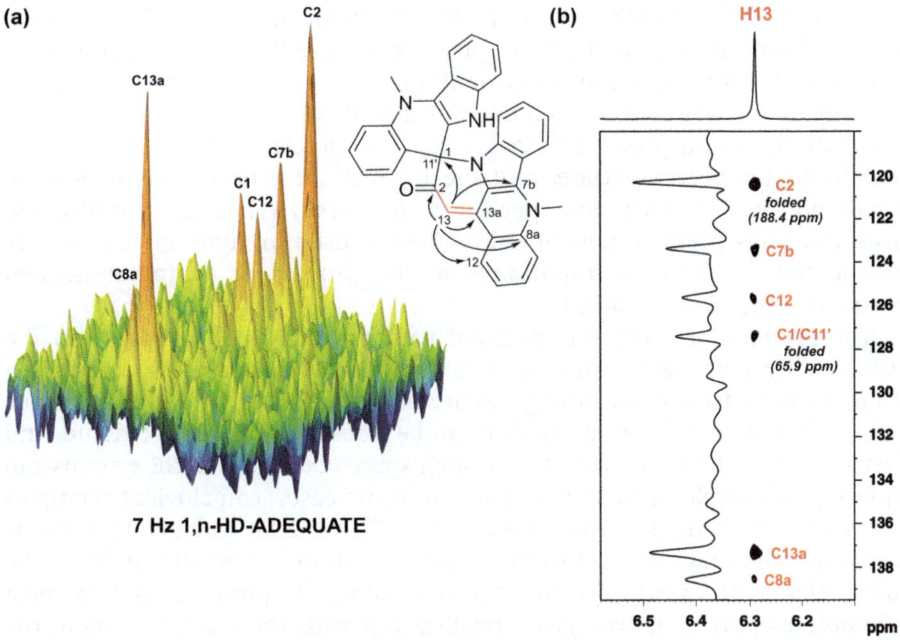

Figure 1.16 (a) Stack plot showing correlations in a 7 Hz optimized 1,*n*-ADEQUATE spectrum of the alkaloid cryptospirolepine.[33–35] (b) Contour plot from the 7 Hz optimized 1,*n*-ADEQUATE spectrum of cryptospirolepine. The data contained in both presentations are identical but there is little argument that the presentation in (b) is far easier to interpret than the stack plot shown in (a).

of the pure shift spectra such as PSYCHE-TOCSY.[41] Although there have been demonstrations of the use of unsymmetrical or generalized indirect covariance processing to calculate ^{13}C–^{15}N[42] or ^{13}C–^{31}P[43] correlation plots, as noted earlier, the former can now be obtained directly using the HCNMBC experiment.[6] What new ways there may be of utilizing advanced NMR data processing techniques remains to be seen.

Since the inception of NMR as a structure elucidation technique, reference data resulting from the analyses have been published in the literature. In the early days of the technique, NMR assignments were, of course, limited to proton NMR. With the advent of heteronuclear NMR, specifically ^{13}C NMR, the availability of reported data allowed other databases to be assembled. By the mid-1990s, one of the authors was investigating the purchase of a ^{13}C NMR database containing just over 10 000 compounds, running on a Unix workstation, averaging a cost of almost \$2 per compound for 1 year of access. Technology changes quickly, and within 2 years the author had installed a PC-based ^{13}C NMR shift prediction package and database containing 10 000 compounds for less than \$0.1 per compound and with a perpetual license. By the mid-2000s, reference data extracted from the literature were optionally available *via* a web interface for a number of nuclei, including ^{1}H, ^{13}C,

^{15}N, ^{19}F, and ^{31}P, together with prediction algorithms assembled from the data collections. Currently, it is possible to predict NMR spectra on a number of websites, on an iPad or iPhone. Large collections of assigned NMR data are available as open data for download and repurposing. One of the authors was a product manager for commercial NMR predictors and databases for over a decade, and the majority of expectations relative to prediction performance and speed were delivered. What was not foreseen was how NMR prediction would ultimately be used for automated structure verification[44,45] and its importance in the process of computer-assisted structure elucidation (CASE).

The promise of CASE was initiated with the DENDRAL Project[46] in the 1960s. Fifty years later, with the availability of high-resolution mass spectrometry to assist in determining a molecular formula, and with the enormous array of NMR techniques available to probe direct and long-range homo- and heteronuclear through-bond and through-space couplings, CASE systems can now ingest complex arrays of data and, in some cases, can elucidate complex chemical structures in a few seconds[20,26] (Figure 1.17, Table 1.1). CASE is actually in its infancy in terms of adoption, with only a small number of laboratories in the world utilizing the technology. At present, CASE is most valuable as part of a synergistic relationship with the scientist, where the scientist contributes as much detail as possible in terms of class of compound, fragments identified in the mass spectrum, partially assigned spectra, *etc.* However, as the amount of data extracted from the literature expands and finds its way into the multinuclear NMR prediction databases, and the knowledge base of molecular fragments grows from these data, then CASE is likely to become less dependent on a scientist's input in the majority of cases. The greatest challenges for CASE to elucidate a chemical structure successfully are good peak-picking, specifically within the 2D-NMR spectra, and in a related manner, determining the bond order of correlations within the 2D spectra. There has been significant progress in improving both of these areas in recent years, especially with the advent of pure shift techniques[3,35] and experiments to identify correlations of a specific order.[18] The time is nearing when structures *will* be automatically elucidated on the instrument using CASE techniques, and the primary questions of most chemists will be answered directly – is this compound what I think it is and, if not, what is it? Bruker BioSpin is already taking steps in this direction with their CMC-se program package,[47] and it will be interesting to watch progress in this area of natural product structure elucidation.

What technologies that are being investigated now may lead to new leaps in NMR sensitivity, additional resolving power *via* manipulation of coherences across spin systems, or the ability to task general structure elucidation to software algorithms using historical data, thereby reducing the burden of a scientist to duplicate the work of others, is very difficult to even speculate. Work continues unabated to move the technologies forward on these fronts, and others, and we have attempted to offer here our views on some of the great promise that lies ahead.

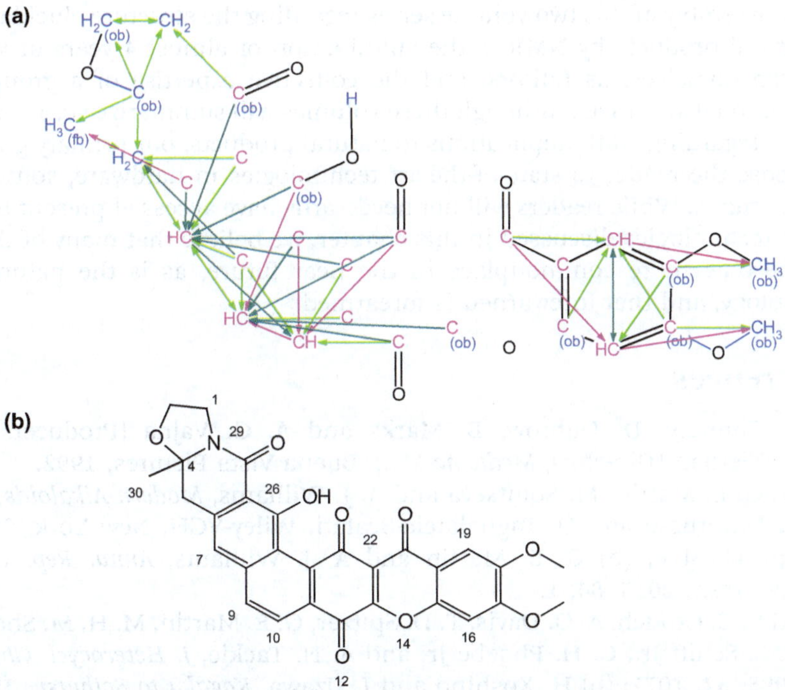

Figure 1.17 (a) Molecular connectivity diagram (MCD) taken from the Structure Elucidator CASE data for a study of the impact of various long-range heteronuclear chemical shift correlation data on structure generation times for the xanthone antibiotic cervinomycin A_2.[26] (b) Structure of cervinomycin A_2.[22] The study demonstrated that the availability of very long-range (*e.g.* $\geq{}^4J_{CH}$) can have a profound impact on both the number of structures generated as well as the generation times (see Table 1.1).

Table 1.1 Results obtained from various Structure Elucidator CASE program computation runs for various sets of input data for the xanthone antibiotic cervinomycin A_2 (see Figure 1.17b for the structure). As can be readily seen from the first two rows of the table, restricting the input data file to data that are likely to have primarily ${}^2J_{CH}$ and ${}^3J_{CH}$ correlations with perhaps only sparse ${}^4J_{CH}$ correlations (rows 1 and 2) leads to lengthy computation runs. However, when 2 Hz optimized LR-HSQMBC data, which can contain ${}^4J_{CH}$–${}^6J_{CH}$ correlations (rows 3 and 4), are included in the data input file, computation times shorten precipitously and the number of structures generated is also significantly reduced.[26]

COSY HSQC	^1H–^{13}C HMBC 8 Hz	^1H–^{13}C HMBC 4 Hz	^1H–^{13}C LR-HSQMBC 4 Hz	^1H–^{13}C LR-HSQMBC 2 Hz	Structure generation time	No. of structures generated
+	+	+			49 h	314
+	+	+	+		37 h	4
+	+	+		+	150 s	7
+	+	+	+	+	104 s	1

The assembly of this two-volume series regarding the structure elucidation of natural products by NMR is the culmination of almost 4 years of work between ourselves, as Editors, and the collective expertise of a group of international scientists. Although these volumes will summarize some of the history regarding NMR applications to natural products, our primary goal is to expose the reader to state-of-the-art technologies in hardware, software, and methods. While readers will not necessarily have access at present to all of the technologies discussed in this chapter, we believe that many of these will become more commonplace in the near future, as is the nature of technology, and that forewarned is forearmed.

References

1. S. Connery, D. Dubrow, B. Marks and A. G. Vajna (Producers), J. McTiernan (Director), *Medicine Man*, Buena Vista Pictures, 1992.
2. (a) G. E. Martin, M. Solntseva and A. J. Williams, *Modern Alkaloids*, ed. E. Fattorusso and O. Taglialatela-Scafati, Wiley-VCH, New York, 2007, pp. 411–476; (b) G. E. Martin and A. J. Williams, *Annu. Rep. NMR Spectrosc.*, 2015, **84**, 1.
3. (a) R. C. Crouch, A. O. Davis, T. D. Spitzer, G. E. Martin, M. H. M. Sharaf, P. L. Schiff Jr., C. H. Phoebe Jr. and A. N. Tackie, *J. Heterocycl. Chem.*, 1995, **32**, 1077; (b) H. Koshino and J. Uzawa, *Kagaku to Seibutsu*, 1995, **33**, 252.
4. (a) R. W. Adams, *eMagRes*, 2014, **3**, 1; (b) K. Zangger, *Prog. NMR Spectrosc.*, 2015, **86–87**, 1; (c) L. Castañar and T. Parella, *Magn. Reson. Chem.*, 2015, **53**, 399.
5. (a) R. T. Williamson, A. V. Buevich, G. E. Martin and T. Parella, *J. Org. Chem.*, 2014, **79**, 3887; (b) R. T. Williamson, A. V. Buevich and G. E. Martin, *Tetrahedron Lett.*, 2014, **55**, 3365.
6. (a) S. Cheatham, P. Gierth, W. Bermel and Ē. Kupče, *J. Magn. Reson.*, 2014, **247**, 38; (b) S. Cheatham, M. Kline and Ē. Kupče, *Magn. Reson. Chem.*, 2015, **53**, 363.
7. (a) G. E. Martin, *Annu. Rep. NMR Spectrosc.*, 2011, **74**, 215; (b) G. E. Martin, M. Reibarkh, A. V. Buevich, K. A. Blinov and R. T. Williamson, *eMagRes*, 2014, **3**, 215.
8. (a) R. C. Crouch and G. E. Martin, *J. Nat. Prod.*, 1992, **55**, 1343; (b) R. C. Crouch and G. E. Martin, *Magn. Reson. Chem.*, 1992, **30**, 66.
9. (a) G. E. Martin, J. E. Guido, R. H. Robins, M. H. M. Sharaf, P. L. Schiff Jr. and A. N. Tackie, *J. Nat. Prod.*, 1998, **61**, 555; (b) G. E. Martin, R. C. Crouch and A. P. Zens, *Magn. Reson. Chem.*, 1998, **36**, 551; (c) C. E. Hadden and G. E. Martin, *J. Nat. Prod.*, 1998, **61**, 969; (d) G. E. Martin, in *Encyclopedia of Nuclear Magnetic Resonance*, ed. D. M. Grant and R. K. Harris, Wiley, New York, 2002, vol. 9, pp. 98–112; (e) G. E. Martin, *Annu. Rep. NMR Spectrosc.*, 2005, **56**, 1–99; (f) F. C. Schroeder and M. Gronquist, *Angew. Chem., Int. Ed.*, 2006, **45**, 7122; (g) G. E. Martin, in *Encyclopedia of Nuclear Magnetic Resonance*, ed.

R. K. Harris and R. A. Wasylishen, Wiley, New York, online, 2011, DOI: 1002/9780470034590.emrstm1300.
10. (a) G. E. Martin, in *Encyclopedia of Nuclear Magnetic Resonance*, ed. D. M. Grant and R. K. Harris, Wiley, New York, 2002, vol. 9, pp. 33–35; (b) G. E. Martin, D. J. Russell, K. A. Blinov, M. E. Elyashberg and A. J. Williams, *Ann. Magn. Reson.*, 2003, **2**, 1.
11. (a) D. J. Russell, C. E. Hadden, G. E. Martin, A. A. Gibson, A. P. Zens and J. L. Carolan, *J. Nat. Prod.*, 2000, **63**, 1047; (b) B. D. Hilton and G. E. Martin, *J. Nat. Prod.*, 2010, **73**, 1465.
12. Y. Liu, M. D. Green, R. Marques, T. Pereira, R. Helmy, W. R. T. Williamson, W. Bermel and G. E. Martin, *Tetrahedron. Lett.*, 2014, **55**, 5450.
13. T. F. Molinski, *Curr. Opin. Drug Discovery Dev.*, 2009, 197(b) D. S. Dalisay and T. F. Molinski, *J. Nat. Prod.*, 2009, **72**, 739; (c) T. F. Molinski, *Curr. Opin. Biotechnol.*, 2010, **21**, 819; (d) T. F. Molinski, *Nat. Prod. Rep.*, 2010, **27**, 321.
14. (a) T. D. W. Claridge, *High-Resolution NMR Techniques in Organic Chemistry*, Pergamon Press, Amsterdam, 1999, pp. 239–240; (b) L. Paudel, R. W. Adams, P. Király, J. A. Aguilar, M. Foroozandeh, M. J. Cliff, M. Nilsson, P. Sándor, J. P. Waltho and G. A. Morris, *Angew. Chem., Int. Ed.*, 2013, **52**, 11616.
15. A. Bax and M. F. Summers, *J. Am. Chem. Soc.*, 1986, **108**, 2093.
16. (a) W. F. Reynolds, *Encyclopedia of Magnetic Resonance*, John Wiley & Sons, Ltd., 2010, DOI: 10.1002/9780470034590.emrstm1176; (b) W. Schoefberger, J. Schlagnitweit and N. Müller, *Annu. Rep. NMR Spectrosc.*, 2011, **72**, 1–60; (c) J. Furrer, *Annu. Rep. NMR Spectrosc.*, 2011, **74**, 293–354; (d) W. F. Reynolds and D. C. Burns, *Annu. Rep. NMR Spectrosc.*, 2012, **76**, 1–21; (e) J. Furrer, *Concepts Magn. Reson.*, 2012, **40A**, 101; (f) J. Furrer, *Concepts Magn. Reson.*, 2012, **40A**, 149; (g) J. Furrer, *Concepts Magn. Reson.*, 2015, **43A**, DOI: 10.1002/cmr.a.21317.
17. P. Sakhaii, B. Haase and W. Bermel, *J. Magn. Reson.*, 2013, **228**, 125.
18. M. Reibarkh, R. T. Williamson, G. E. Martin and W. Bermel, *J. Magn. Reson.*, 2013, **236**, 126.
19. R. B. Woodward, M. P. Cava, W. D. Ollis, A. Hunger, H. U. Daeniker and K. Schenker, *J. Am. Chem. Soc.*, 1954, **76**, 4749.
20. M. E. Elyashberg, A. J. Williams and G. E. Martin, *Prog. NMR Spectrosc.*, 2008, **53**, 1–104.
21. (a) T. F. Molinski and B. I. Morinaka, *Tetrahedron*, 2012, **68**, 9307; (b) P. Ralifo and P. Crews, *J. Org. Chem.*, 2004, **69**, 9025.
22. S. Ōmura, Y. Iwai, K. Hinotozawa, Y. Takahashi, J. Kato, A. Nakagawa, A. Hirano, H. Shimizu and K. Handea, *J. Antibiotics*, 1982, **35**, 645.
23. M. M. Senior, R. T. Williamson and G. E. Martin, *J. Nat. Prod.*, 2013, **76**, 2088.
24. A. V. Buevich, R. T. Williamson and G. E. Martin, *J. Nat. Prod.*, 2014, **77**, 1942.
25. K. Furihata and H. Seto, *Tetrahedron Lett.*, 1995, **36**, 2817.

26. K. A. Blinov, A. V. Buevich, R. T. Williamson and G. E. Martin, *Org. Biomol. Chem.*, 2014, **12**, 9505.
27. (a) T. Domke, *J. Magn. Reson.*, 1991, **95**, 174; (b) R. C. Crouch, A. O. Davis and G. E. Martin, *Magn. Reson. Chem.*, 1995, **33**, 889; (c) C. E. Hadden, G. E. Martin, J.-K. Luo and R. N. Castle, *J. Heterocycl. Chem.*, 1999, **36**, 553.
28. G. E. Martin and R. C. Crouch, *J. Nat. Prod.*, 1991, **54**, 1.
29. A. Bax, R. Freeman and S. P. Kempsell, *J. Am. Chem. Soc.*, 1980, **102**, 4849.
30. (a) R. T. Williamson, A. V. Buevich and G. E. Martin, *Org. Lett.*, 2012, **14**, 5098; (b) K. Kövér and P. Forgo, *J. Magn. Reson.*, 2004, **166**, 47; (c) C. M. Thiele and W. Bermel, *Magn. Reson. Chem.*, 2007, **45**, 889.
31. S. Glöggler, J. Colell and S. Appelt, *J. Magn. Reson.*, 2013, **235**, 130.
32. (a) H. Barkhuijsen, R. de Beer, W. M. M. J. Bovée and D. van Ormondt, *J. Magn. Reson.*, 1985, **65**, 465; (b) D. S. Stephenson, *Prog. NMR Spectrosc.*, 1988, **20**, 512; (c) J. C. Hoch and A. S. Stern, *NMR Data Processing*, 1996, Wiley-Liss, New York; (d) P. Koehl, *Prog. NMR Spectroscopy*, 1999, **34**, 257.
33. A. N. Tackie, G. L. Boye, M. H. M. Sharaf, P. L. Schiff Jr., T. D. Spitzer, R. L. Johnson, J. Dunn, D. Minick and G. E. Martin, *J. Nat. Prod.*, 1993, **56**, 553.
34. G. E. Martin, C. E. Hadden, D. J. Russell, B. D. Kaluzny, J. E. Guido, W. K. Duholke, B. B. A. Stiemsma, T. J. Thamann, R. C. Crouch, K. A. Blinov, M. Elyashberg, E. R. Martirosian, S. G. Molodtsov, A. J. Williams and P. L. Schiff Jr., *J. Heterocyc. Chem.*, 2002, **39**, 1241.
35. J. Saurí, W. Bermel, A. V. Buevich, M. H. M. Sharaf, P. L. Schiff Jr., T. Parella, R. T. Williamson and G. E. Martin, *Angew. Chem., Int. Ed.*, 2015, 54, DOI: 10.1002/anie.201502540.
36. (a) J. C. J. Barna, E. D. Laue, M. R. S. Mayger, J. Skilling and S. J. P. Worrall, *J. Magn. Reson.*, 1987, **73**, 69–77; (b) J. C. J. Barna and E. D. Laue, *J. Magn. Reson.*, 1987, **75**, 384; (c) A. D. Schuyler, M. W. Maciejewski, A. S. Stern and J. C. Hoch, *J. Magn Reson.*, 2015, **254**, 121.
37. (a) R. Brüschweiler and F. Zhang, *J. Chem. Phys.*, 2004, **120**, 5253; (b) D. A. Snyder and R. Brüschweiler, *eMagRes*, 2007, DOI: 10.1002/9780470034590.emrstm1098; (c) D. A. Snyder and R. Brüschweiler, in *Multidimensional NMR Methods for the Solution State*, ed. G. A. Morris and J. W. Emsley, Wiley, New York, 2010, pp. 97–104; (d) M. Jaeger and R. L. E. G. Aspers, *Annu. Rep. NMR Spectrosc.*, 2014, **83**, 271–349.
38. (a) F. Zhang and R. Brüschweiler, *J. Am. Chem. Soc.*, 2004, **126**, 13180; (b) K. A. Blinov, N. I. Larin, M. P. Kvasha, A. Moser, A. J. Williams and G. E. Martin, *Magn. Reson. Chem.*, 2005, **43**, 999.
39. (a) K. A. Blinov, N. I. Larin, A. J. Williams, M. Zell and G. E. Martin, *Magn. Reson. Chem.*, 2006, **44**, 107; (b) K. A. Blinov, N. I. Larin, A. J. Williams, K. A. Mills and G. E. Martin, *J. Heterocyc. Chem*, 2006, **44**, 163; (c) G. E. Martin, B. D. Hilton, P. A. Irish, K. A. Blinov and A. J. Williams, *J. Nat. Prod.*, 2007, **70**, 1393.
40. D. A. Snyder and R. Brüschweiler, *J. Phys. Chem. A*, 2009, **113**, 12898.

41. (a) G. A. Morris, J. A. Aguilar, R. Evans, S. Haiber and M. Nilsson, *J. Am. Chem. Soc.*, 2010, **132**, 12770; (b) M. Foroozandeh, R. W. Adams, M. Nilsson and G. A. Morris, *J. Am. Chem. Soc.*, 2014, **136**, 11867.
42. (a) G. E. Martin, P. A. Irish, B. D. Hilton, K. A. Blinov and A. J. Williams, *Magn. Reson. Chem.*, 2007, **45**, 624; (b) G. E. Martin, B. D. Hilton, K. A. Blinov and A. J. Williams, *J. Nat. Prod.*, 2007, **70**, 1966.
43. E. R. Zartler and G. E. Martin, *J. Biomol. NMR*, 2011, **51**, 357.
44. S. S. Golotvin, E. Vodopianov, B. A. Lefebvre, A. J. Williams and T. D. Spitzer, *Magn. Reson. Chem.*, 2006, **44**, 524.
45. S. S. Golotvin, E. Vodopianov, R. Pol, B. A. Lefebvre, A. J. Williams, R. D. Rutkowske and T. D. Spitzer, *Magn. Reson. Chem.*, 2007, **45**, 803.
46. R. K. Lindsay, B. G. Buchanan, E. A. Feigenbaum and J. Lederberg, *Applications of Artificial Intelligence for Organic Chemistry: The DENDRAL Project*, McGraw-Hill, New York, 1980.
47. G. E. Martin, R. T. Williamson, T. Kühn and S. Groscurth, Structure Elucidation of Proton-Deficient Natural Products by NMR Spectroscopy, poster presentation, ASP meeting, Oxford, MS, August 3–5, 2014, Poster PN6; *Planta Med.*, **80** (10) ASP meeting abstracts issue.

CHAPTER 2

NMR Magnets: A Historical Overview

RAZVAN TEODORESCU

Bruker BioSpin Corporation, Billerica, MA 01821, USA
Email: razvan.teodorescu@bruker.com

2.1 Introduction

This chapter presents a historical review of NMR magnet development focusing on milestones in superconducting magnets over the past four decades. These developments have contributed to significant advances in both fundamental and applied research fields, including natural products.

The evolution of magnets for NMR spectroscopy has been driven by several important factors for improving the performance and quality of research when using NMR spectroscopy, to open up new applications, and to allow more scientists to take advantage of the benefits of NMR in numerous fields of research.

The needs driving NMR magnet development include increasing the NMR sensitivity and resolution of spectra, improving the magnetic field homogeneity, maximizing the magnetic field stability, minimizing magnetic stray fields, mitigating external magnetic field disturbances, reducing the physical size and weight of magnets, and minimizing the cryogenic consumption rates. Significant progress in meeting these needs has been enabled by the development of superconducting wire technologies, novel magnet coil and cryostat designs, and refrigeration technologies.

The development of NMR magnets operating at higher and higher fields and leading to increases in both sensitivity and resolution (chemical

Modern NMR Approaches To The Structure Elucidation of Natural Products,
Volume 1: Instrumentation and Software
Edited by Antony J. Williams, Gary E. Martin and David Rovnyak

Published by the Royal Society of Chemistry, www.rsc.org

dispersion) has been essential for the natural product research applications. The secondary metabolites, which constitute the bioactive components desired from marine or terrestrial natural product sources, are typically isolated in very small quantities (<1 mg) early in the discovery process. By acquiring high-quality NMR data early in the discovery process, de-replication of known metabolites can proceed rapidly and structure elucidation for novel metabolites can begin early. Higher field NMR magnets aid in this process by providing enhanced sensitivity to allow the analysis of very small quantities for early de-replication. The enhanced resolution afforded by these high-field magnets then assists the researcher in the structure elucidation of these often structurally complex molecules by simplifying the analysis of the observed coupling coherences that would be overlapped at lower field strengths. As a result, natural product researchers seek data from the highest field strength magnet available to enhance their research process with regard to speed and accuracy in the discovery of novel metabolites.

2.2 Field Strength and NMR Sensitivity and Resolution

Increasing the sensitivity and resolution of NMR spectra continues to be the driving force for developing new, higher field NMR magnets.

The first superconducting commercial NMR magnets were introduced in the late 1960s and early 1970s.[1] These were operated at resonance frequencies of 180 and 270 MHz for protons (^{1}H), using single-filament niobium–titanium (NbTi) superconductors. These were soon followed by 400 MHz (^{1}H) magnets operating at 9.4 T, and using multi-filament NbTi superconductors, which improved field stability.

The next step was the development in the late 1970s of 500 MHz magnets operating at 11.7 T. Since this field strength was beyond the critical field of NbTi superconductors at the operating temperature of 4.2 K, the use of more complex and more expensive niobium–tin (Nb_3Sn) superconductors was required, owing to their higher critical field. The niobium–tin superconductors were able to push operational field strengths further to 14.1 T, allowing the introduction of commercial 600 MHz NMR magnets in the late 1980s.

The next generation of magnets operating beyond 17 T was made possible by two significant developments. One was the introduction of niobium–tantalum–tin [$(NbTa)_3Sn$] and niobium–tantalum–titanium–tin [$(NbTaTi)_3Sn$] superconductors with increased electrical and mechanical performance, solving the challenges of achieving higher fields and sustaining greatly increased stress levels caused by significantly higher forces. The other was the sub-cooling cryostat technology for operating magnets at temperatures below 4.2 K (*i.e.* below the boiling temperature of helium at ambient pressure), pioneered by Bruker in the early 1990s.[2] These two developments led to a broad variety of ultra-high-field NMR magnets over the last two decades, ranging from 750 MHz (17.6 T) to 1 GHz (23.5 T).

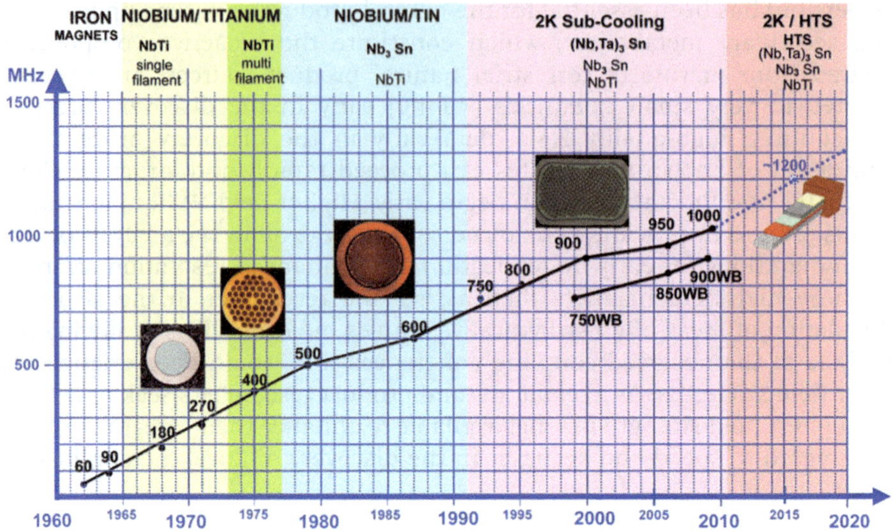

Figure 2.1 Historical milestones for superconductors and NMR magnet field strength.

However, once again, the critical field of Nb_3Sn superconductors prevents them from being used to reach higher fields. Now current efforts are being focused on nesting an insert coil inside a conventional NMR magnet. The use of an insert coil wound with high-temperature superconductors, featuring an increased critical field, is very promising for reaching fields beyond 1 GHz.

The historical milestones for superconductors and NMR magnet field strength are illustrated in Figure 2.1.

2.3 Magnetic Field Homogeneity

Optimum field homogeneity is critical to NMR spectral quality. The very early NMR spectrometers were based on permanent magnets or iron-core resistive magnets and were limited in field homogeneity, which depended on the shape of the poles and was affected by the saturation and the inhomogeneities in the iron. The development of superconducting NMR magnets has enabled much improved field homogeneity to be achieved.

NMR superconducting magnets are based on solenoid coils with graded conductors, meaning that thicker superconductors are in the windings closer to the center of the coil and thinner conductors comprise the outer windings, leading to a higher current density on the outside, which is possible owing to the decay of the field strength in the radial direction.

NMR solenoid coil designs incorporate "notches," which are regions with windings having a reduced current density with respect to the rest of the section to improve the magnetic field homogeneity. Superconducting shim coils are additional means to correct residual field inhomogeneities prior to the final corrections using room temperature shims. The superconducting shims are also known as cryo-shims; these have currents that are set once during the magnet system installation. The room temperature shims are adjusted regularly for each NMR sample prior to performing an NMR experiment.

2.4 Magnetic Field Stability

The NMR magnetic field must be extremely stable and decay by no more than a few parts per billion per hour, which translates to a few hertz (^{1}H NMR) per hour.

The rate of electrical current decay in a solenoid magnet coil is given by the equation

$$\frac{\mathrm{d}I}{\mathrm{d}t} \approx \frac{-(R_{\text{wire}} + R_{\text{joints}})I}{L} \tag{2.1}$$

where I is the operational current, L is the coil inductance, and R_{wire} and R_{joints} are the residual electrical resistances of the superconducting wire and the superconducting joints (splices), respectively, needed in order to connect the various coil sections.

Although in the early years of superconducting NMR magnets it was desirable to have a lower current and larger inductance in order to minimize the field drift, the residual resistance of the superconducting joints remained the primary cause of field drift. Modern NMR magnets are based on a high current and small inductance coil design and are able to reduce drift instead by incorporating advanced superconducting joints operating at high current and maintaining a residual resistance of 10^{-12} Ω or lower, critical to achieving field drift rates of less than 10 ppb per hour.

2.5 Minimizing Stray Magnetic Fields

Siting older non-shielded superconducting NMR magnets with very large stray magnetic fields required a very large footprint for the NMR laboratory while impacting the adjacent spaces, including both above and below the NMR room. Iron shielding plates on the ceiling or under the floor of the magnet room were sometimes used to reduce the very large vertical fringe fields in the rooms above and below such non-shielded magnets.

Actively shielded NMR magnets became available in the mid-1990s, and are now standard in the industry. This has been a major breakthrough, allowing many NMR users to take advantage of small spaces for siting their NMR instrument or to site multiple NMR systems in one room. Also, siting NMR magnets on upper floors is no longer uncommon these days owing to the small magnetic footprint.

The basic operational principle of actively shielded magnets is shown in Figure 2.2. The superconducting coil consists of primary inner coil sections colored in blue, producing the main field for NMR, and shielding outer coil

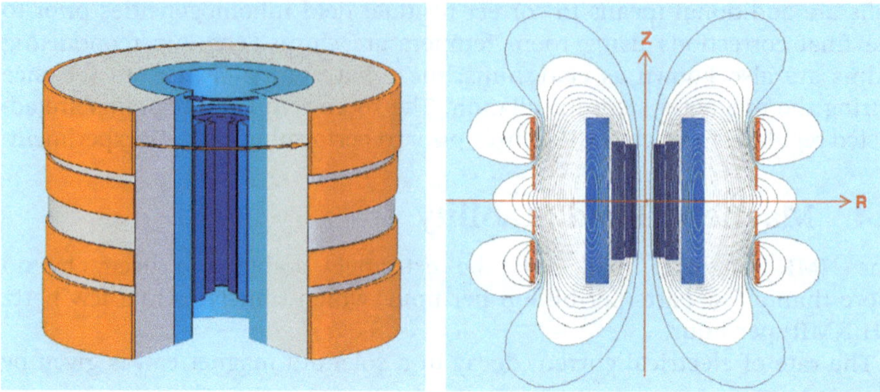

Figure 2.2 Schematic of coil arrangements in an actively shielded magnet. Left: the coils generating the main field are shown in blue and the shielding coils in red. Right: the field geometry resulting from the coil arrangement shown.

sections, colored in red, producing an opposing magnetic field in order to reduce the stray magnetic field. This is achieved by connecting the inner and outer coil sections in series but in such a way that the electrical current in the shielding sections flows in the opposite direction to the current in the primary sections.[3]

Over a period of 15 years, Bruker has developed three generations of actively shielded NMR magnets known as UltraShield, UltraShield Plus, and Ascend (Figure 2.3–2.5). Following the successful introduction of UltraShield magnets that had already reduced the 5 G stray magnetic field volume by 90% compared to non-shielded magnets, further advances in superconductors and magnet design enabled the development of second and third generations of actively-shielded magnets of reduced physical size and weight while reducing the stray magnetic fields even further. In fact, the 5 G line of most Ascend magnets these days is enclosed within the posts supporting the magnet.

An additional benefit of the Ascend magnets is the ability to site them in environments with external magnetic field disturbances, previously considered impossible for an NMR laboratory. Further details are presented in the next section.

2.6 Mitigating External Magnetic Field Disturbances

Older non-shielded magnets are based on a typical solenoid coil design that responds to an external magnetic field disturbance by a self-induced electrical current that opposes the external disturbance. This phenomenon gave such magnets a partial intrinsic insensitivity to external disturbances. The typical screening efficiency of non-shielded magnet coils against external disturbances is ~70%, hence the residual shift of the B_0 central field at the

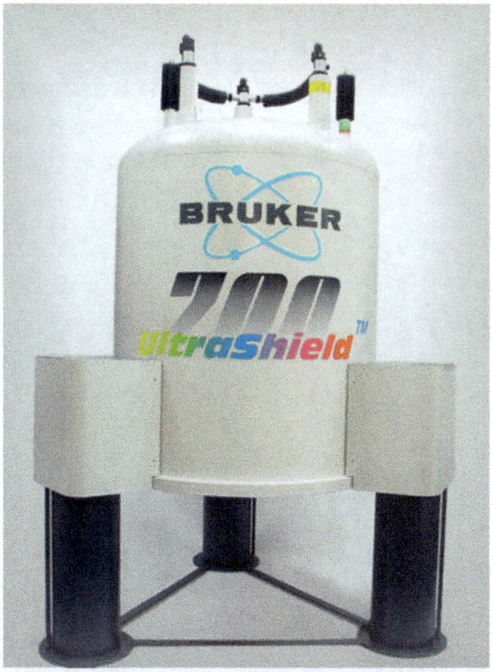

Figure 2.3 UltraShield, Bruker's first-generation actively shielded magnets.

magnet center (with the NMR lock off) would be $\sim$30% of the magnetic field disturbance outside the magnet.

However, the introduction of actively shielded magnets presented a new challenge given that the currents in the inner coil sections and the outer shielding sections circulate in opposing directions. As such, an actively shielded magnet based only on the basic design principle of reversed current sections to reduce the stray magnetic fields will not screen an external disturbance, unless it is fitted with an additional technology to reduce the disturbance.

Suppressing disturbances in actively shielded magnets, in addition to reducing the magnetic footprint of the magnets, is important since it allows these magnets to be sited in closer proximity to sources of external magnetic field disturbances such as outside traffic, elevators, and power lines.

This problem has been addressed with clinical magnetic resonance imaging (MRI) magnets at sites with large external magnetic field disturbances. Active compensation solutions using external Helmholtz coils built into the magnet room (typically on the RF shielding enclosure) have been used at some difficult MRI sites to mitigate external magnetic field interferences.

For NMR magnets, Bruker has developed a proprietary EDS technology for external disturbance suppression that is integrated within the NMR magnet coil system and does not require an outside power source, unlike the Helmholtz coils solution. As mentioned before, without particular measures,

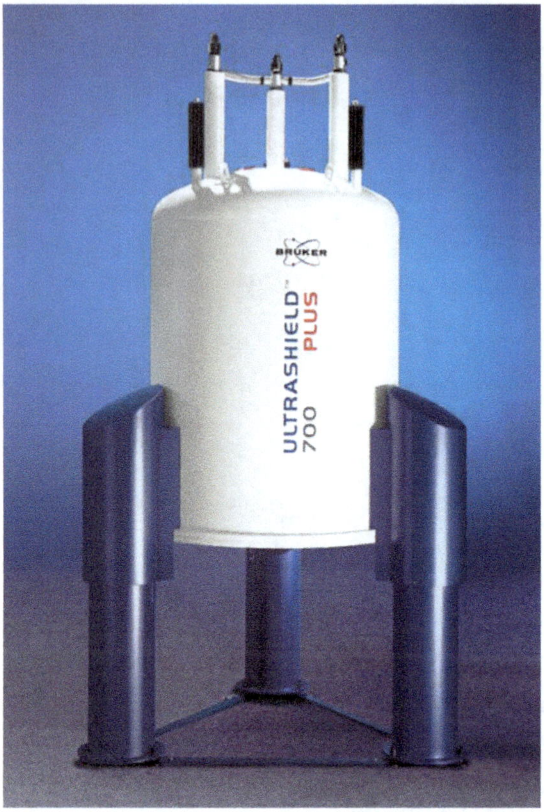

Figure 2.4 UltraShield Plus, Bruker's second-generation actively shielded magnets.

an actively shielded magnet would not have the ability to suppress external disturbances because the main coils and the shielding coils work against each other. The EDS technology circumvents this flaw by introducing additional current loops in the magnet coil system. Careful adjustment of the different current loops is necessary to achieve optimal performance.

The original introduction of EDS technology resulted in a screening efficiency of 90%. The latest generation of EDS (which is integrated in the Ascend magnet coils), however, is capable of suppressing both DC and AC external magnetic field disturbances, typically by 99%. This has allowed for successful installations of NMR systems at sites that have previously been considered extremely problematic, such as those in proximity to subway or tram lines where the level of electromagnetic disturbances is usually fairly high.[4]

2.7 Reducing the Physical Size and Weight

The developments in superconductor technologies leading to an increase in current-carrying capabilities have enabled higher operational currents

Figure 2.5 Ascend, Bruker's third-generation actively shielded magnets.

to be used, which in turn meant less wire and reduced coil mass. Along with novel magnet designs, these advances have contributed significantly to the manufacture of compact magnet systems of reduced physical size and weight for a given field strength compared with their predecessors.

The reduction in the physical size and weight of the magnets has provided siting flexibility benefits to NMR users in terms of a reduced physical footprint in the laboratory, less ceiling height clearance requirements, reduced floor loading, and less complex/costly rigging.

An additional key benefit of the reducing the magnet and cryostat size is the significant reduction in cryogen consumption because of the reduced radiation surface of smaller physical size magnets combined with less conductive heat load through the neck-tubes supporting a reduced weight of the cryostat's vessels.

The side-by-side comparison of the three generations of actively shielded 700 MHz magnets shown in Figure 2.6 illustrates the reduction in physical size, weight, and helium consumption.

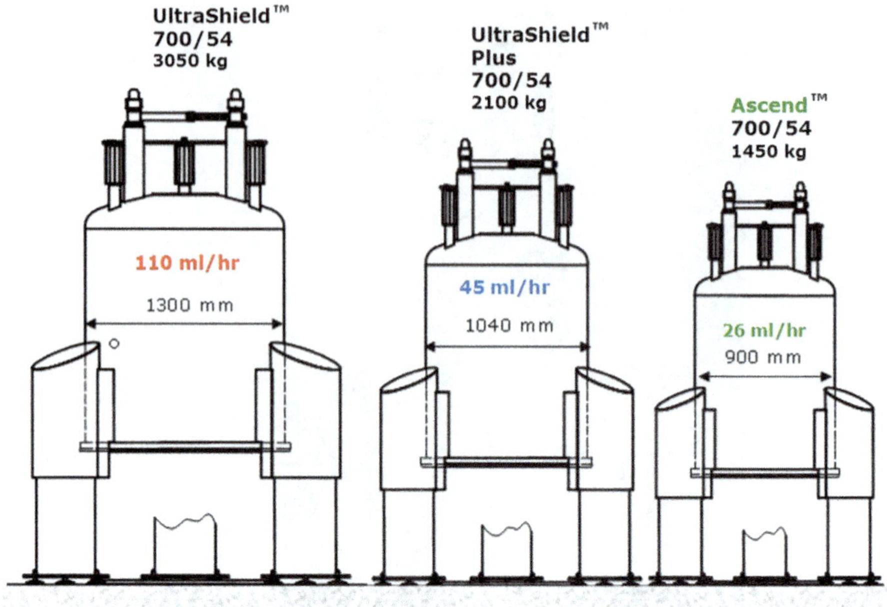

Figure 2.6 The comparison of three generations of actively shielded 700 MHz NMR magnets.

2.8 Cryogen Conservation and Future Outlook

In parallel with lowering the cryogen consumption by reducing the physical size and weight of magnets, there have been other developments of various refrigeration technologies aimed at conserving cryogens.

In the late 2000s, Bruker introduced a nitrogen liquefier as an optional accessory that could be added to any system with a CryoProbe and the latest generation of CryoPlatform without adding extra infrastructure. A few years later, Bruker also introduced a standalone nitrogen liquefier as an optional accessory that could be added to an Ascend or UltraShield Plus magnet that is not equipped with a CryoProbe. Both of these nitrogen liquefaction solutions eliminate the need for regular nitrogen fills, thus conserving liquid nitrogen and increasing the user convenience by eliminating the regular downtimes associated with nitrogen fills and reducing the overall requirements for regular maintenance.

Helium remains much more critical of the two types of cryogens used in traditional NMR magnets, given that the global helium supply has been decreasing and the costs have increased in recent years due to the shortage. Given that the future outlook for helium availability and costs does not look promising, the NMR community has begun seeking out helium conservation solutions.

Some NMR facilities have been able to take advantage of existing helium liquefiers associated with physics research laboratories and have implemented helium gas recovery lines in NMR cryostats. Such liquefier systems are capable

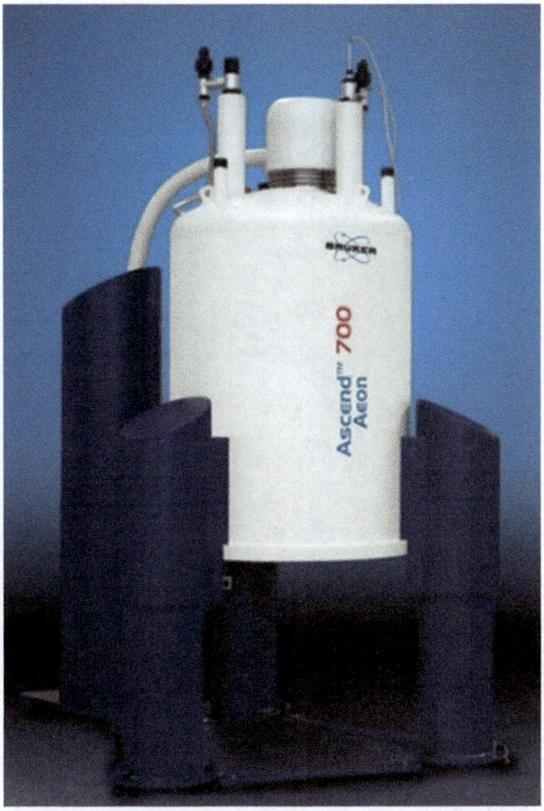

Figure 2.7 Bruker's Ascend Aeon 700 MHz NMR magnet.

of producing several hundred liters of helium per day but tend to be expensive in terms of infrastructure and associated costs. These systems require large helium gas storage balloons, a compressing station, compressed gas cylinder storage, and purifiers, ending with the actual liquefier.

Some cryogenic companies have recently started to offer smaller helium liquefiers rated to produce 22 liters or even less per day. Although these may be suitable for NMR laboratories with multiple systems, the same infrastructure chain (*i.e.* gas bags, compressors, gas cylinders, purifier, and liquefier) would still be required because of the high losses experienced during the helium refills and the need to capture the helium gas during these periodic events.

Bruker has been very active in developing an integrated and active refrigeration technology for NMR magnets. Although such a technology has been available since the early 2000s for their horizontal bore superconducting magnets for MRI and Fourier transform mass spectrometry (FTMS), it has been difficult to apply this technology to vertical bore NMR magnets owing to their higher susceptibility to vibrations, which may cause artifacts in NMR spectra. Resolving the vibration issues has been an engineering challenge that took many years of development before Bruker was ready to introduce its Ascend Aeon magnet product line in 2013 (Figure 2.7 and 2.8).

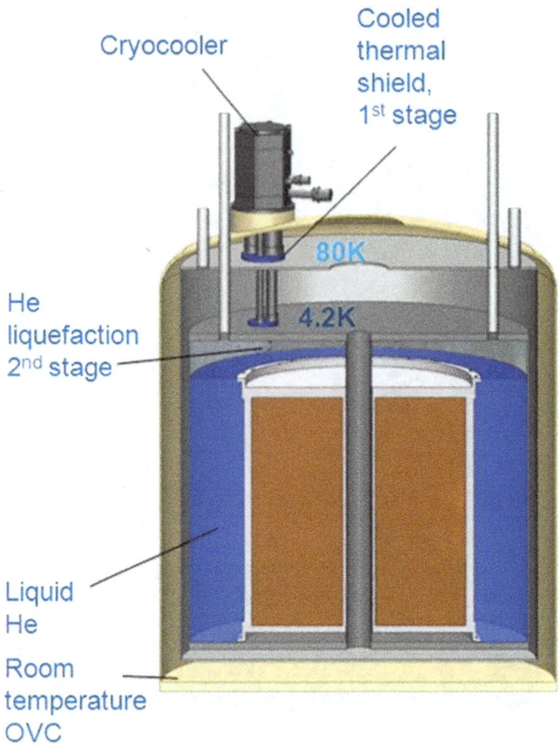

Figure 2.8 Cross-section through a Bruker Ascend Aeon magnet with a two-stage cryocooler.

The new Ascend Aeon magnets feature integrated active refrigeration using pulse-tube cryocoolers, are nitrogen free, and permit long-term care-free operation without the need for user maintenance and without compromising the NMR performance (Figure 2.9).

The pulse-tube cryocooler is a closed-loop device with an oscillating helium gas pressure at one end generating an oscillating helium gas flow in the rest of the system, which in turn removes heat from the point being cooled (also known as the cooling stage).

Pulse-tube cryocoolers are well suited to NMR magnets because they do not rely on moving parts at the cold end, thus reducing the vibration levels, which is essential for NMR spectral quality. At the same time, these systems are more robust and reliable, with regular maintenance needed every 2 years, compared with traditional Gifford–McMahon (G-M) cryocoolers, which require annual maintenance.

In summary, two priorities will likely continue to define the future of NMR magnet development. One is the development of ultra-high-field NMR magnets beyond 1 GHz using insert coils wound with superconducting materials of higher critical fields compared with the traditional Nb_3Sn superconductors. The other area is expanding the active refrigerated

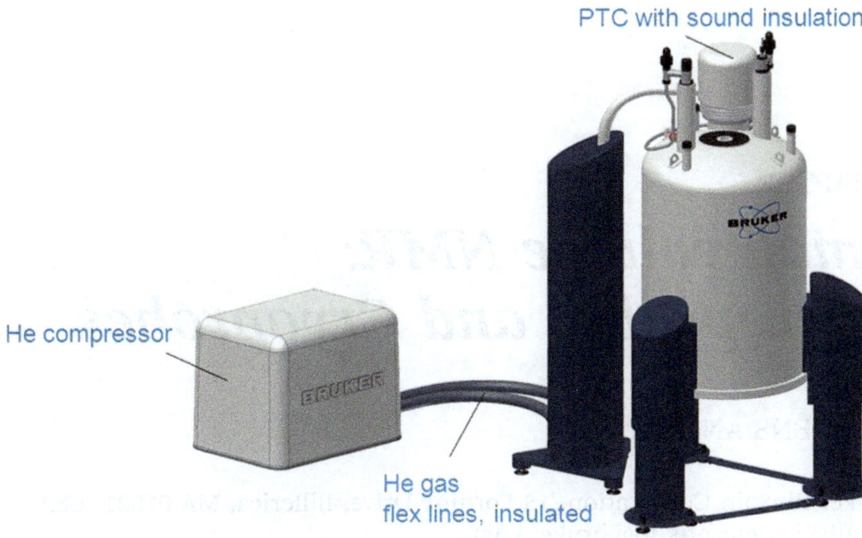

Figure 2.9 Complete active refrigeration system including the NMR magnet, pulse-tube cryocooler (PTC), He gas lines, and He compressor.

magnet product line and further perfecting the technology to improve its efficiency.

Acknowledgements

I would like to acknowledge the contributions of several of my Bruker colleagues with whom I have interacted and worked over many years: Robert Schauwecker, Daniel Baumann, Pierre-Alain Bovier, Riccardo Tediosi, Agnès Glémot, Daniel Eckert, René Jeker, Claus Hanebeck, Gerhard Roth, and Werner Maas.

References

1. D. D. Laukien and W. H. Tschopp, *Concepts Magn. Reson.*, 1993, **6**, 255.
2. G. Roth, *Bruker Spin Report*, 2003, **152/153**, 14.
3. G. Roth, *Bruker Spin Report*, 2005, **156**, 33.
4. R. Teodorescu, D. Baumann, J. Guo and A. Makriyannis, *ENC Poster Session*, 2007, **140**.

CHAPTER 3

Small-volume NMR: Microprobes and Cryoprobes

CLEMENS ANKLIN

Bruker BioSpin Corporation, 15 Fortune Drive, Billerica, MA 01821, USA
Email: clemens.anklin@bruker.com

3.1 Introduction

Generous quantities of material have rarely been available to the natural product chemist, and this limitation often made the acquisition of NMR data an extremely difficult and time-consuming task. From the early days of NMR spectroscopy, attempts were made to optimize the instrumentation and the experimental conditions in a way that would guarantee the highest sensitivity for these mass-limited samples. Diluting a small amount of a precious compound in a large volume of solvent is certainly not a way to achieve optimum results. With this in mind, a variety of ways to limit the sample volume have been introduced over the years. However, for data collection, most of these reduced-volume samples were still inserted into standard 5 mm NMR probes. It was not until the early 1990s that smaller diameter probes became widely available in the form of the Nalorac 3 mm probe.[1,2] This probe was followed a few years later with the 1.7 mm probe from Nalorac.[3] In 2003, Bruker introduced a 1 mm conventional probe and 3 years later, at the 2006 Experimental NMR Conference (ENC), reintroduced the 1.7 mm probes to the NMR community. The most significant advance in sensitivity came in the late 1990s with the commercial introduction of cryogenically cooled probes. The first models were presented in 1995 at the

Modern NMR Approaches To The Structure Elucidation of Natural Products,
Volume 1: Instrumentation and Software
Edited by Antony J. Williams, Gary E. Martin and David Rovnyak

Published by the Royal Society of Chemistry, www.rsc.org

ENC in Boston. The first installations in customer laboratories followed in 1999. Initially, the majority of cryogenically cooled probes were used for biomolecular NMR. However, natural products chemists quickly discovered the benefits of these probes and used them to examine ever smaller quantities. The introduction of cryogenically cooled probes optimized for ^{13}C observation was a welcome addition to the tools of the NMR spectroscopist. The development of a 1 mm cryogenic probe using high-temperature superconducting materials for its coils[4] at the University of Florida in 2006 was followed in 2008 by the commercial introduction of the 1.7 mm cryogenically cooled probe. This led to new horizons in sensitivity of NMR probes. The sensitivity is equal to that of a conventional 5 mm probe at less than 1/15th of the volume. Early applications were again based in protein NMR or protein–ligand screening, but shortly thereafter this probe was used with great success in the NMR analysis of natural products. This highest sensitivity probe allowed the easy collection of NMR data on quantities of 1 nmol of material or less.[5] At the 2011 ENC, Bruker BioSpin introduced the first probe in which the rf coil was cooled by liquid nitrogen. Owing to the lower costs of purchase and operation, this probe is likely to make cryogenic technology available to a wider circle of spectroscopists.

3.2 Theoretical and Practical Aspects of Small-volume Probes

The overall sensitivity of the NMR experiment has increased by well over three orders of magnitude in the 60 years since the early days of NMR spectroscopy. Using the signal of a sample of 0.1% ethylbenzene in deuterated chloroform as a reference, a Fourier transform NMR spectrometer in the early 1960s would produce a signal-to-noise ratio (S/N) of just over 10 : 1. A modern 900 MHz instrument equipped with a cryogenically cooled probe would provide an S/N of over 10 000 : 1. This increase can be attributed to several major factors. NMR probe design is one important aspect and will be discussed in more detail below. The other important factors are magnetic field strength, radiofrequency technology and digital signal processing. It is generally accepted that the sensitivity of the NMR experiment is proportional to the 3/2 power of the increase in field strength.[6,7] On doubling the field strength, this results in a factor of approximately 2.8 increase in S/N. For the factor of 10 increase in field strength from 90 to 900 MHz, this would result in a gain of a factor of over 30.

Improvements in the electronics of NMR spectrometers have led to further gains in sensitivity. Better, more advanced components, miniaturization, receivers with lower noise figures and higher dynamic range and also the introduction of digital signal processing[8] are contributors to these gains. These gains are reflected in the comparison of quoted S/N values for 0.1% ethylbenzene over the years at a constant field. Starting at an S/N of 180 : 1 in 1979 when 500 MHz spectrometers were introduced, and reaching 900 : 1 for a modern instrument, this corresponds to a fivefold increase. A proportion of the improvement has to be attributed to probe design, but a significant part results from improved console hardware. Another important increase in overall sensitivity of NMR is the result of newly introduced experiments. For direct detected heteronuclei, the introduction of polarization transfer experiments such as INEPT[9] and DEPT[10] delivered a sizeable increase in sensitivity. In 2D-NMR, the inverse-detected experiments HMQC,[11] HSQC,[12] and HMBC[13] resulted in the largest increases in sensitivity. Last but not least, the use of pulsed-field gradients[14–20] led to, amongst other approaches, a reduction in artifacts, which also helped to lower the detection limits.

In addition to the factors mentioned above, the characteristics and the design of the NMR probe are important factors in determining the overall sensitivity of the spectrometer. The S/N of an NMR probe can be described by the following equations:[21]

$$\mathrm{S/N} = \frac{k_0 (B_1/i) V_\mathrm{s} N_\mathrm{s} \gamma (h/2\pi)^2 I(I+1)({\omega_0}^2)}{3\sqrt{2} V_\mathrm{noise} k_\mathrm{B} T} \tag{3.1}$$

$$\mathrm{S/N} \propto \frac{{\omega_0}^2 (B_1/i) V_\mathrm{s}}{V_\mathrm{noise}} \tag{3.2}$$

According to these equations, the signal should show a dependence that is proportional to the square of the Larmor frequency ${\omega_0}^2$, but the field dependence of the noise, as shown in eqn (3.3),[22] counteracts this. The noise voltage V_noise is given by

$$V_\mathrm{noise} = \sqrt{4 k_\mathrm{b} T_\mathrm{c} R_\mathrm{noise} \Delta f} \tag{3.3}$$

where R_noise is the resistance of the entire probe circuit:

$$R_\mathrm{noise} = \frac{1}{p}\sqrt{\frac{\mu \mu_0 \omega_0 \rho(T_\mathrm{c})}{2}} \tag{3.4}$$

Eqn (3.1) and (3.2) show that the S/N is proportional to the sample volume V_s, the number of spins N_s, the spin quantum number I, and the gyromagnetic ratio γ, with the term B_1/i described below along with other factors. These equations also show the inverse proportional relationship of sensitivity and temperature. We will see later that this is exploited in cryogenically cooled probes.

The NMR S/N is proportional to B_1/i, the magnitude of the magnetic field induced in the coil per unit current. The induced magnetic field is dependent on the coil geometry as described in eqn (3.5) for a saddle or Helmholtz coil,[23]

$$\frac{B_1}{i} = \frac{n \mu_0 \sqrt{3}}{\pi}\left[\frac{2dh}{(d^2+h^2)^{3/2}} + \frac{2h}{d\sqrt{d^2+h^2}}\right] \tag{3.5}$$

where n is the number of turns, d is the diameter, and h is the height of the coil. With everything else constant, the induced magnetic field would increase as a function of $1/d$. This inverse proportional relationship between the diameter and the intrinsic sensitivity was the driving force leading to the development of small-diameter NMR probes. The S/N of solenoid coils has a different dependence on their geometry.

As shown in eqn (3.3) and (3.4), the temperature dependences of the noise itself and resistance of the coil are the major contributors to the higher sensitivity of cryogenically cooled probes. An additional reduction in noise level is also achieved by cooling the preamplifier circuit. These gains are offset by a slightly lower filling factor due to the required insulation between the NMR sample and the cold coils. At coil temperatures of ~20 K and preamplifier temperatures of ~77 K, an overall increase of a factor of 4–6 over conventional probes can be obtained. The S/N for a cryogenic probe can also be expressed as

$$\mathrm{S/N} \propto \frac{1}{\sqrt{1 + \alpha \dfrac{R_s}{R_c}}} \tag{3.6}$$

where R_s is the resistance of the sample, R_c is the resistance of the coil and α is a function of the noise temperatures of the coil (T_c), the sample (T_s) and the preamplifier (T_a) according to the equation (Figure 3.1)

$$\alpha \propto \frac{T_s + T_a}{T_c + T_a} \tag{3.7}$$

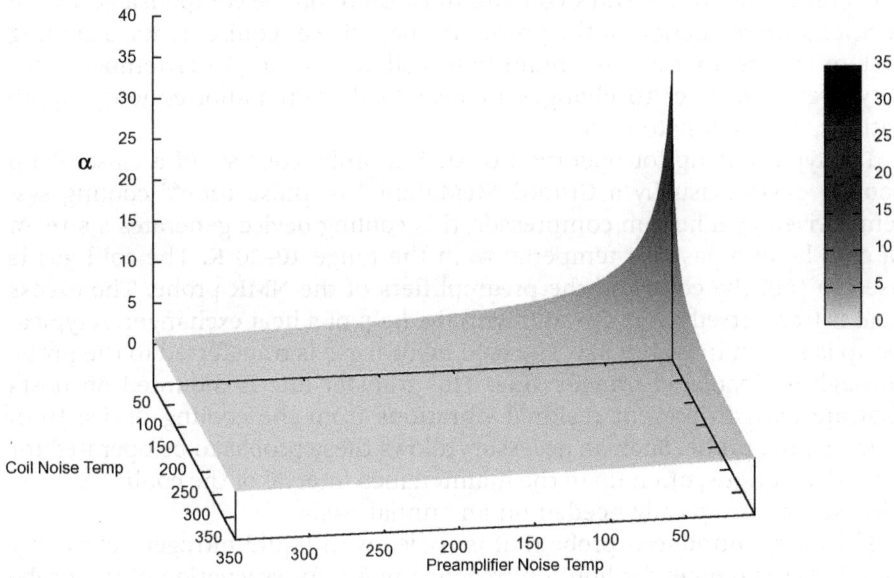

Figure 3.1 Dependence of α in eqn (3.7) as a function of preamplifier noise, temperature, and coil noise temperature.

According to eqn (3.7), only the cooling of both the preamplifier and the coil to very low temperatures leads to significant gains in sensitivity. For a room-temperature probe α is near 1, but for a cryogenically cooled probe it is about 7.8. This results from a coil noise temperature T_c of near 20 K and a preamplifier noise temperature T_a of ~15 K. Narrow-band low-noise preamplifiers cooled to 77 K have a noise figure near 0.1 dB. According to the equation

$$T_a = 290\left[10^{(NF/10)-1}\right] \tag{3.8}$$

this results in a noise temperature T_a of 15 K or less. Cooling just the coil leads to a value of α of only 3.8 when using a preamplifier with a noise figure of 1 dB or better. Cooling both the coil and the preamplifier to liquid nitrogen temperature also leads to a value of $\alpha > 3.5$ or an overall sensitivity enhancement of a factor of 2 or more. Bruker introduced just such a probe at the ENC in 2011.

The technical realization of keeping the coil at ~20 K, the preamplifier at 77 K, and the sample at room temperature poses an engineering challenge. These temperatures have to be maintained in a very narrow range to guarantee the stability required for advanced NMR experiments. Deviations of a few tenths of a degree can result in degradation of the spectroscopic performance. To optimize the filling factor, the insulation is only a few millimeters thick and in many cases the samples would not stay at room temperature unless actively kept warm by the variable temperature control accessory of the spectrometer. Applying radiofrequency currents can alter the temperature of the NMR coil and must therefore be compensated by the temperature regulation of the probe. To meet these requirements, a cooling system is used that can maintain both coil and preamplifier temperatures very precisely, react to changes in heat load from radiofrequency applications, and is easy to operate.

The typical setup for operation of such a probe consists of a closed-loop cooling device, usually a Gifford–McMahon[24] or pulse–tube[25] cooling system. Driven by a helium compressor, this cooling device generates a stream of cold helium gas at a temperature in the range 10–20 K. The cold gas is used to cool the coils and the preamplifiers of the NMR probe. The excess heat is transferred to air or water with the help of a heat exchanger. A typical setup is shown in Figure 3.2. The cold helium gas is transferred to the probe through an insulated transfer line. This transfer line is mounted on posts that are used to prevent residual vibrations from the cooling device from reaching the probe. Such an accessory allows these probes to be operated for extended periods, often up to the maintenance interval of the cooling device. This service is typically needed on an annual basis.

The newly introduced probe that is cooled with liquid nitrogen relies only on a larger container for liquid nitrogen, a pump for evacuation of the probe body, and control electronics. The lower sensitivity, compared with cryoprobes cooled with helium, is balanced by lower operating and maintenance costs.

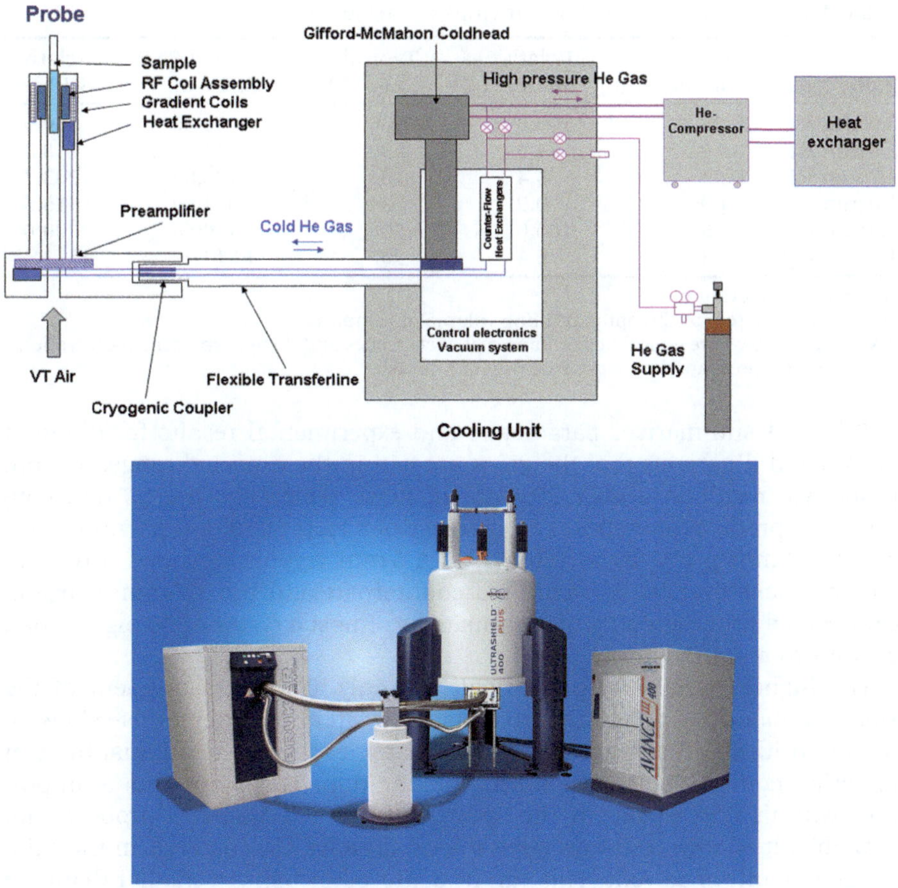

Figure 3.2 Top: schematic setup for operation of a cryogenically cooled NMR probe. Bottom: typical setup of an instrument.

The comparison of the performances of different probes is a confusing subject and requires a clear definition of the sensitivity. NMR sensitivity can be defined in terms of mass or concentration sensitivity. Concentration sensitivity compares S/Ns obtained for samples of equal concentration whereas mass sensitivity bases the comparison on results obtained with a constant mass of material. When comparing concentration sensitivities, the results have to be scaled based on either total or active volume of the probes. This can be a difficult task as information about the coil height is not always readily available. For example, a result of an S/N of 1200 : 1 for 0.1% ethylbenzene obtained in a 5 mm tube with ~0.6 mL of sample is equivalent to 400 : 1 in a 3 mm tube as the volume is only 0.2 mL assuming equal coil height. Probes with very small coil diameters typically also have shorter coil lengths. When measuring the 3 mm sample in a 3 mm probe, one can expect an S/N of over 500 : 1 owing to the increased mass sensitivity of this probe.

Table 3.1 Comparison of the sensitivities of various probes.

Probe type[a]	Total volume[b] (mL)	Relative volume (%)	Typical SNR[c] at 500 MHz	Scaled SNR[d] SNR/vol.	Relative SNR (%)
5.0 mm RT	0.55	100.0	900	900.0	100.0
3.0 mm RT	0.19	34.5	430	1244.7	138.3
1.7 mm RT	0.03	5.4	100	1833.3	203.7
1.0 mm	0.005	0.9	34	3740.0	415.5
5.0 mm cryo	0.55	100.0	4500	4500.0	500.0
1.7 mm cryo	0.035	6.4	900	14 142.9	1571.4

[a]RT, room temperature.
[b]Total volume refers to the optimal filling volume of samples.
[c]SNR values shown are typical performances of such probes and do not represent specifications.
[d]Scaled SNR represents the mass sensitivity of a probe.

Table 3.1 summarizes parameters and experimental results for different probes and illustrates that the gains are within the expected range. A 3 mm probe will result in a 40% increase in mass sensitivity, the 1.7 mm conventional probe demonstrates an increase of a factor of 2 over a 5 mm probe and the 1 mm probe by a factor of 4. A 5 mm cryoprobe shows a fivefold enhancement over its conventional counterpart and the 1.7 mm cryogenically cooled probe shows a mass sensitivity almost a factor of 20 higher than a 5 mm room-temperature probe.

The highest sensitivity probes will not only allow measurement of the smallest quantities of natural products; they will unfortunately also show all the impurities and contaminants. NMR samples of very small quantities of material require special care during preparation. Trace amounts of impurities will also be visible in the spectra. If a small-diameter probe is not available, it is almost always advisable to dissolve the material in the minimum amount of solvent. This can be achieved by using a smaller diameter tube in a larger probe. Using a tube with 3 mm outer diameter will reduce the required volume of the solution from typically 0.6 to 0.2 mL and consequently increase the concentration of the solute by a factor of 3. This also means that the ratio of solute to contaminants also improves by the same factor, at least for those introduced by the solvent or liquid handling process. On progressing to 1.7 mm diameter tubes, the volume is now reduced to about 6–7% of the original volume. Again, undesired signals, such as solvent, water, and contaminants, are further reduced and the overall S/N is improved by ~50%. In Figure 3.3, the contaminant signals around 1.0 ppm are clearly visible in the top spectrum whereas they almost disappear in the noise in the bottom spectrum. The ^{13}C satellites of the residual solvent signal also clearly exceed the solute signals in the more dilute 5 mm sample. For particularly small sample amounts, the ^{13}C satellites of the solvent signals can exceed the signals of the sample even in 1.7 mm probes. ^{13}C decoupling of proton spectra is occasionally used in these cases to avoid overlap of the ^{13}C satellites with the solute signals.

Reducing the sample volume in the same diameter probe usually does not lead to a significant improvement of the S/N but the relative intensity of the

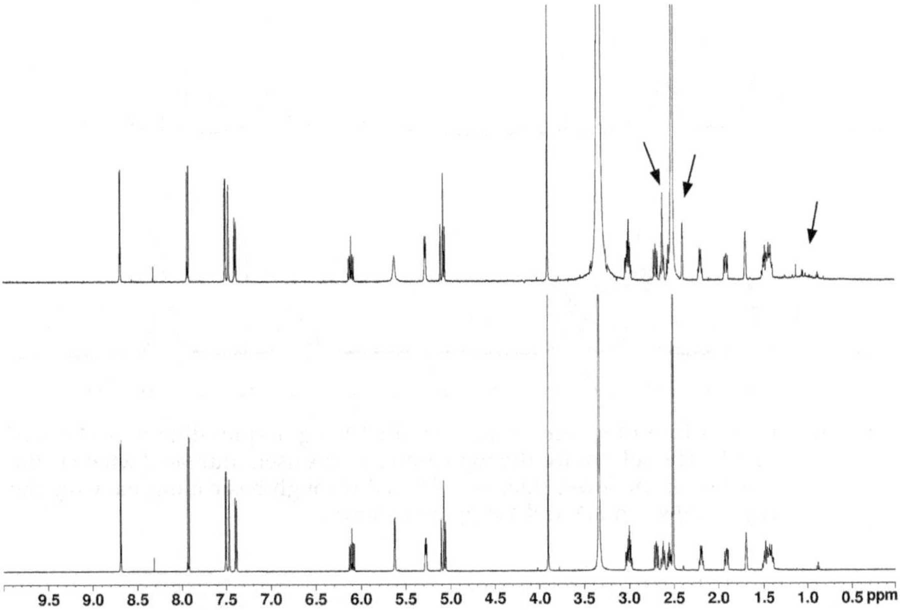

Figure 3.3 100 μg of quinidine in 0.55 mL of DMSO-d_6 in a 5 mm tube (top) compared with the same quantity in 0.035 mL of DMSO-d_6 in a 1.7 mm tube (bottom). Contaminants and ^{13}C satellite signals of DMSO-d_6 are indicated by arrows.

contaminant signals is greatly reduced. Of course, it is advisable to avoid the introduction of contaminants in the first place. When working with low quantities and volumes, it is essential to use the highest purity solvents. In many cases, it is recommended to use a fresh batch of solvent. Most solvent vendors offer small quantities of solvents in ampoules. Special care has to be taken with sensitive compounds. Acid-sensitive materials can easily be affected by the hydrochloric acid present in $CDCl_3$. Filtration of the solvent through basic alumina will typically remove the acid. As an example, spectra of 100 μg of quinidine in 0.55 mL of $CDCl_3$ are shown in Figure 3.4.

The effect of traces of HCl are clearly visible in the top spectrum. CD_2Cl_2 can be used as an alternative solvent, as it does not develop HCl upon storage. Other critical solvents include DMSO-d_6, whose water content can cause problems. DMSO is very hygroscopic and will absorb atmospheric water fairly quickly. Solvent or samples left open to the air will show an ever-increasing water peak with time. Some sources of DMSO-d_6 show elevated D_2O contents that are used to mask high water contents. Should a high water content be suspected, it is worth acquiring both proton and deuterium NMR spectra of the pure solvent to establish the overall H_2O and D_2O content.

NMR tubes are usually very clean when new and rinsing new tubes is, in most cases, not necessary. On the other hand, it is advisable to keep the outside of an NMR tube clean at all times. Touching the part of the tube

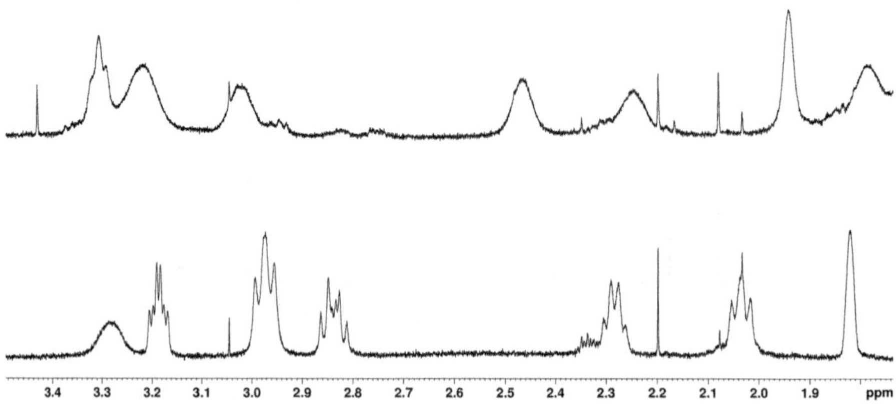

Figure 3.4 Comparison of spectra acquired with 100 μg of quinidine in 0.55 mL of $CDCl_3$. The solvent for the top spectrum was used untreated whereas the solvent for the lower trace was filtered through basic alumina. Only the region between 3.5 and 1.8 ppm is shown.

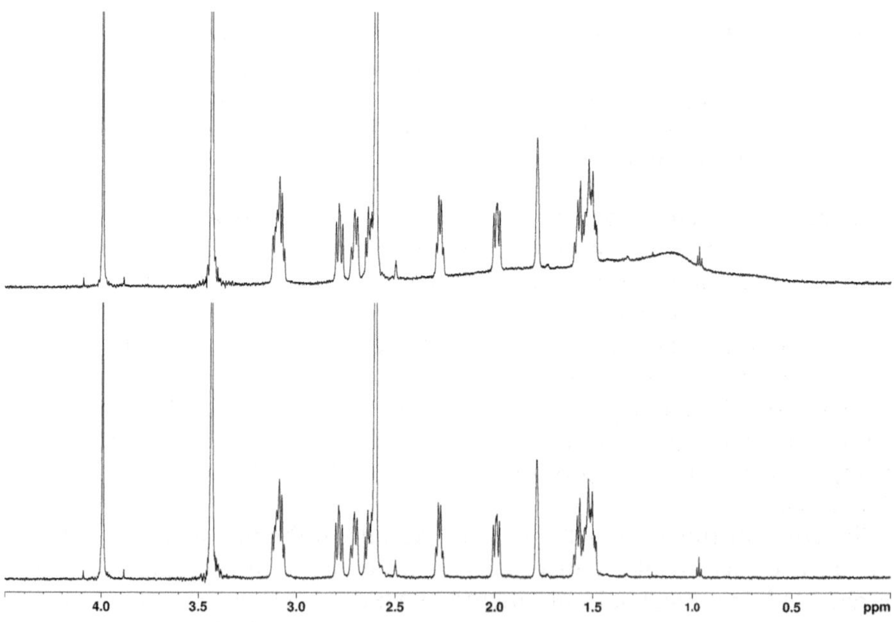

Figure 3.5 Contamination of sample with a fingerprint. The top spectrum is the data obtained with 100 μg of quinidine in a 1.7 mm tube where the tube has been touched. The bottom spectrum shows the same sample after cleaning of the tube. The broad signal between 2.5 and 0.5 ppm originates from the lipids in the fingerprint.

where the sample is contained can lead to clearly visible signals in a spectrum. Typically, these signals present themselves in the aliphatic region as they originate from lipids left on the tube in the fingerprints (Figure 3.5).

3.3 Conventional Small-volume Probes

Before small-volume or micro NMR probes became routinely available, many different approaches for limitation of the sample volume were utilized. The earliest forms included the use of cylindrical or spherical sample cells inserted in a 5 mm tube or the vertical limitation of the sample volume with plugs.

Figure 3.6 shows a collection of such sampling cells. These modes of volume limitation suffered from bad lineshape and resolution due to the susceptibility effects introduced by the materials. Shimming was very difficult until susceptibility-matched materials were used. Based on an idea by Zens,[26] Doty introduced a series of susceptibility-matched plugs for 3 mm, 5 mm, and larger NMR tubes. The use of materials matched to the susceptibility of different common NMR solvents allowed the restriction of the sample volume to equal to or less than the active volume of the probe, thus reducing the required solvent volume by 50–70%. After many years of success in the field of protein NMR with the tubes matched for water, Shigemi also introduced tubes matched for $CDCl_3$, DMSO, and MeOD, solvents more commonly used in the NMR of natural products. These tubes are available in

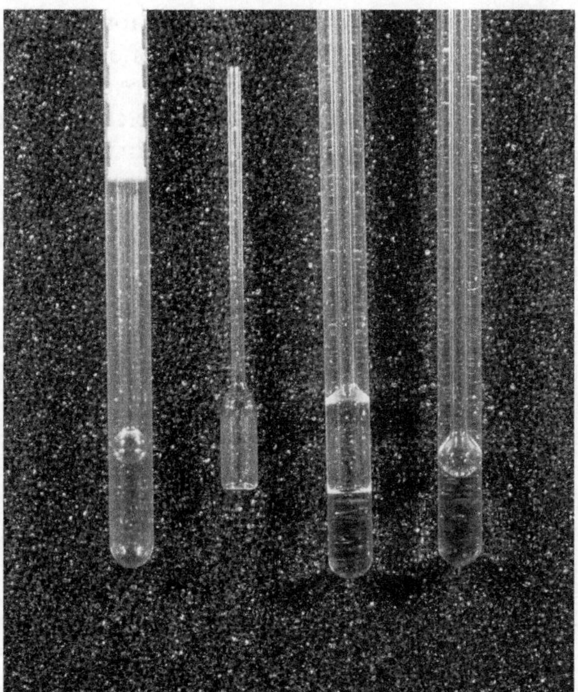

Figure 3.6 Sampling cells for small volumes. From the left: cylindrical insert in 5 mm tube, showing Teflon holder; cylindrical insert for 5 mm tube; 5 mm tube with cylindrical cavity; and 5 mm tube with spherical cavity. All tubes from Wilmad Glass.

a variety of diameters and, in the same way as the matched plugs, allow restriction of the sample volume to less than the active volume of the coil without detrimental effects on lineshape and resolution. The use of these devices, for example, allows the easy restriction of the volume from the typical 0.6 mL of a 5 mm tube to approximately 70 μL when using a 3 mm tube with the volume matched to the coil size. This corresponds to a sevenfold increase in concentration when using the same mass.

The simplest mode of volume limitation is the use of tubes with a smaller diameter than the probe. Without vertical sample limitation, this method does not offer the optimal volume limitation. For example, a 3 mm tube in a 5 mm probe requires only ~180 μL compared with the 550–600 μL for a 5 mm tube. These samples are typically much easier to shim. The radial inhomogeneity contributions disappear with smaller tube diameters in a larger probe. The use of smaller diameter tubes in a larger probe also allows the use of a shorter solvent column. Whereas a 5 mm tube has to be filled to ~40 mm, a 3 mm tube can be shimmed well with only a 30 mm filling height.

Smaller diameter probes were available for iron magnets for many years, as winding a smaller solenoid coil is far easier than the design of a Helmholtz or saddle coil. However, at the fields typically available for iron magnets, the overall sensitivity was not sufficient for NMR experiments on very small quantities. Several manufacturers offered smaller diameter probes for superconducting magnets, but these were often hard to shim and difficult to use. Probes with 2.5 mm sample diameter in indirect and direct observe configurations were available from Bruker in the late 1980s and early 1990s. They found limited acceptance in the NMR community with one of the main reasons being the tubes that were available. Whereas 5 and 3 mm tubes were available in a standard 7 in length, the 2.5 mm tubes were constructed as a 5 mm tube tapered down to 2.5 mm in the bottom part. These tubes never reached the quality of the 5 or 3 mm tubes and led to inferior results.

The development of a high-quality, easy to use, small-diameter conventional NMR probe started in the early 1990s with a 3 mm inverse-detection probe built by Nalorac. This probe was launched in 1992 and the first experimental results with this probe were published by Crouch and Martin.[1,2] Typically, these probes displayed about a 30–40% increase in mass sensitivity, leading to a factor of almost 1.7 in time savings. Later, a probe for direct ^{13}C observation was also developed with the same sample diameter. The success of the 3 mm probe prompted the development of probes with further reduced sample diameters and led in 1998 to the introduction of the 1.7 mm sub-micro NMR probes, again by Nalorac and in collaboration with Martin *et al.*[3]

In 2001, Bruker introduced to the NMR market a 1 mm proton observe probe with ^{13}C and ^{15}N decoupling. Although the initial use of this probe was for the screening of large compound libraries in the pharmaceutical industry,[27] micro-scale protein structure determination,[28] and in metabolomics,[29] the general utility of this probe for many types of mass-limited

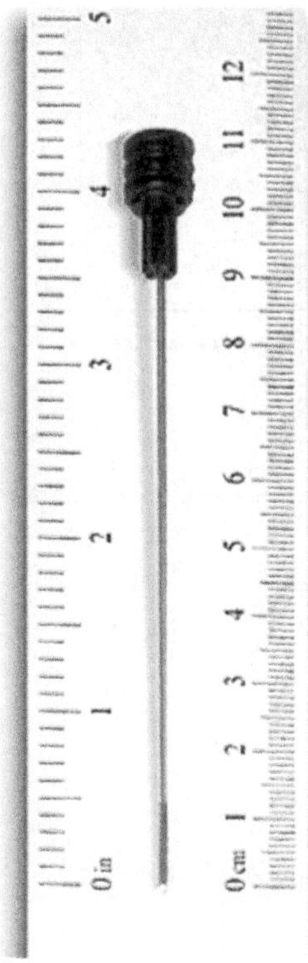

Figure 3.7 Tube for Bruker 1 mm microprobe. Total sample volume is 5 μL.

samples was quickly recognized. This probe only requires 5 μL of solution (Figure 3.7) and provides a mass sensitivity that is a factor of 4 higher than that of a standard 5 mm probe. In other words, the performance is comparable to that of cryogenically cooled probes with 5 mm diameter.

The 1 mm probe permits the collection of all essential experiments for the elucidation of the structure of natural products when only a few tens of micrograms are available. The spectra in Figure 3.8 were acquired in a couple of hours on a sample of 25 μg of jaspamide at 700 MHz.

Simple 1D spectra and homonuclear 2D spectra of a compound can be acquired on even smaller quantities, such as the spectra of 0.852 μg or 1 nmol of taxol shown in Figure 3.9.

In 2006, Bruker reintroduced the 1.7 mm probe to the NMR market, responding to multiple requests for this probe diameter. This probe offered an added benefit over the previous version from Nalorac as, in addition to the expected high proton sensitivity, it also provides a high carbon sensitivity. With an optimal volume of 30–35 μL, this probe allows the measurement of carbon spectra on only a few hundred micrograms of material. Compared with the popular 5 mm multinuclear probes, this probe provides a gain in mass sensitivity for carbon of a factor of 1.3–1.5. This is only surpassed by the performance of carbon-optimized cryogenically cooled probes.

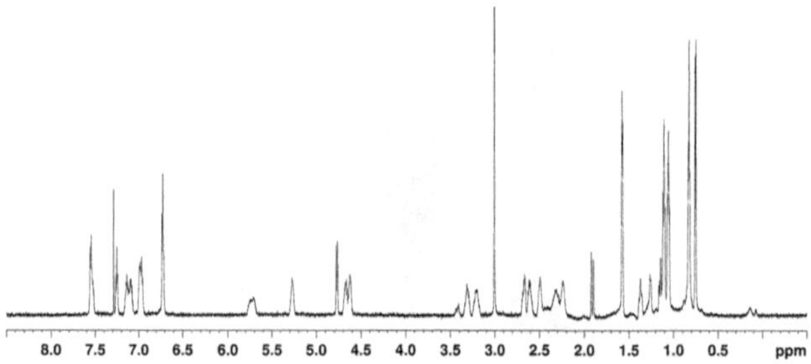
8.0 7.5 7.0 6.5 6.0 5.5 5.0 4.5 4.0 3.5 3.0 2.5 2.0 1.5 1.0 0.5 ppm

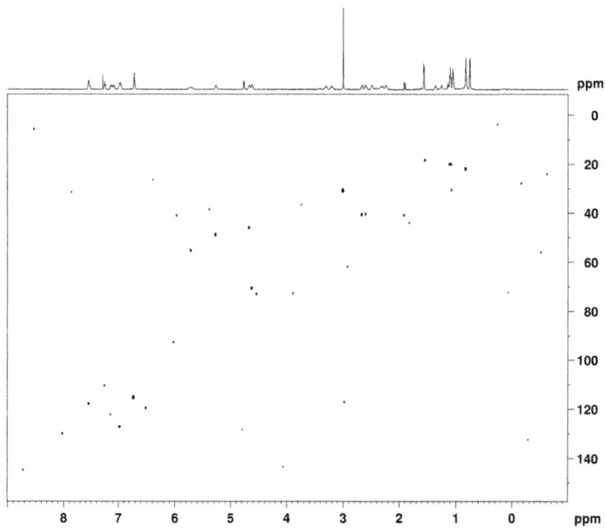
ppm
0
20
40
60
80
100
120
140
8 7 6 5 4 3 2 1 0 ppm

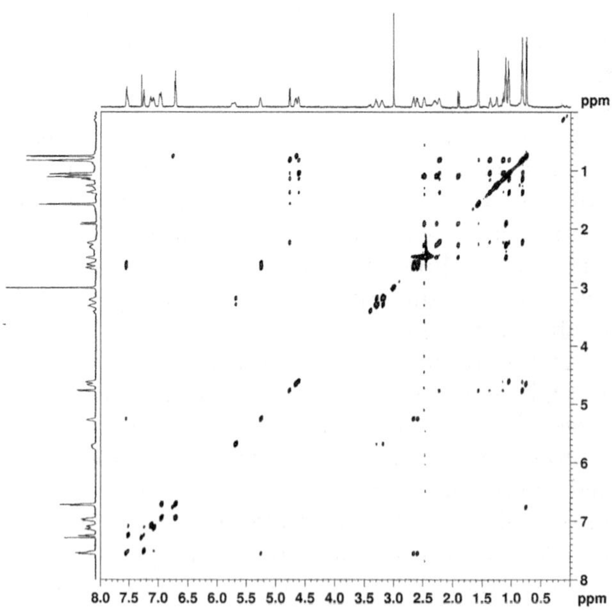
ppm
1
2
3
4
5
6
7
8
8.0 7.5 7.0 6.5 6.0 5.5 5.0 4.5 4.0 3.5 3.0 2.5 2.0 1.5 1.0 0.5 ppm

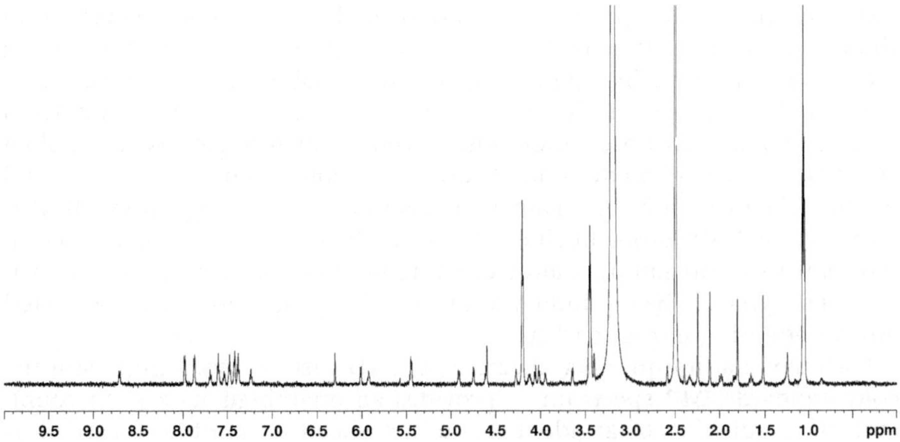

Figure 3.9 1D proton spectrum of 1 nmol (0.852 μg in 5 μL) of taxol in DMSO-d_6, 700 MHz 1 mm triple resonance microprobe, 1024 scans, 1.5 h experimental time.

A different approach for small-volume probes was taken by Varian with the Nano Probe.[30,31] In this probe, a sample with a volume of 40 μL is aligned along the magic angle and rotated at speeds exceeding 2 kHz. This eliminates susceptibility effects from the limited sample volume.[32] In addition to the use of small sample amounts with these probes, a similar development derived from solid-state probes termed HR-MAS found use with samples coming from solid-phase synthesis.[33,34] The addition of a pulsed-field gradient[35] made these probes much more useful for all of the above applications.

3.4 Cryogenically Cooled Small-volume Probe

Some of the earliest work on cryogenically cooled probes[36] dates back to the first half of the 1980s. The commercial development of cryogenically cooled probes began in the first half of the 1990s and culminated with the first introduction of these probes at the 36th ENC in Boston in 1995. This development was carried out independently in two collaborations, one between Conductus and Varian and the other between General Electric Central Research and Development Laboratories and Bruker. The Conductus/Varian probe was based on high-temperature superconducting materials and the GE-CRD/Bruker probe used cold-wire technology. Both of these probes were

Figure 3.8 Top: 1D proton spectrum of 25 μg of jaspamide in $CDCl_3$ acquired at 700 MHz with a 1 mm HCN triple resonance probe. 128 scans, 7.5 min experimental time. Middle: multiplicity-edited HSQC spectrum of 25 μg of jaspamide in $CDCl_3$ acquired in 4 h with 96 scans and 100 complex increments. Bottom: TOCSY spectrum of 25 μg of jaspamide in $CDCl_3$ acquired in 2 h with a 100 ms mixing time, 32 scans and 100 complex increments.

dedicated probes for proton observation. Only a few basic spectra were shown at the time. It took another 4 years for commercialization of the probes and the first installations in customer laboratories to take place. These first probes were designed as triple resonance inverse probes for 5 mm samples for proton observation with carbon and nitrogen as decoupling channels. They were predominantly used in biomolecular NMR, and natural product chemists only later gained access to this technology. Later, Bruker acquired the NMR probe business from Conductus and continued work on probes with 3 mm sample diameter using high-temperature superconducting coils. Typically being limited to two fixed nuclei, these probes attracted limited attention in the market.

For many natural product chemists, the 1D carbon spectrum is still the gold standard NMR spectrum. It provides an unambiguous carbon count, and the acquisition of an identical carbon spectrum is often regarded as proof of a successful synthesis of a compound. However, carbon-13 NMR spectroscopy requires larger quantities of material. Solubility considerations tend to preclude the use of micro-volume probes for this application. Cryogenically cooled probes optimized for carbon observation provide the highest sensitivity. The S/N as measured on standard samples such as 10% ethylbenzene or 60% C_6D_6 in dioxane typically show an enhancement of a factor of 5–6 over conventional dual-frequency or multinuclear probes. S/N values approaching 3000 : 1 have been achieved at 600 MHz. For comparison, a multinuclear probe at the same field would typically result in an S/N of ~600 : 1. This can result in a 30-fold reduction in experiment times and permits the acquisition of carbon spectra for medium-sized molecules in quantities as low as 100 μg in a few hours. With this sensitivity, it is possible to observe the carbon satellites in the spectrum of 10% ethylbenzene in a single scan (Figure 3.10).

Such a probe was used in the collection of an INADEQUATE[37] spectrum of only 1 mg of a fractionally (3–4%) ^{13}C-enriched sample of karlotoxin-2[38] (Figure 3.11) or in the acquisition of the carbon spectrum of the final product in the synthesis of amphidinolides by Carter and co-workers.[39]

Whereas early inverse detection cryogenic probes only used cold ^{1}H and lock preamplifiers, the second generation of these probes have also been equipped with cooled ^{13}C preamplifier electronics and typically show about double the sensitivity of conventional probes for carbon or about half the sensitivity of the dedicated carbon observe cryogenic probes. This modification facilitated direct carbon detection with higher sensitivity to a larger community, since these probes could be more easily shared between different research groups.

The combined success of conventional small-volume NMR probes and cryogenically cooled probes led to the idea of combining the two and creating a small-volume cryoprobe. The first such probe, a 1 mm cryogenically cooled probe, was introduced in early 2006 in a collaboration between the University of Florida, the National High Magnetic Field Laboratory and Bruker BioSpin.[4] This probe was based on high-temperature

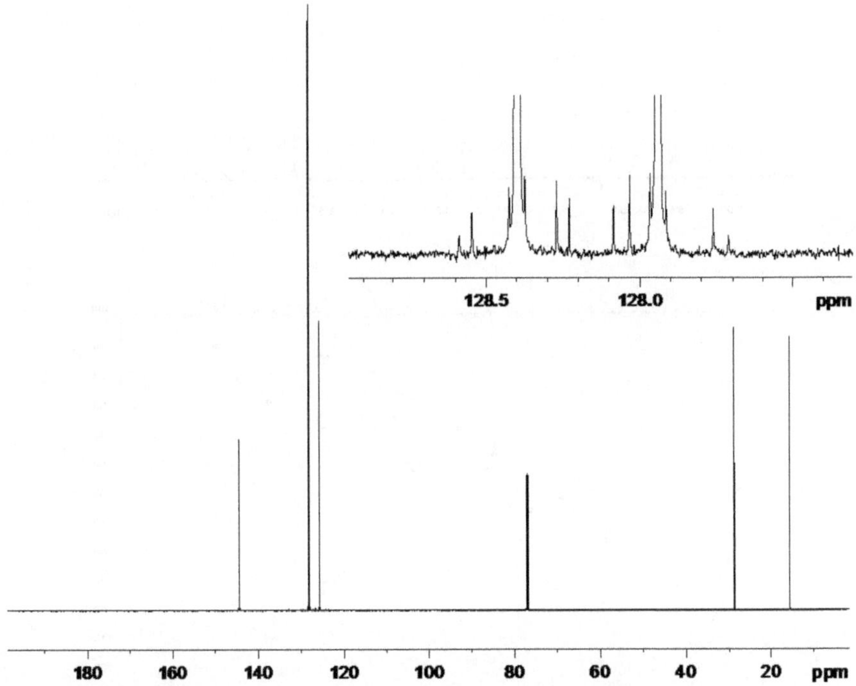

Figure 3.10 Single-scan carbon spectrum of 10% ethylbenzene in $CDCl_3$. The inset shows the carbon satellites of the aromatic signals.

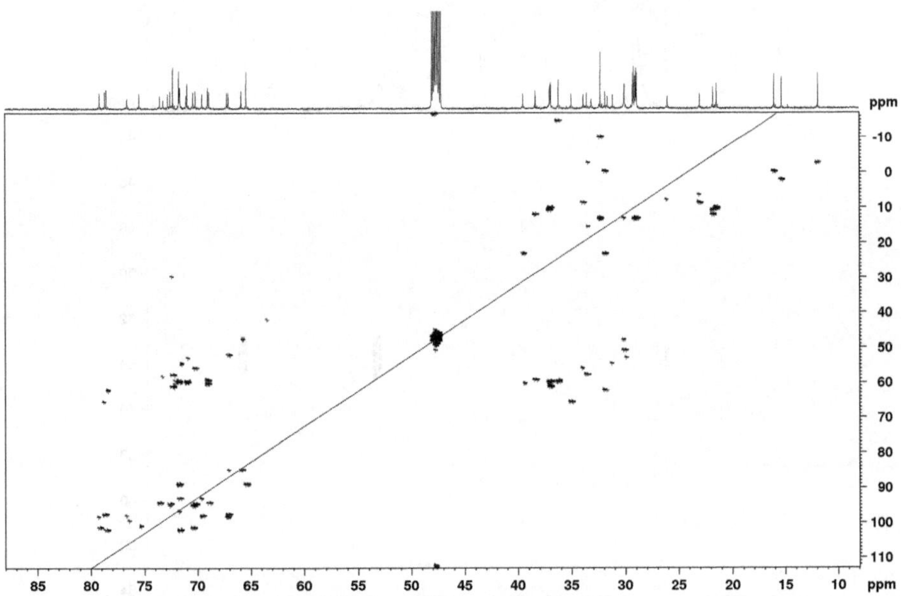

Figure 3.11 2D INADEQUATE spectrum of 1 mg of karlotoxin-2 ($\sim$3–4% ^{13}C enriched) in CD_3OD. Acquisition time $\sim$60 h, 600 MHz carbon optimized 5 mm cryogenically cooled probe, 384 scans and 128 complex increments.

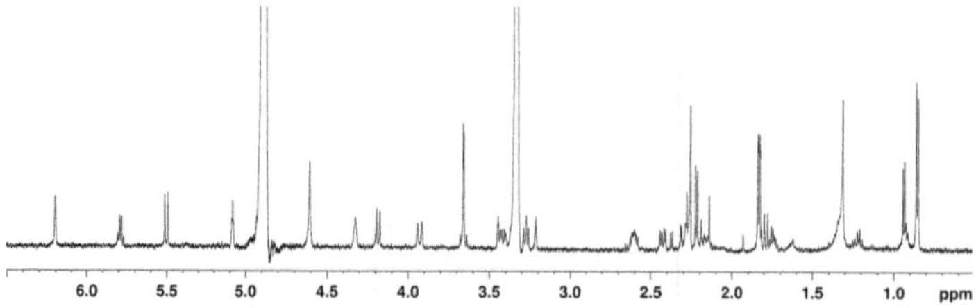
6.0 5.5 5.0 4.5 4.0 3.5 3.0 2.5 2.0 1.5 1.0 ppm

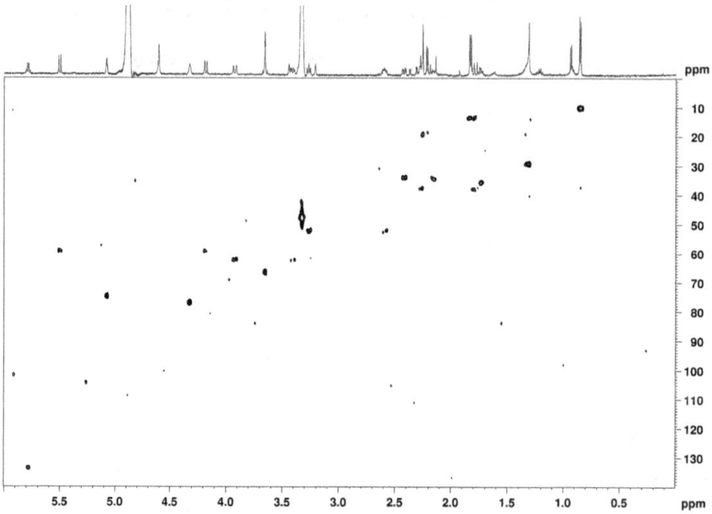
ppm
10 20 30 40 50 60 70 80 90 100 110 120 130
5.5 5.0 4.5 4.0 3.5 3.0 2.5 2.0 1.5 1.0 0.5 ppm

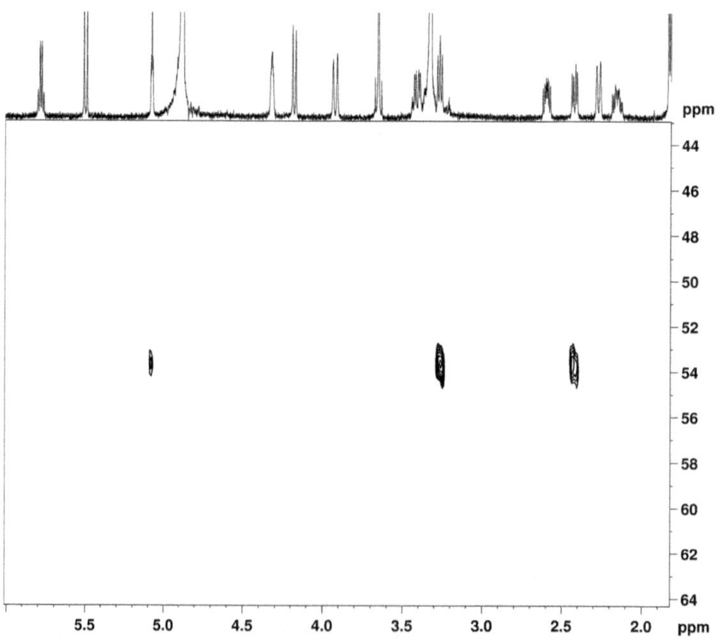
ppm
44 46 48 50 52 54 56 58 60 62 64
5.5 5.0 4.5 4.0 3.5 3.0 2.5 2.0 ppm

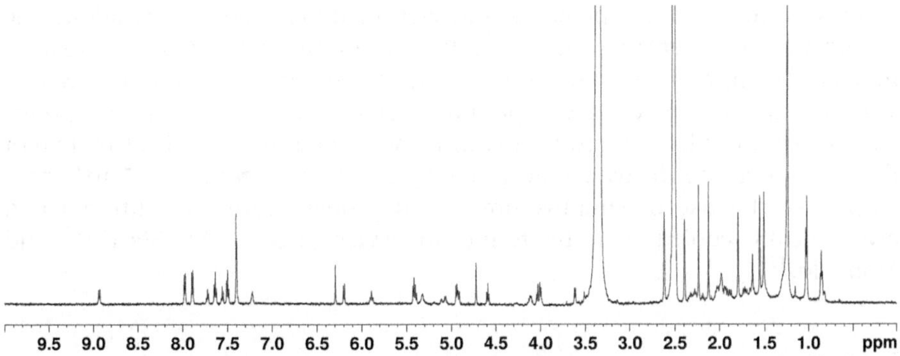

Figure 3.13 1D proton spectrum of 1 nmol (0.852 μg in 30 μL) of taxol in DMSO-d_6, 700 MHz 1.7 mm triple resonance microcryoprobe, 64 scans and 4.5 min experimental time. This corresponds to 1/20th of the time it took to acquire the spectrum in Figure 3.9.

superconductors and was built with four sets of HTS coils for ^{1}H, ^{13}C, ^{15}N, and ^{2}H lock. An early exemplary application of this probe is the examination of the chemical composition of defensive secretions from walking stick insects.[40] The secretion from a single insect could be collected and analyzed with this 1 mm HTS probe.

Bruker BioSpin later engaged in the development of a 1.7 mm cryogenically cooled probe. This probe was introduced at the ENC in 2007 as a triple resonance probe with proton detection and carbon and nitrogen decoupling, and attracted immediate interest from a variety of NMR users. In addition to applications in the field of proteins and nucleic acids, several groups in the field of natural product research made use of the sensitivity of this probe to study extremely small quantities of material. An early example of such experiments is the spectra collected on 19 μg of retrorsine for ^{1}H/^{13}C correlation experiments and 190 μg for ^{1}H/^{15}N HMBC data (Figure 3.12).

The results obtained with 1 mm conventional probes could easily be surpassed and the acquisition of NMR data of 1 nmol of material in just a few minutes had become a reality, as demonstrated by the spectrum of 0.853 μg of taxol in Figure 3.13.

The impact of this high-sensitivity probe on the experimental limits of small-sample NMR has been investigated by Hilton and Martin.[5] In addition to the high sensitivity for proton detection, this probe also exhibits very short pulses for carbon and nitrogen. Of special importance for natural product

Figure 3.12 Top: proton spectrum of 19 μg of the alkaloid retrorsine in CD_3OD acquired in 2.75 min at 600 MHz with a 1.7 mm triple resonance microcryoprobe. Middle: ^{1}H/^{13}C HSQC experiment on 19 μg of retrorsine in CD_3OD acquired in 5.5 h at 600 MHz with a 1.7 mm triple resonance microcryoprobe. Bottom: ^{1}H/^{15}N HSQC experiment on 190 μg of retrorsine in CD_3OD acquired in 12 h at 600 MHz with a 1.7 mm triple resonance microcryoprobe.

studies is the short 90° pulse for nitrogen of under 25 μs. This allows the observation of the entire chemical shift range of nitrogen, which can span as much as 600 ppm, in a single experiment, as demonstrated by Martin *et al.*[41] The structures of several compounds isolated from the marine sponge *Phorbas* sp. in only microgram quantities were determined with the help of the 1.7 mm cryogenically cooled probe by Molinski's group.[42,43] Further examples of the use of small-volume NMR probes, both conventional and cryogenically cooled, can be found in review articles by Martin[44] and Molinski.[45]

References

1. R. C. Crouch and G. E. Martin, *J. Nat. Prod.*, 1992, **55**, 1343.
2. R. C. Crouch and G. E. Martin, *Magn. Reson. Chem.*, 1992, **30**, 66.
3. G. E. Martin, R. C. Crouch and A. P. Zens, *Magn. Reson. Chem.*, 1998, **36**, 551.
4. W. W. Brey, A. S. Edison, R. Nast, J. Rocca, S. Saikat Saha and R. S. Withers, *J. Magn. Reson.*, 2006, **179**, 290.
5. B. D. Hilton and G. E. Martin, *J. Nat. Prod.*, 2010, **73**, 1465.
6. A. Abragam, *The Principles of Nuclear Magnetism*, Oxford University Press, Oxford, 1961, p. 82.
7. H. D. W. Hill and R. E. Richards, *J. Phys. E: Sci. Instrum. Ser. 2*, 1968, **1**, 977.
8. D. Moskau, *Concepts Magn. Reson.*, 2002, **15**, 164.
9. G. A. Morris and R. Freeman, *J. Am. Chem. Soc.*, 1979, **101**, 760.
10. M. R. Bendall, D. M. Doddrell and D. T. Pegg, *J. Am. Chem. Soc.*, 1981, **103**, 4603.
11. A. Bax, R. H. Griffey and B. L. Hawkins, *J. Magn. Reson.*, 1983, **55**, 301.
12. G. Bodenhausen and D. J. Ruben, *Chem. Phys. Lett.*, 1980, **69**, 185.
13. A. Bax and M. F. Summers, *J. Am. Chem. Soc.*, 1986, **108**, 2093.
14. P. Barker and R. Freeman, *J. Magn. Reson.*, 1985, **64**, 334.
15. I. M. Brereton, S. Crozier, J. Field and D. M. Doddrell, *J. Magn. Reson.*, 1991, **93**, 54.
16. M. von Kienlin, C. T. W. Moonen, A. van der Thorn and P. C. M. van Zijl, *J. Magn. Reson.*, 1991, **93**, 423.
17. R. E. Hurd, *J. Magn. Reson.*, 1990, **87**, 422.
18. A. Bax, P. G. de Jong, A. F. Mehlkopf and J. Smidt, *Chem. Phys. Lett.*, 1980, **69**, 567.
19. C. H. Sotak, D. M. Freeman and R. E. Hurd, *J. Magn. Reson.*, 1988, **78**, 355.
20. A. L. Davis, E. D. Laue, J. Keeler, D. Moskau and J. A. B. Lohman, *J. Magn. Reson.*, 1991, **94**, 637.
21. D. I. Hoult and R. E. Richards, *J. Magn. Reson.*, 1976, **24**, 71.
22. F. D. Doty, T. J. Connick, X. Z. Ni and M. N. Clingan, *J. Magn. Reson.*, 1988, 77, 536.

23. D. Doty, G. Entzminger and Y. A. Yang, *Concepts Magn. Reson.*, 1998, **10**, 156.
24. W. E. Gifford and H. O. McMahon, *Proceedings 10th International Congress of Refrigeration*, 1959, vol. 1.
25. E. I. Mikulin, A. A. Tarasov and M. P. Shkrebyonock, *Adv. Cryo. Eng.*, 1984, **31**, 629.
26. P. Zens, Controlled Susceptibility Plugs, *U.S. Pat.*, No. 4,549,136, 1985.
27. G. Schlotterbeck, A. Ross, R. Hochstrasser, H. Senn, T. Kühn, D. Marek and O. Schett, *Anal. Chem.*, 2002, **74**, 4464.
28. J. M. Aramini, P. Rossi, C. Anklin, R. Xiao and G. T. Montelione, *Nat. Methods*, 2007, **4**, 491.
29. J. L. Griffin, A. W. Nicholls, H. C. Keun, R. J. Mortishire-Smith, J. K. Nicholson and T. Kuehn, *Analyst*, 2002, **127**, 582.
30. J. N. Shoolery, *Prog. Nucl. Magn. Reson. Spectrosc.*, 1995, **28**, 37.
31. P. A. Keifer, L. Baltusis, D. M. Rice, A. A. Tymiak and J. N. Shoolery, *J. Magn. Reson.*, 1996, **119A**, 65.
32. T. Barbara, *J. Magn. Reson.*, 1994, **109A**, 265.
33. W. L. Fitch, G. Detre, C. P. Holmes, J. N. Shoolery and P. A. Keifer, *J. Org Chem.*, 1994, **59**, 7995.
34. R. C. Anderson, M. A. Jarema, M. J. Shapiro, J. P. Stokes and M. Ziliox, *J. Org. Chem.*, 1995, **60**, 2650.
35. W. E. Maas, F. H. Laukien and D. G. Cory, *J. Am. Chem. Soc.*, 1996, **118**, 13085.
36. P. Styles, N. F. Soffe, C. A. Scott, D. A. Crag, F. Row, D. J. White and P. C. J. White, *J. Magn. Reson.*, 1984, **60**, 397.
37. A. Bax, R. Freeman and S. P. Kempsell, *J. Am. Chem. Soc.*, 1980, **102**, 4849.
38. J. Peng, A. R. Place, W. Yoshida, C. Anklin and M. T. Hamann, *J. Am. Chem. Soc.*, 2010, **132**, 3277.
39. L. Lu, W. Zhang and R. G. Carter, *J. Am. Chem. Soc.*, 2008, **130**, 7253.
40. A. T. Dossey, S. S. Walse, J. R. Rocca and A. S. Edison, *ACS Chem. Biol.*, 2006, **1**, 511.
41. G. E. Martin, B. D. Hilton, D. Moskau, N. Freytag, K. Kessler and K. Colson, *Magn. Reson. Chem.*, 2010, **48**, 935.
42. D. S. Dalisay and T. F. Molinski, *Org. Lett.*, 2009, **11**, 1967.
43. D. S. Dalisay, B. I. Morinaka, C. K. Skepper and T. F. Molinski, *J. Am. Chem. Soc.*, 2009, **131**, 7552.
44. G. E. Martin, *Annu. Rep. NMR Spectrosc.*, 2005, **56**, 1.
45. T. F. Molinski, *Curr. Opin. Drug Discovery Dev.*, 2009, **12**, 197.

CHAPTER 4

Cryogenically Cooled NMR Probes: a Revolution for NMR Spectroscopy

KIMBERLY L. COLSON

Bruker BioSpin Corporation, Billerica, MA 01821, USA
Email: kim.colson@bruker.com

4.1 Introduction

Advances in NMR probe technology over the past two decades have revolutionized the capabilities of NMR, including a broad expansion of its applications. The speed of acquisition of spectra went from being the rate-limiting step to being so rapid that the time required to analyze data has become the major concern. The most notable advancement has been the development and commercialization of cryogenically cooled NMR probes. Now common in many NMR facilities, cryogenically cooled NMR probes enable researchers to obtain data considered impossible to obtain less than two decades ago.

4.2 Historical Perspective

In the early to mid-1980s, NMR spectroscopists were hampered by the inherent low sensitivity of NMR spectroscopy. Long data collection times were the norm and booking enough NMR time was always a concern. The overnight run was often reserved for a single experiment. This valuable 16 h block of time would reward the NMR spectroscopist with only a 1D ^{1}H

Modern NMR Approaches To The Structure Elucidation of Natural Products,
Volume 1: Instrumentation and Software
Edited by Antony J. Williams, Gary E. Martin and David Rovnyak

Published by the Royal Society of Chemistry, www.rsc.org

spectrum, provided that the sample had a limited quantity (<1 mg) of a small molecule of medium complexity (~800 amu). With more sample (~5–10 mg), the NMR spectroscopist could use this precious block of time to acquire an NMR spectrum on an inherently insensitive nucleus, such as ^{13}C or ^{15}N. Two-dimensional NMR experiments were also becoming commonplace and the overnight run would grant the user a COSY and HETCOR or a COLOC experiment on this 5–10 mg sample.

During the 1980s, pioneers in cryogenically cooled NMR probes, including the groups of Peter Styles[1] and Daniel Marek,[2] started to experiment with increasing the sensitivity of NMR technology by reducing the noise factor in the NMR electronics by cooling the rf coils and also the preamplifiers. Hoult and Richards,[3] a decade earlier, had proposed that the sensitivity of a NMR spectrometer, which is defined by the signal-to-noise ratio (S/N), could be enhanced by lowering the coil temperature. The inverse relationship between the signal (S), temperature (T) and resistance (R) of the NMR coil (Figure 4.1) served as the foundation for the developments in cryogenically cooled NMR probe technology. This generalized equation reflects the theoretical signal-to-noise obtainable. To establish the actual S/N achievable for a liquid-state NMR experiment, the total resistance factor and the temperature must be considered from the overall signal path. In an elegant review article by Kovacs *et al.*,[4] the resistance and temperature of the sample, preamplifier, and coil are considered and described as shown in Figure 4.2. The resistance

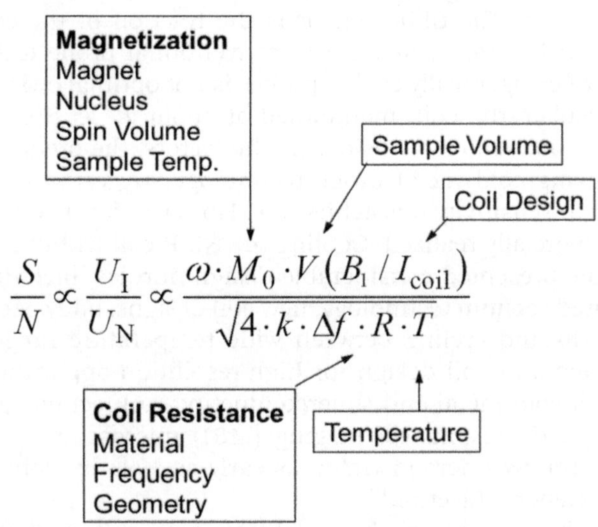

Figure 4.1 A general signal-to-noise ratio equation and corresponding NMR parameters.[3] Prior to the advent of the cryogenically cooled NMR probe, sensitivity enhancements resulted primarily from increases in the magnetic strength (M). Signal (S) has an inverse relationship to the temperature (T) and coil resistance (R). Noise (N) decreases with temperature and resistance. Note: preamplifier noise and sample loss is not accounted for in this equation.

$$\frac{S}{N} \propto \frac{1}{\sqrt{4k_B \times \Delta f \times \left[R_c\left(T_c + T_a\right) - R_s\left(T_s + T_a\right)\right]}}$$

Figure 4.2 The signal-to-noise ratio equation presented by Kovacs *et al.*[4] considers the resistance of the coil (R_c) and sample (R_s) and the temperature of the coil (T_c), sample (T_s) and preamplifier (T_a).

factor (R) includes resistance from coils (R_c) and the sample (R_s), reflecting the inductive coupling between the sample and the coil. While the resistance and temperature of the coil (T_c) are low, the resistance and temperature of the sample (T_s), being maintained near room temperature, are high. The conductivity, or ionic strength, of the sample solution, particularly buffered aqueous solvents used for measurements of proteins, provides a significant source of resistance and consequently may markedly reduce the S/N achievable s as shown by Kovacs *et al.*[4] and Voehler *et al.*[5] To reduce the consequence of ionic strength and enhance the S/N, smaller sample tubes and shaped NMR tubes with susceptibility-matched glass are often used.[6] Fortunately, natural product samples are typically acquired under low ionic conditions, hence resistance from the sample is small relative to protein and RNA/DNA applications. Therefore, the use of smaller tubes and low conductivity solvents is typically not needed for natural products to achieve optimal sensitivity as a result of the sample resistance factor.

Another factor that significantly influences the probe sensitivity is the probe filling factor. The filling factor is the fraction of the coil detection volume filled with sample. Relative to conventional probe technology, the filling factor[4] of cryogenically cooled probes is not optimal owing to the need to thermally isolate the coil, maintained at about 25 K, from the sample, maintained at about 300 K. Considering the factors, the predicted S/N gains (Figure 4.3) using cold metal probe technology suggested that a fourfold enhancement in sensitivity was achievable. However, it was years before this could be commercially realized. Cooling the NMR coil within millimeters of a liquid sample presented considerable design hurdles, including the need for sophisticated vacuum technology, new coil designs, and special materials that could withstand cycling between wide temperature ranges. Research and developments in coil design for high-resolution applications explored the use of both cold metal and superconducting coil technology.[7–9] A basic setup for magnetic resonance imaging (MRI) microscopy using a superconducting rf coil was demonstrated as early as 1993 by Robert Black and co-workers at General Electric.[10]

In 1996, the first commercially available cryogenically cooled NMR probe was released by Conductus. This probe featured two nuclei (typically ^{1}H and ^{2}H), a cryogenically cooled coil, and a manually operated cooling unit. Inverse detected heteronuclear experiments, gradient technology and three-dimensional NMR had become routine by the 1990s, and early adopters of the first commercial cryogenically cooled probe design sacrificed these advanced methods for a probe having high sensitivity. In 1997, the

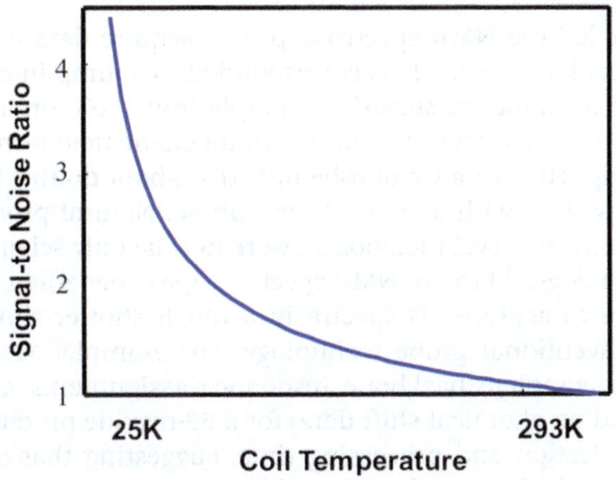

Figure 4.3 Typical S/N enhancements of a cold metal high-resolution NMR probe as a function of coil temperature.

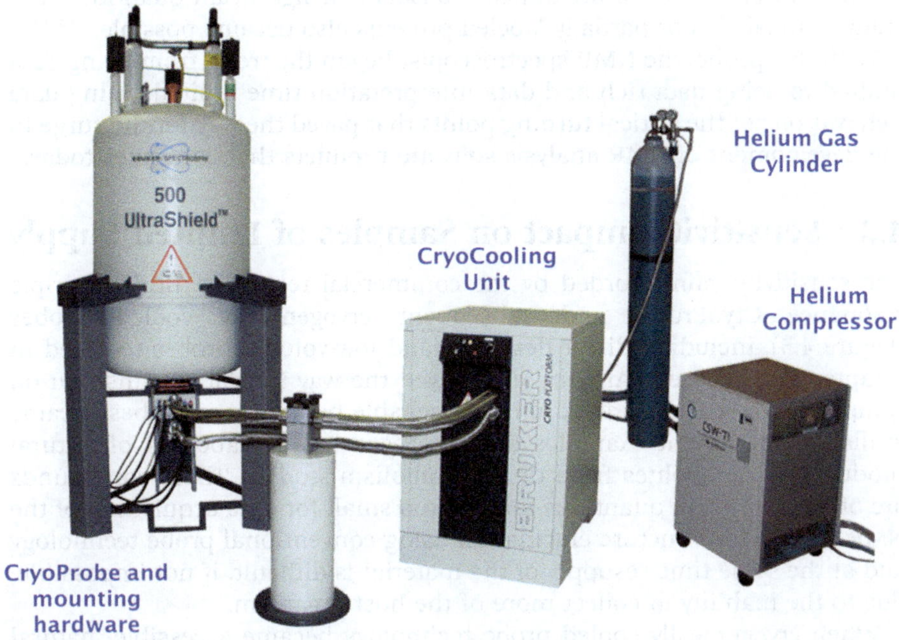

Figure 4.4 Bruker's 500 MHz 5 mm triple resonance *z*-gradient CryoProbe system released in 1999.

Conductus NMR division was purchased by Bruker Instruments (now Bruker BioSpin). The revolutionary sensitivity gains were first made available to the general NMR community starting in 1999, when Bruker released a 5 mm triple resonance (^{1}H, ^{2}H, ^{13}C, ^{15}N) inverse detected *z*-gradient CryoProbe with a fully automated close looped CryoCooling Unit (Figure 4.4). This probe delivered a fourfold sensitivity enhancement in ^{1}H sensitivity and, as a result, enabled the NMR spectroscopist to acquire data in 1 h that previously took overnight to run. This corresponded to a jump in the S/N for the 0.1% ethylbenzene standard sensitivity sample from 1100 on a conventional probe to 4400 on a CryoProbe. To put this in magnetic field terms, a 500 MHz instrument equipped with a CryoProbe provided about double the sensitivity of an 800 MHz system with a conventional probe. Natural product chemists and those studying low-level metabolites were not the only scientists adopting CryoProbe technology.[11] Protein NMR spectroscopists benefited tremendously from being able to acquire 3D spectra in a much shorter time frame than when using conventional probe technology. For example, Monleon *et al.*[12] achieved nearly complete backbone resonance assignments and secondary structures (based on chemical shift data) for a 59-residue protein in less than 30 h of data collection and processing time, suggesting that a 10-fold time saving in data acquisition may be achievable under some conditions. Perhaps even more significant was the ability to acquire data on proteins at concentrations less than 1 mM, opening the door to studying proteins that aggregate at high concentrations or are difficult to isolate in significant quantities. The study of unlabeled or partially labeled proteins also became possible.[4,13,14]

With this probe, the NMR spectroscopist began the trend from being data limited to being data rich and data interpretation time limited. Being data rich was one of the critical turning points that paved the way for the surge in the development of NMR analysis software products that continues today.

4.3 Sensitivity Impact on Samples of Limited Supply

The sensitivity gain afforded by the commercial release of Bruker's triple resonance CryoProbe, and subsequent cryogenically cooled probes (Figure 4.5), including direct detection and low-volume probes (covered in Chapter 3 by Clemens Anklin), also paved the way for data acquisition on samples that were considered unapproachable before these probes became available. Prominent examples are the secondary metabolites of natural products or metabolites from drug metabolism studies. These compounds are often isolated in quantities that are too small for data acquisition of the NMR spectra for structure elucidation using conventional probe technology and at the same time resupply of the material is difficult, if not impossible, due to the inability to collect more of the host organism.

When cryogenically cooled probe technology became accessible, natural product researchers were able to determine the chemical structures of materials available in sub-milligram quantities with relative ease. In some cases, sample-limited material was stored for decades awaiting new technological breakthroughs to allow the structure elucidation of these precious samples.[15–17] One particular example is a single specimen of a sea slug collected in 1990 by Tadeusz Molinski. Isolation fractions from this sea slug were frozen for 18 years until a 1.7 mm MicroCryoProbe was available to his laboratory for the structure elucidation of this prized material.[15] Data acquisition on these small sample sizes was not possible prior to the

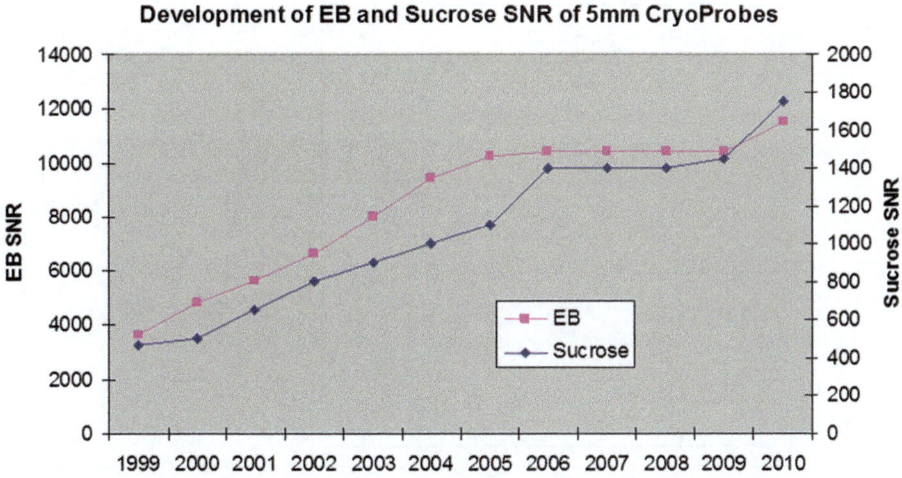

Figure 4.5 Development of 5 mm CryoProbes at new field strengths and with improved technology as a function of time and the resulting maximum available S/R on ethylbenzene (EB) and sucrose available.

cryogenically cooled probe technology because even a 1D proton experiment was time prohibitive, let alone obtaining the 2D data that were essential for the structure elucidation. Cryogenically cooled probe technology allowed these samples to be analyzed at last. Tadeusz Molinski of UC San Diego, a leader in marine natural products chemistry and an early adopter of cryogenically cooled NMR technology, utilized this technology for a new nanomole natural products discovery initiative. He credits this technology, and its sensitivity gains, as key to the discovery of many new natural product compounds unapproachable using conventional NMR probe technology, including muironolide A isolated from a marine sponge in a quantity of only 90 µg (Figure 4.6).[15–22] Additional information on small-volume NMR probes can be found in Chapter 3.

4.4 Experimental Options Expand

The accessibility to cryogenically cooled NMR technology has redefined the limits of detection for NMR spectroscopy and made experiments accessible that were previously considered unapproachable. Gary Martin of the Merck Research Laboratories conducted the much needed study to redefine the amount of sample and time required for typical experiments using the very sensitive low-volume 1.7 mm MicroCryoProbe.[23] Through this work, Hilton and Martin[24] discovered that the 1,1-ADEQUATE experiment, a rarely used ^{1}H detection experiment due to its inherent insensitivity, is now achievable even on 870 µg of strychnine using this technology. ^{13}C detection of natural products material is also possible. Shimba *et al.*[25] demonstrated that the combination of cryogenically cooled NMR probe technology and pulse

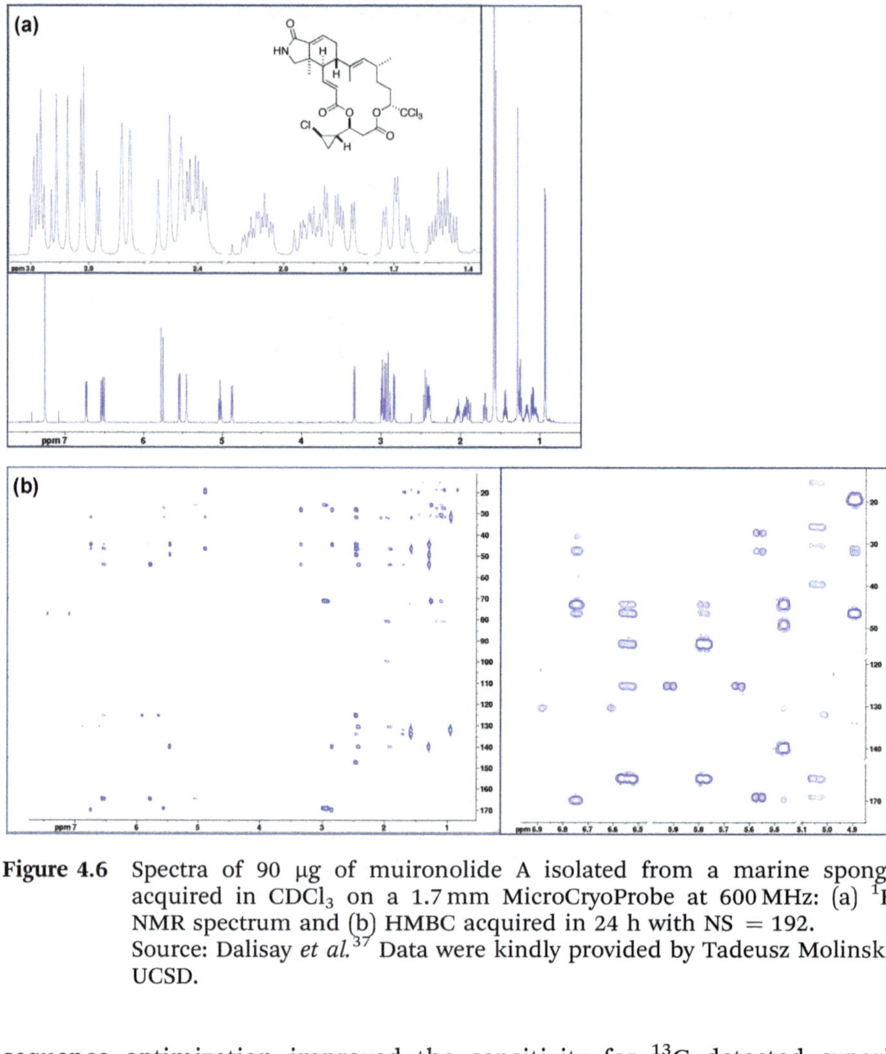

Figure 4.6 Spectra of 90 μg of muironolide A isolated from a marine sponge acquired in $CDCl_3$ on a 1.7 mm MicroCryoProbe at 600 MHz: (a) ^1H NMR spectrum and (b) HMBC acquired in 24 h with NS = 192. Source: Dalisay *et al.*[37] Data were kindly provided by Tadeusz Molinski, UCSD.

sequence optimization improved the sensitivity for ^{13}C detected experiments. Kovacs *et al.* demonstrated the carbon detection capabilities of a direct detection CryoProbe with a ^{13}C–^{13}C INADEQUATE experiment on a 9.8 mg sample of quinine in an overnight run on a 500 MHz system.[4]

4.5 Magnetic Resonance Imaging

While the NMR spectroscopists who were focused on high resolution were the first to adopt cryogenically cooled NMR probes, the technology expanded to imaging applications and studies on metabolism as well.[26–29] Researchers have reduced scan times by 75% without sacrificing image quality. They have also doubled the spatial resolution without sacrificing scan time.[30] Typical results far exceed conventional technology, as shown in Figures 4.7 and 4.8.

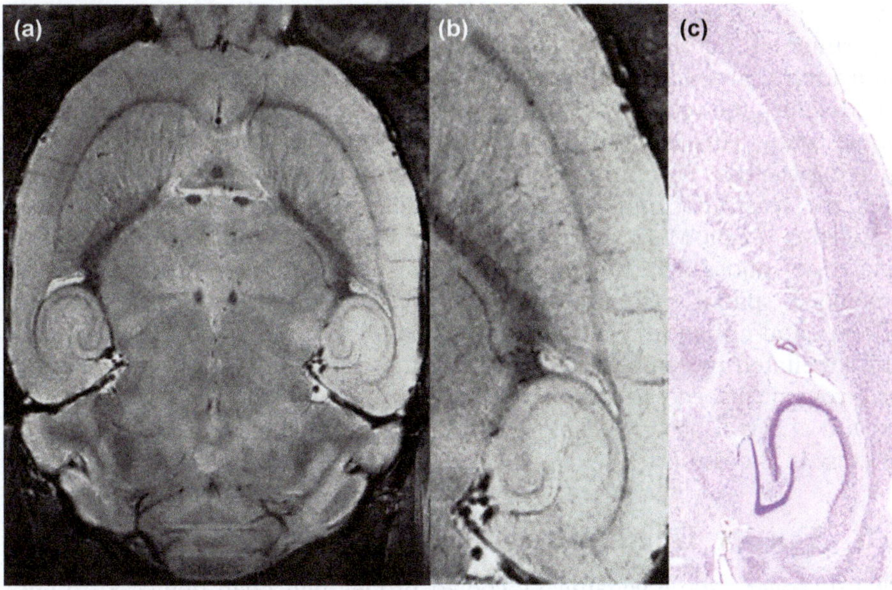

Figure 4.7 Mouse brain high-resolution RARE 15.2 T scan verses a histology plate. (a) Full field of view coronal RARE image, (b) expanded view of the hippocampal area, and (c) a corresponding Nissl stained plate. Acquisition details: matrix, 660×660; field of view, 1.9 cm^2; TR, 3.5 s; TE, 25 ms; echoes, 6; slices, 7.
Source: Bruker BioSpin MRI GmbH, Ettlingen, Germany.

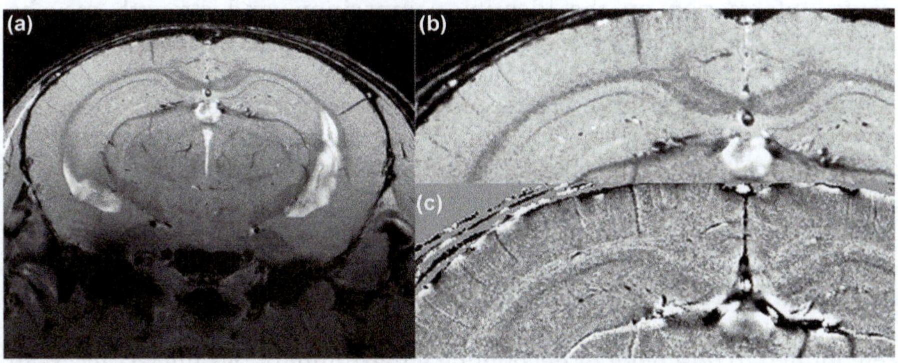

Figure 4.8 *In vivo* magnetic resonance microscopy images acquired in 21 min on mouse brain using a Bruker MRI CryoProbe at 15.2 T, 19.5 μm^2 in-plane and 150 μm slice thickness. (a) FLASH image; (b) an expanded area; (c) phase image. Acquisition details: matrix, 768×768; field of view, 1.5 cm^2; TR, 550 ms; TE, 4.4 ms; slices, 7.
Source: Bruker BioSpin MRI GmbH, Ettlingen, Germany.

MRI using a cryogenically cooled probe was applied for non-invasive phenotyping of mouse brains that are used for mouse models of human disease used in biomedical research.[31] Applications to brain imaging have shown inflammatory infiltrates in detail in even the early stages of an experimental autoimmune encephalomyelitis model.[31] This is a clear indication that the technology is at the edge of imaging very small objects such as Purkinje cell layer *in vivo* with MRI while still being able to image the whole brain. This is novel for the biologist and adds value to the existing imaging toolbox. Similarly, a study conducted by Wagenhaus *et al.*[32] evaluated the feasibility and benefits of cardiac magnetic resonance in mice employing a 400 MHz cryogenic rf surface coil, compared with a conventional mouse heart coil array operating at room temperature. The enhanced spatial resolution afforded better delineation of myocardial borders and enhanced the depiction of papillary muscles and trabeculae – and facilitated more accurate cardiac chamber quantification. Applications of cryogenically cooled probe technology to MRI will surely continue to expand and will possibly explore organisms producing natural products of interest in the future.

4.6 Future Developments

By 2013, over 1700 cryogenically cooled probes had been delivered to customers worldwide by two of the major suppliers of NMR equipment. Further advances are expected in this technology to increase capabilities and applications, increase accessibility to users, and increase sensitivity. Expanded capabilities and applications will continue to drive research and development efforts. For high-resolution NMR, a very new application that benefits from CryoProbe technology is evaluation of nutraceuticals and medicinal plants. Information on the identity, composition, and strength may be accessed using this technology. The strength of nutraceutical products is typically reflected by the concentration of a single component or a short list of specific components. Nutraceuticals are often complex mixtures (Figure 4.9) composed of metabolites with a wide range of concentrations of the individual components. The complex mixture provides a wealth of information, including species identification and quantification of key metabolites and even farm location.[33–36] Improved S/N and the ability to accelerate data acquisition benefit the evaluation of these materials. Broadband nuclear capabilities (Bruker standard and cost-effective Prodigy CryoProbes) recently expanded the previously available configurations that were limited to observing or decoupling a maximum of five nuclei. Access to additional nuclei is particularly important for academic laboratories that support a wide range of research initiatives and the materials industry. For imaging applications, the implementation of multichannel phased array coils in a cryogenically cooled design will continue to provide further enhanced sensitivity over a large dynamic range and permit parallel imaging applications, greatly expanding current cryo-MRI technology.

Applications of cryogenically cooled technology are expected in the solid-state NMR magic angle spinning (MAS) field. The S/N in solid-state spectra is hindered by inherently broad lines. As a result, even gains of 2–3-fold in S/N

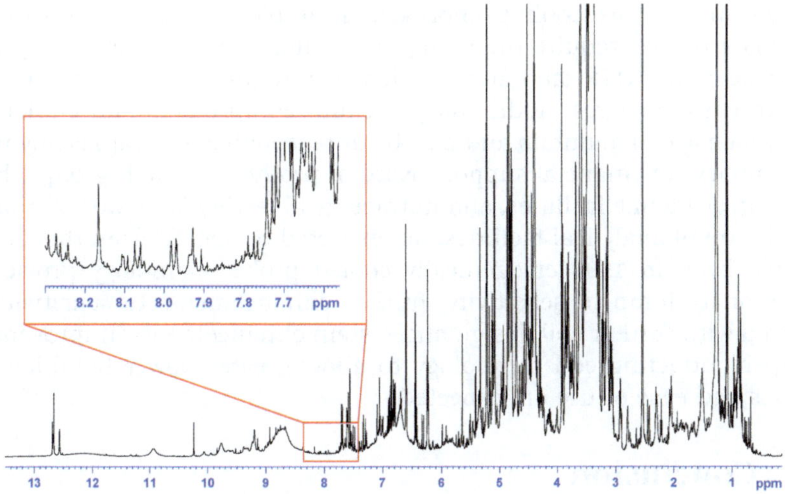

Figure 4.9 NMR spectrum of blueberry leaf (*Vaccinium angustifolium*) extract dissolved in DMSO-d_6. Data acquired on a Bruker 600 MHz TCI CryoProbe. This crude mixture shows a wide dynamic range in metabolite concentrations.
Source: Bruker BioSpin, Billerica, MA, USA.

will be advantageous to those researching biosolids, materials, and polymorphs. Developments in this area are more challenging than for the high-resolution forerunner. The most significant hurdle is the need for the NMR coils to withstand very high power pulses and long decoupling cycles. At low temperature, the coils are more efficient and it becomes easier to arc a probe at the same voltage as a room temperature probe. With the solid-state NMR probe requiring as much as three times the power of a high-resolution probe, careful attention to power handling necessitates significant design and development efforts. Another challenge to developments in this area is the need to spin samples at high speeds, where the expected demand is for spinning ranges from 1 to 50 kHz. Although a cryogenically cooled MAS probe is available (Doty Scientific) and will satisfy many solid-state NMR users, some research may require additional sensitivity gains that may be achieved through dynamic nuclear polarization (DNP) technology.

Already, increased accessibility to cryogenically cooled probes is being realized as cryogenically cooled probes that utilize an open-loop cooling system are now being sold commercially, as in the case of the Bruker Prodigy CryoProbe. In this design, the probes are cooled by liquid nitrogen boil-off rather than a closed-loop helium gas design, reducing maintenance costs and infrastructure needs that are required by a helium compressor-equipped CryoProbe. Although the Prodigy CryoProbe has about half the sensitivity of its bigger cousin, the ability to place the open-loop liquid nitrogen probe in most laboratories without significant siting restrictions or infrastructure changes makes this cooling approach desirable in many cases. Helium gas-cooled probes require the siting of a helium compressor and optional chiller near the NMR laboratory, which may require potentially expensive infrastructure changes. Additionally, helium compressors and chillers are often expensive to maintain and may be unreasonable in facilities relying on grant money for financial support. Natural product research groups, being mainly in academic facilities, and nutraceutical testing laboratories, that are typically small analytical facilities, are expected to benefit from this option.

While back in 1999 cryogenically cooled probe technology provided a revolutionary jump in sensitivity, further enhancements to sensitivity are expected with further coil design and system enhancements. Improvements to superconducting coil technology to allow greater power handling capabilities and ease of use are expected.

4.7 Conclusion

Two decades ago, NMR sensitivity gains were mainly accomplished through increases in field strength. Cryogenically cooled probe technology changed that and made it possible to obtain data on low- and mid-range field systems with the sensitivity that was previously reserved for very high-field magnets. Limits of NMR detection have been redefined and applications have been expanded. Experiments that were considered too insensitive have found their way into the NMR spectroscopist's toolbox as a result of this technology. The future gains from this ground-breaking technology will continue to reward scientists with additional new horizons to explore.

Acknowledgements

This chapter is dedicated to the memory of Detlef Moskau, my gracious colleague and friend, who gave so much to many within Bruker and many customers worldwide. His warm smile, eager eyes and "can-do" spirit will always be remembered and serve as inspiration for me throughout the remainder of my life.

I am grateful for the contributions of many of my Bruker colleagues with whom I have worked with over the years, including Werner Maas, Detlef Moskau, Helena Kovacs, Oskar Schett, Daniel Marek, Klemens Kessler, Urs Seehofer, Daniel Oberli, Tim Wokrina, Mat Brevard, Pavel Kostikin, Rich Withers and Clemens Anklin. Thanks are also due to David Rovnyak for his very helpful suggestions for this chapter.

References

1. P. Styles, N. F. Soffe, C. A. Scott, D. A. Cragg, D. J. White and P. C. J. White, *J. Magn. Reson.*, 1984, **60**, 397.
2. D. Marek and co-workers, Bruker Instruments, unpublished data.
3. D. I. Hoult and R. E. Richards, *J. Magn. Reson.*, 1976, **24**, 71.

4. H. Kovacs, D. Moskau and M. Spraul, *Prog. Nucl. Magn. Reson. Spectrosc.*, 2015, **46**, 131.
5. M. W. Voehler, G. Collier, J. K. Young, M. P. Stone and M. W. Germann, *J. Magn. Reson.*, 2006, **183**, 102.
6. M. Takeda, K. Hallenga, M. Shigezane, M. Waelchli, F. Löhr, J. L. Markley and M. Kainosho, *J. Magn. Reson.*, 2011, **209**, 167.
7. V. Kotsubo and R. Nast, in *Advances in Cryogenic Engineering, a Cryogenic Engineering Conference Publication*, ed. P. Kittel, Springer, New York, 1996, vol. 41, pp. 1857–1864.
8. W. W. Brey, A. S. Edison, R. E. Nast, J. R. Rocca, S. Saha and R. S. Withers, *J. Magn. Reson.*, 2006, **179**, 290.
9. W. A. Anderson, W. W. Brey, A. L. Brooke, B. Cole, K. A. Delin, L. F. Fuks, H. D. W. Hill, M. E. Johanson, V. Kotsubo, R. Nast, R. S. Withers and W. H. Wong, *Bull. Magn. Reson*, 1995, **17**, 98.
10. R. Black, T. A. Early, P. B. Roemer, O. M. Mueller, A. Mogro-Campero, L. G. Turner and G. A. Johnson, *Science*, 1993, **259**, 793.
11. K. L. Colson, *Mod. Drug Discovery*, July 2003, 47.
12. D. Monleon, K. Colson, H. N. B. Moseley, C. Anklin, R. Oswald, T. Szyperski and G. T. Montelione, *J. Struct. Funct. Genetics*, 2002, **2**, 93.
13. P. Selenko, Z. Serber, B. Gadea, J. Ruderman and G. Wagner, *Proc. Natl. Acad. Sci. U. S. A.*, 2006, **103**, 11904.
14. P. J. Barrett, J. Chen, M.-K. Cho, J.-H. Kim, Z. Lu, S. Mathew, D. Peng, Y. Song, W. D. Van Horn, T. Zhuang, F. D. Sönnichsen and C. R. Sanders, *Biochemistry*, 2013, **52**, 1303.
15. D. S. Dalisay, E. W. Rogers, A. S. Edison and T. F. Molinski, *J. Nat. Prod.*, 2009, **72**, 732.
16. D. S. Dalisay, B. I. Morinaka, C. K. Skepper and T. F. Molinski, *J. Am. Chem. Soc.*, 2009, **131**, 7552.
17. D. S. Dalisay and T. F. Molinski, *Org. Lett.*, 2009, **11**, 1967.
18. T. F. Molinski, *Curr. Opin. Drug Discov. Dev.*, 2009, **12**, 197.
19. D. S. Dalisay and T. F. Molinski, *J. Nat. Prod.*, **72**, 739.
20. D. S. Dalisay, E. W. Rogers, A. Edison and T. F. Molinski, *J. Nat. Prod.*, 2009, **72**, 732.
21. T. F. Molinski, *Curr. Opin. Biotechnol.*, 2010, **21**, 819.
22. T. F. Molinski, *Nat. Prod. Rep.*, 2010, **27**, 321.
23. G. Martin, *Encyclopedia of Magnetic Resonance Online*, John Wiley & Sons, Ltd., June 15, 2012, and references therein.
24. B. D. Hilton and G. E. Martin, *J. Nat. Prod.*, 2010, **73**, 1465.
25. N. Shimba, H. Kovacs, A. S. Stern, A. M. Nomura, I. Shimada, J. C. Hoch, C. S. Craik and V. Dötsch, *J. Biomol. NMR*, 2004, **30**, 175.
26. L. Darrasse and J.-C. Ginefri, *Biochimie*, 2003, **85**, 915.
27. C. Baltes, N. Radzwill, S. Bosshard, D. Marek and M. Rudin, *NMR Biomed.*, 2009, **22**, 834.
28. A. Seuwen, A. Schroeter and M. Rutin, *ISMRM*, 2013, 0860.
29. T. Wokrina, M. Gottschalk, S. R. Hermann, M. Sacher, T. Fitze and D. Marek *ISMRM*, 2012, 3233.

30. I. Vernikouskaya, A. Bornstedt and V. Rasche, *theresonance*, February 19, 2013.
31. H. Waiczies, J. M. Millward, S. Lepore, C. Infante-Duarte, A. Pohlmann, T. Niendorf and S. Waiczies, *PLoS One*, 2012, 7, e32796.
32. B. Wagenhaus, A. Pohlmann, M. A. Dieringer, A. Els, H. Waiczies, S. Waiczies, J. Schulz-Menger and T. Niendorf, *PLoS One*, 2012, 7, e42383.
33. J. M. Hicks, A. Muhammad, J. Ferrier, A. Saleem, A. Cuerrier, J. T. Arnason and K. L. Colson, *J. AOAC Int.*, 2012, **95**, 1406.
34. M. A. Markus, S. M. Luchsinger, J. Yuk, J. Ferrier, A. Muhammad, J. M. Hicks, K. B. Killday, F. Berrue, C. Kirby, K. Knagge, T. Goedecke, B. Ramirez, G. Pauli, I. Burton, J. T. Arnason and K. L. Colson, *Planta Med.*, 2014, **80**, 732.
35. J. Ferrier, E. H. Chen, M. Markus, J. T. Arnason, and K. L. Colson, *Practical Applications of NMR in Industry Conference (PANIC)*, 2015, San Diego, CA, USA. Retrieved from http://www.panicnmr.com.
36. J. Yuk, K. L. McIntyre, C. Fischer, J. Hicks, K. L. Colson, E. Lui, D. Brown and J. T. Arnason, *Anal. Bioanal. Chem.*, 2013, **405**, 4499.
37. D. S. Dalisay, B. I. Morinaka, C. K. Skepper and T. F. Molinski, *J. Am. Chem. Soc.*, 2009, **131**, 7552.

CHAPTER 5

Application of LC-NMR to the Study of Natural Products

MANFRED SPRAUL,* ULRICH BRAUMANN, MARKUS GODEJOHANN, CRISTINA DAOLIO AND LI-HONG TSENG

Bruker BioSpin GmbH, Silberstreifen, D-76287 Rheinstetten, Germany
*Email: manfred.spraul@bruker.com

5.1 Introduction

NMR spectroscopy has been applied for many years to the structure elucidation of pure compounds. Therefore, it was necessary, prior to NMR analysis, to separate mixtures by means of extraction and preparative chromatography. Such procedures required larger amounts of material and a chromatographic separation good enough to produce a more or less pure compound, a situation that often needed multiple chromatographic steps. In addition NMR sensitivity required milligram amounts in order to be able to run 2D heteronuclear experiments, the cornerstone of structure elucidation. Over the years, NMR sensitivity has been enhanced by improved probehead technology and increased magnetic field strength. With the introduction of cryogenic probes,[1] a major enhancement in signal-to-noise ratio (S/N) was achieved, commonly a factor of 4 in most solvents used. It is now possible to run the relevant experiments for structure elucidation by NMR in the low microgram range.

In the 1970s, another approach to the analysis of compounds in mixtures started with the first on-line (on-flow) liquid chromatography (LC)-NMR

Modern NMR Approaches To The Structure Elucidation of Natural Products,
Volume 1: Instrumentation and Software
Edited by Antony J. Williams, Gary E. Martin and David Rovnyak

Published by the Royal Society of Chemistry, www.rsc.org

experiments, reported by Watanabe and Niki.[2] They used a laboratory-built device that employed a Teflon capillary in a conventional NMR tube. Sensitivity was limited with this approach and consequently stopped-flow experiments had to be performed. Stopping the flow increases the time window for the NMR measurement, since in the on-flow mode only a few scans can be accumulated before the LC peak leaves the NMR flow cell. Another drawback was the need to use normal phase separations with proton-free solvents such as carbon tetrachloride. The situation improved with the introduction of NMR probes with dedicated flow cells,[3] allowing an optimized filling factor with regard to distance to the receiver coils. These probes also afforded resolution and lineshapes comparable to those with conventional tube probes.

Lineshape is an important factor also when confronted with the need to perform solvent suppression, which is often the case for reversed-phase separations, where typically a gradient of water and organic solvent (methanol or acetonitrile) is used. The organic solvent would be too expensive in its deuterated form, and this is the alternative to data acquisition without solvent suppression. As a result, solvent suppression schemes were developed to remove the solvent signals efficiently, allowing the full dynamic range of the receiver systems to be utilized.[4–6] It was obvious that on-flow LC-NMR, where repeated short NMR experiments are run during the separation, had severe sensitivity limitations, as explained later in the technical section, and LC-NMR interfaces were developed where the peaks were traced by UV detection. UV detection allowed both stopped-flow[7] and loop collection.[8,9] However, the real breakthrough for LC-NMR sensitivity came with the introduction of post-column solid-phase extraction[10] and the introduction of dedicated cryogenic flow probes,[11] and later cryoprobes with a flow insert. Another important addition to the LC-NMR hardware configuration was the integration of mass spectrometry (MS), where the MS information could be used to determine which peaks to use for NMR analysis. Such an LC-SPE-NMR/MS system could be operated with the highest selectivity on the trapped peaks. With all of these tools in place, LC-NMR became an important player in the detection and structure elucidation of new natural products.

5.2 LC-NMR Technology

In this section, the LC-NMR technology is explained in detail, starting with on-flow LC-NMR and ranging through to LC-SPE-NMR/MS. The first operational mode discussed is on-flow NMR and its relative requirements.

5.2.1 On-flow LC-NMR

As can be seen from Figure 5.1, in on-flow LC-NMR the flow stream leaving the LC column is guided through a non-NMR detector first, which can be either a UV or a diode array detector (DAD), before the flow is guided through the NMR flow probe and the flow cell itself.

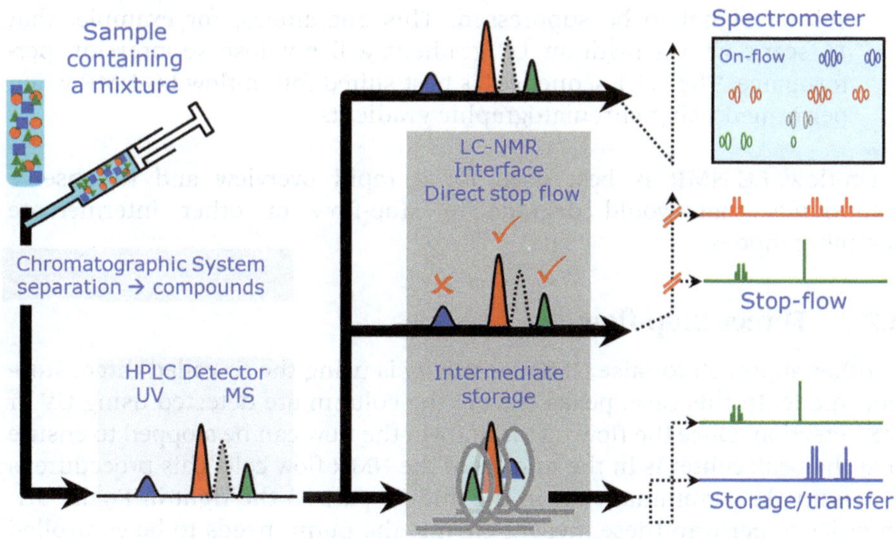

Figure 5.1 A schematic flowchart of LC-NMR operation modes visualizing on-flow (upper part on the right), direct stop-flow (middle part on the right) and loop collection (lower part on the right).

When using a mass spectrometer as an additional detector, a flow splitter has to be used to divert a small fraction (typically 1–5%) of the flow to the mass spectrometer. The splitting ratio addresses the differences in the sensitivity of the two detection methods. On-flow LC-NMR, however, has significant limitations that have led to its use on only rare occasions:

- The number of scans is limited, since when a peak passes through the flow cell at a flow rate of typically 1 mL min^{-1} on an analytical column, a maximum of only 16 scans can be acquired, limiting the sensitivity to detection levels in the upper microgram range. The flow rate can be reduced to allow for more scans, but this increases the run time of the chromatogram correspondingly.
- If gradient elution is used, which is a necessity in reversed-phase chromatography, then the steepness of the gradient has to be restricted because of needs associated with the solvent suppression applied prior to NMR detection. Changes in the solvent ratio lead to changes in the chemical shift position of the solvent signals. Since the NMR spectrometer is typically locked to D_2O during on-flow LC-NMR, the resonance position of the organic solvent moves with the LC gradient. This means that in a series of scans the solvent suppression will degrade, as suppression is set up using a prescan and then transferred to the experiment recorded. This means that during the accumulation, the position of the organic solvent signal is moving. Different solvent suppression modes are available. In on-flow LC NMR, it is best to use a pulse sequence that produces a broader zero excitation field around the

solvent signal to be suppressed. This guarantees, for example, that 16 scans with a medium LC gradient will not lose suppression performance. The WET sequence[5] is best suited for on-flow LC-NMR when performed using chromatographic gradients.

On-flow LC-NMR is best used for a rapid overview and to observe compounds that would degrade in stop-flow or other intermediate sampling modes.

5.2.2 Direct Stop-flow

Another approach to raise NMR sensitivity is using the so-called direct stop-flow mode. In this case, peaks leaving the column are detected using UV or MS detection. Once the flow rate is known, the flow can be stopped to ensure that the peak center is in the middle of the NMR flow cell. This procedure is represented diagrammatically in the middle part on the right in Figure 5.1. In order to perform these measurements, the pump needs to be controlled directly through the LC-NMR software. The reason for this is that in normal LC operation a stop-flow is interpreted as being at the end of the chromatography run and the pump will readjust the starting conditions when flow continues. This is not possible in direct stop-flow LC-NMR. After a peak is placed in the flow cell, and the measurement is finished, the pump must then continue with the same eluent mixture as when it was stopped. Optimally, a valve is used that switches flow from the pump before the LC column to flush to waste and only then turns back on when the flow is stable and completely returned back to the conditions when the flow was stopped. This approach is required if more than one peak within one separation has to be measured by NMR. The advantage over on-flow is clear: the NMR measurement can be performed for a much longer time and both long 1D acquisitions and 2D experiments can also be performed. As there is no flow during the time of the NMR measurement, solvent suppression experiments can also be used to suppress the solvent lines selectively and leave other resonances only a few hertz away from the presaturation frequency unperturbed. A disadvantage of the direct stop-flow approach is that peaks which are still on the column, or in the transfer line to the NMR instrument, can undergo diffusion while waiting for the pump to restart. This diffusion is partly refocused for peaks still experiencing some residual time on-column after the pump restart, before moving into the NMR flow cell. In principle, this can be understood as a second short column through which the peak is moving after flow restart.

5.2.3 Loop Collection

To overcome the chromatographic resolution problem of multiple stop-flow runs, a loop collection system was introduced.[21] In its first iteration, the system had a collection valve and 12 sample loops attached (see Figure 5.1, lower part on the right). Typically, UV detection or DAD was used to identify the peak positions and to determine when to switch the valve for peak collection. The loops used were adapted to the size of the NMR flow cell. Typically, 4 mm LC probes were used with an active volume of 120 μL and a total volume of 200 μL. The transfer time was set up in a way that placed the peak center exactly in the middle of the flow cell. At the end of the separation, the loop contents were transferred sequentially into the NMR flow cell under full automation. The information stored for each peak includes the chromatographic conditions, the retention time of the peak, and the solvent ratio when the loop collection took place. To optimize the elution from the loop, the solvent composition at the pump, being different from the composition in which the peak was eluting, was taken into account. Therefore, in gradient elutions, the composition of the solvents was readjusted accordingly before transfer into the NMR flow cell. This unit established LC-NMR as a broadly applicable technology. An improved loop collection system was introduced a few years later and today defines the state-of-the-art in loop collection. In this new system, loops are placed in a removable cassette containing 36 sample loops as shown in Figure 5.2.

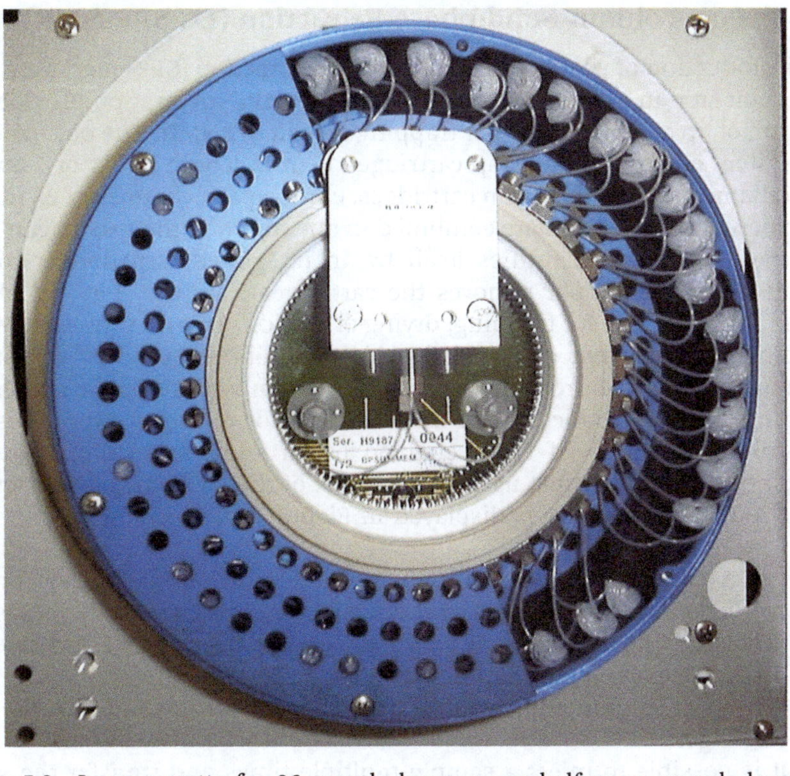

Figure 5.2 Loop cassette for 36 sample loops, cover half open; sample loops are visible on the outer part of the ring and a memory board sits in the center of the cassette.

The advantages of the system in Figure 5.2 compared with the 12-loop system are a threefold increase in the number of sample loops, faster access to the loops, a memory board on the cartridge to store all relevant information of the separation and the peaks, transportability, and the use of multiple cassettes with different volumes, if necessary, thereby allowing the use of 3 and 4 mm probeheads with one loop collection system. The cassette in this case is placed in a temperature-controlled compartment allowing cooling and thereby reducing diffusion in the loops. In this way, the best NMR sensitivity can be guaranteed. Another advantage of the cassette-based system is that it allows for the separation of the loop collection step from the transfer step. This has two consequences:

- During loop collection, the NMR system is free for other tasks.
- In some situations, there may be more than one location and only one NMR cassette will need to be transferred to the NMR location if there are two loop collection systems. All relevant information describing the peaks is stored on the memory board, so the cassette can be operated autonomously.

5.2.4 Post-column Solid-phase Extraction (LC-SPE-NMR)

The introduction of post-column solid-phase extraction (SPE) marks another important step in the success story of LC-NMR. SPE integration provides an S/N gain of up to a factor of 4 per trapping step in the NMR. The post-column SPE system contains small trap cartridges 1 cm in length and 1 or 2 mm in inner diameter. With the 2 mm cartridges, eluting peak volumes of 25 μL can be achieved; 96 cartridges are combined in one tray, and the system can hold two trays. Each tray identifies itself by an integrated transponder to the overall system. A gripper removes the cartridges from the tray and places them in the flow line for trapping, drying or elution. Two flow lines allow the trapping of peaks that elute side-by-side. Dilutors are used to condition, wash and elute peaks from the trap cartridges. Important for the success of trapping at high organic solvent fractions and for multiple trapping is the post-column addition of water enabled by an additional pump. The detection of peaks of interest uses one or more of the UV, DAD or MS signals. The LC-SPE-NMR process is displayed in Figure 5.3.

- Peaks eluting from the column are detected using UV, DAD or MS methods or a combination of them, since a combination increases the probability of detecting all peaks since UV detection is blind to compounds lacking a chromophore.
- With post-column addition of water, the peaks are retained on the SPE cartridges.
- It is possible to inject a sample multiple times and transfer the same peaks to the same trap cartridge to increase sensitivity further.

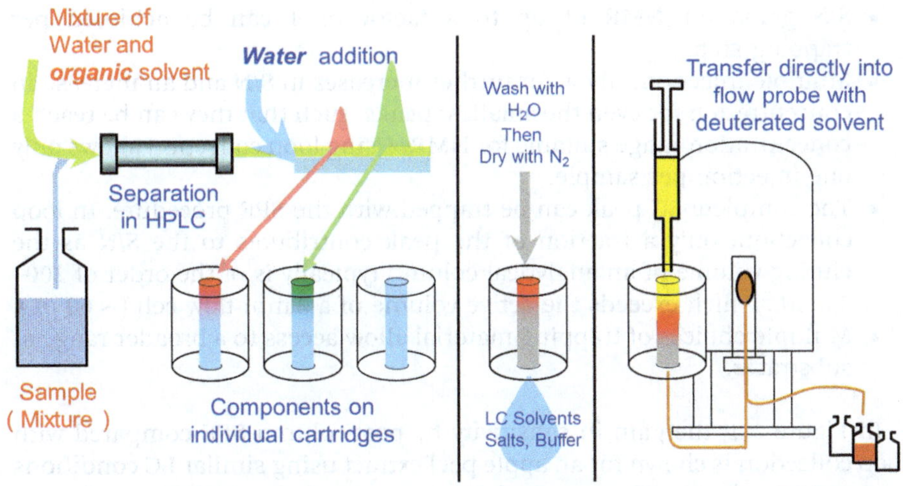

Figure 5.3 A flowchart describing LC-SPE-NMR transfer of trapped peaks into a flow cell. Also possible is the transfer to NMR tubes using a liquid handler.

- Once the separation is finished, the SPE cartridges can be washed with water to remove, for example, any salt content or buffer.
- After washing, the cartridges are dried with nitrogen gas to remove most of the non-deuterated solvent.
- The dried cartridges can then be eluted with pure organic solvent in a small volume into either a flow cell or small-diameter NMR tubes. These NMR tubes can be 1.7, 2, 2.5 or 3 mm in diameter. A 1.7 mm tube needs about 25 µL to perform NMR measurements and is therefore the best match for maximum sensitivity.

Trap cartridges have to be conditioned and cleaned before their first usage to prevent unwanted signals. The whole tray compartment is best flushed with nitrogen constantly in order to avoid the collection of impurities from the laboratory air. For multiple peak trapping, it is best to have a UV flow cell in the outflow of the trap cartridges during a preparation run in order to determine the breakthrough of the compounds. The advantages of post-column SPE over loop collection can be summarized as follows:

- The use of completely non-deuterated solvents during chromatography removes the need for solvent suppression in NMR and also prevents the exchange of exchangeable protons as is always observed with D_2O as one of the solvents.
- The amount of deuterated solvent for elution from the SPE cartridge is much smaller than that needed for a standard 5 mm NMR tube.
- Ionic matrix from buffer systems is removed and therefore variations in chemical shift in the eluted fractions are reduced.

- S/N gains for NMR of up to a factor of 4 can be achieved per trapping step.
- Multiple injections allow for further increases in S/N and an increase in concentration for even the smallest peaks, such that they can be reach a concentration range suitable for NMR. Using loop collection allows only one injection per sample.
- The complete LC peak can be trapped with the SPE procedure. In loop collection, only a fraction of the peak contributes to the S/N as the eluting volume of an analytical column typically is of the order of 200–300 μL, which exceeds the active volume of a 3 mm flow cell (~60 μL).
- Multiple choices of trapping material allow access to a broader range of substrates.

In Figure 5.4, the gain in sensitivity by post-column SPE compared with loop collection is shown for an apple peel extract using similar LC conditions on a RP-18 column. The loop collection was performed with an injection volume of 100 μL, while the LC-SPE injection was only one-fifth of that per single injection with four repetitions. Figure 5.5 shows a comparison between single injection and fourfold injection for the LC-SPE process, and the expected increase in S/N ratio is observed.

In addition to the increase in S/N, thereby allowing the indirect detection 2D NMR proton–carbon correlation experiments generally required for structure elucidation, another advantage of the trapping and elution with deuterated acetonitrile becomes obvious. All exchangeable protons are visible in the spectrum and can be used to assist in structure elucidation. The different solvent systems for the two spectra (acetonitrile–water *versus* acetonitrile) explain the differences in observed chemical shifts.

It should be noted that the SPE approach allows solvent systems to be standardized by using either deuterated methanol or acetonitrile. This is in contrast to LC-NMR, where with gradient elution the solvent system is constantly changing and it is not possible to compare directly spectra obtained at different retention times. Standardization is necessary if spectral databases are to be created using the isolated compounds and to be used later for automated recognition.

5.2.5 Integration of Mass Spectrometric Detection of Peaks of Interest for LC-(SPE)-NMR

When poorly concentrated LC peaks have to be analyzed by NMR for the purpose of structure elucidation, it is necessary to be as specific as possible in the selection of the LC peaks. UV or DAD detection is very unspecific in most cases, and MS therefore plays an important role in both peak selection and structure elucidation, generally delivering precise molecular formulae and molecular fragmentation information.[12–14] In natural products research, it is possible to search selectively, for example, for certain glycosides with interesting structures. In this particular case, one or more conditions

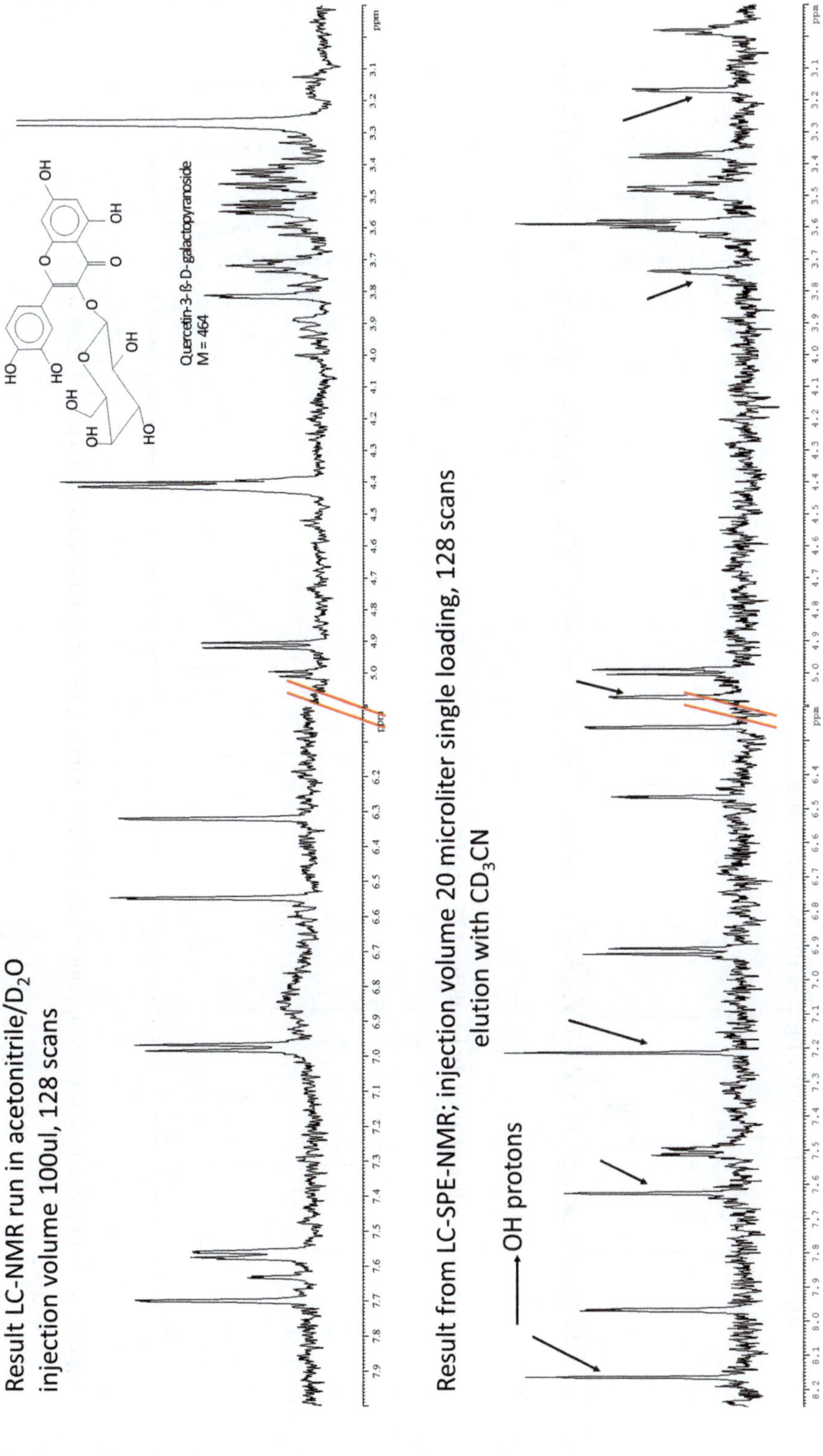

Figure 5.4 A sensitivity comparison of LC-NMR with a 100 µL injection versus fourfold SPE trapping with 20 µL injections each. The spectra shown are for quercitin 3-*O*-galactoside (hyperoside).

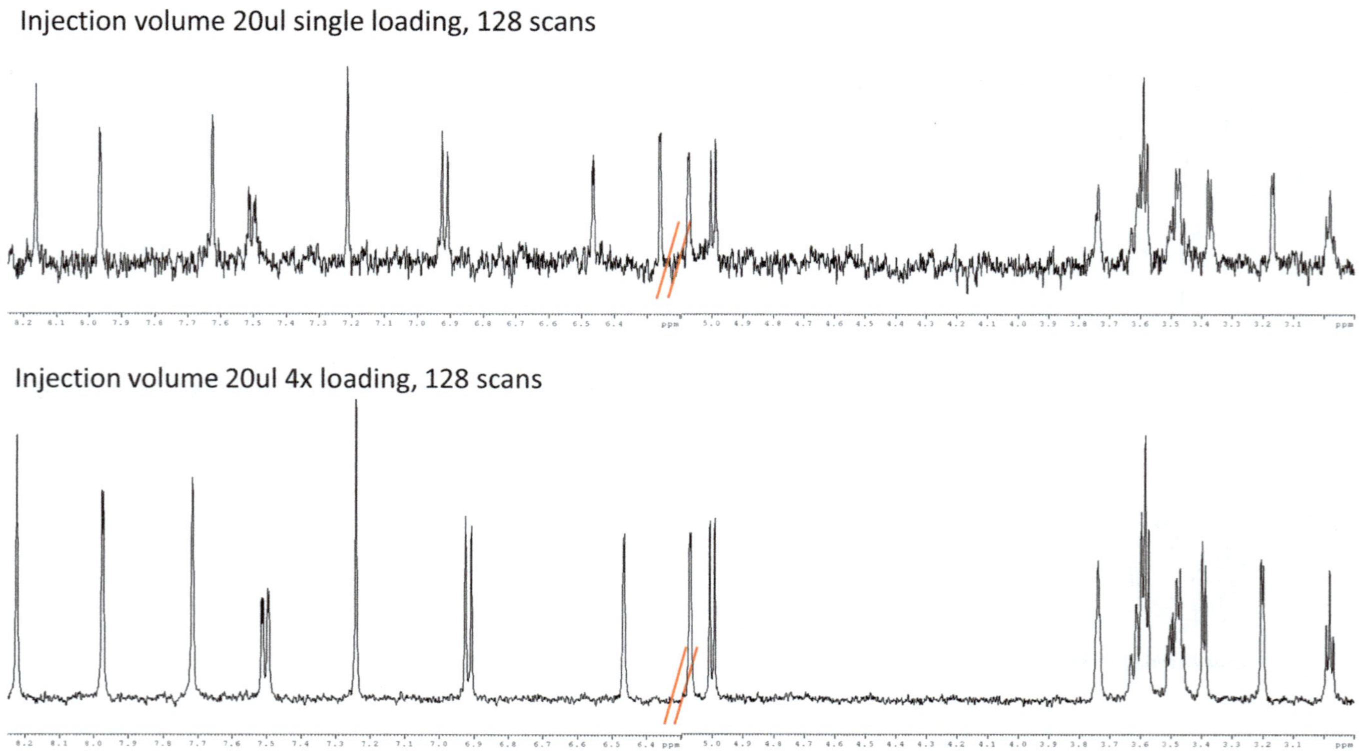

Figure 5.5 Comparison of single trapping *versus* fourfold trapping with the same conditions as in Figure 5.4.

have to be set to define the criteria for when peak collection for NMR should be executed. The simplest way to integrate MS into LC-NMR measurements is using a flow splitter after the LC column. According to the basic sensitivity of NMR and MS, a very small fraction of the flow (typically 5% or less) has to be diverted to the MS detector. The upper panel of Figure 5.6 shows the flowchart of a dedicated LC-NMR/MS interface as used for LC peak selection.

In this case, it is important to have the MS information available before a decision is made whether to collect the peak for NMR analysis using either loop collection or through SPE trapping. The transfer pathway from the splitter to the collection valve must be long enough to allow for both the MS transfer and analysis. It is obvious that, in this case, the line to the MS detector must be as short as possible. If, in addition to the MS, a UV or DAD detector is used, then the transfer capillaries to the individual detectors must be adjusted so that the retention times are identical in the chromatogram display of the software.

If a loop collection device is used, then the LC-NMR/MS interface has a different pathway, as shown in the lower panel of Figure 5.6. In this case, a delay loop is switched in-line on the MS side to delay the transfer until the NMR fraction has reached the flow cell and the main transfer pump of the LC system stops. The MS fraction sitting in the delay loop can now be transferred slowly into the mass spectrometer using a syringe pump, which is part of the interface. The same syringe pump can also be used to dilute the flow to the MS during the peak collection.

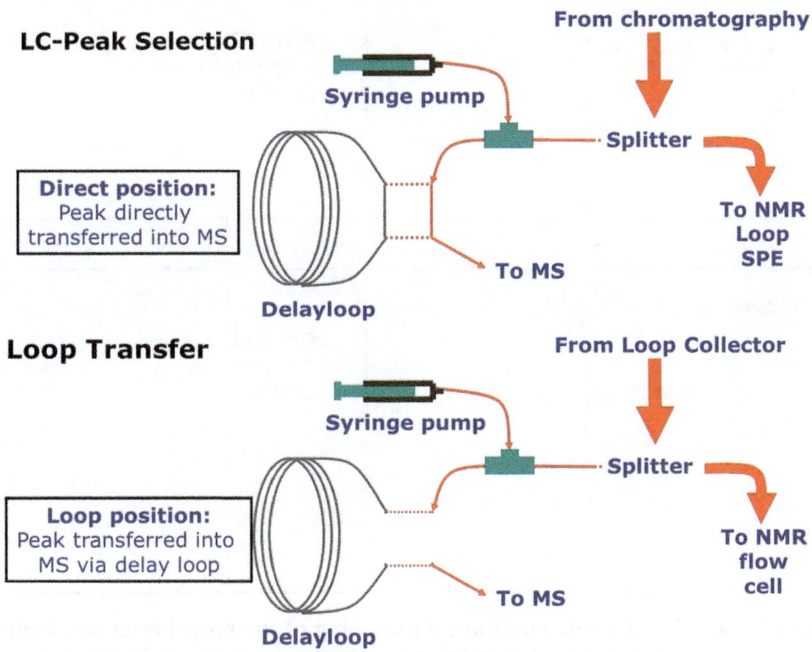

Figure 5.6 LC-(SPE)-NMR/MS interface allowing the use of MS information for peak selection and structure elucidation.

5.2.6 Cryogenic Probes and Their Advantages for LC-(SPE)-NMR

Although SPE post-column collection can generate increases of up to a factor of 4 in sensitivity per trapping step, there is a second tool available to improve NMR sensitivity by a further factor of 4, namely cryogenic probes or cryoprobes. In this case, the sensitivity increase is obtained by cooling the detection coils of the NMR probe to about 20 K using cold helium gas, and by cooling the preamplifiers of the NMR system to roughly 70 K. In modern systems, cryoflow inserts are used in conventional 5 mm cryoprobes and these inserts can have different active volumes depending on the particular application. The typical range is from 120 down to 30 μL active volume. Whereas loop collection requires 60–120 μL, the SPE application works best with small volumes down to 30 μL. In such cases, the filling factor is not fully used in a 5 mm probe. Therefore, the most sensitive approach is to fill small NMR tubes from the SPE cartridges and measure them with an optimal filling factor in a 1.7 mm cryoprobe. In this configuration, an increase in sensitivity by a factor of 6 rather than 4 is possible. In Figure 5.7, the results of both single and fourfold trapping obtained after an injection of 5 μg of the propyl ester of *p*-hydroxybenzoic acid on-column and measurement at 500 MHz using a cryoprobe equipped with a 30 μL active volume are shown.

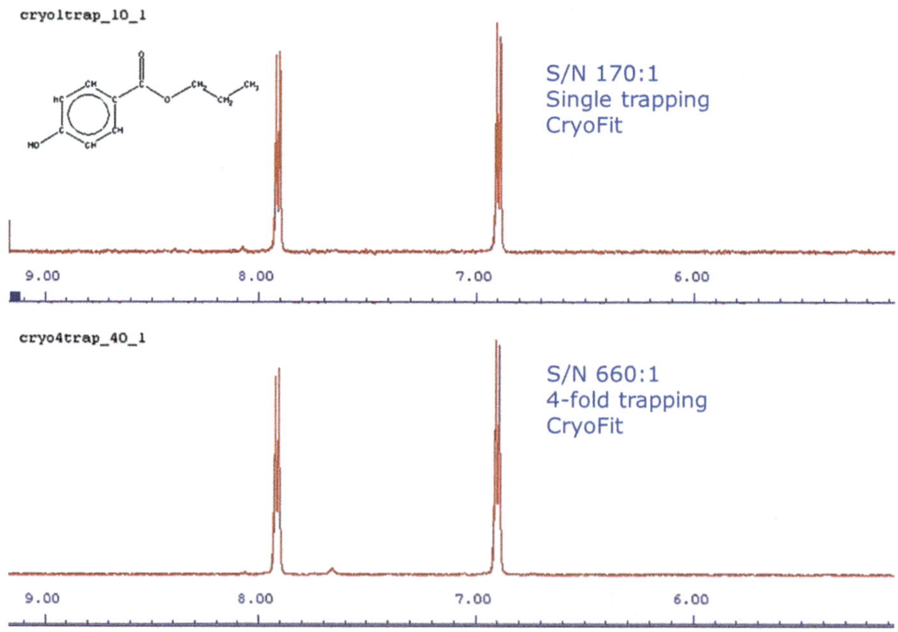

Figure 5.7 Results of single and fourfold trapping of the propyl ester of *p*-hydroxybenzoic acid after a 5 μg injection on-column per trapping and measurement at 500 MHz with 24 scans using a CryoFit (Bruker BioSpin) insert with a 30 μL active volume.

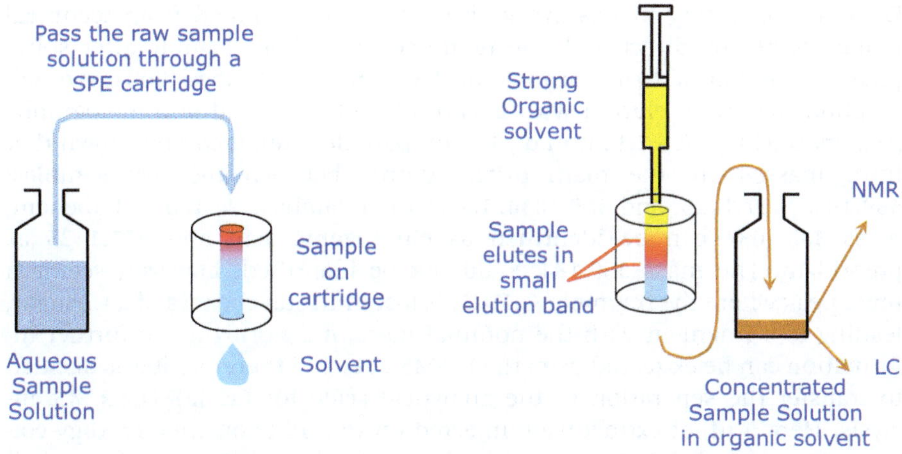

Figure 5.8 Flowchart of an SPE precleaning and enrichment procedure before injection into the LC-SPE-NMR/MS system.

For the same injection with loop collection and a 60 μL conventional probe, an S/N of 23.5 is obtained compared with 660 with fourfold trapping and a cryoprobe.

This result demonstrates the progress made with LC-SPE-NMR in its ultimate configuration. With this setup, it is possible to run even sub-microgram sample quantities and still obtain structurally relevant 2D information.

5.2.7 SPE-LC-SPE-NMR/MS

In order to increase further the performance in LC-SPE-NMR, an SPE enrichment and clean-up step can be added before the LC separation. Depending on the amount of sample available, even larger volumes can be extracted on a robotic system. The flowchart of the precleaning step is shown in Figure 5.8. Such an enrichment step is part of a process that can be called "total analysis." This procedure is described in Section 5.3.2.

5.3 Application Examples from Natural Product-related Samples

5.3.1 Integration of Metabonomics Routines and LC-SPE-NMR/MS

For the quality control of juice samples, it is often necessary to differentiate direct juice from rediluted concentrate. Direct juice is more expensive as it has at least a fivefold larger volume when it is transported and it is considered to be closer to freshly pressed juices as it has undergone fewer processing steps. When comparing LC-MS data from a high-resolution time-of-flight mass spectrometer (*e.g.* a microTOF-Q instrument from Bruker

Daltonics operating in positive ionization mode) obtained from technical replicates of one direct and one rediluted apple juice sample after solid-phase extraction of 10 mL of juice on Baker SDB SPE tubes (200 mg of adsorption material), elution with 2.5 mL of methanol and chromatography (Waters BEH C_{18} 50×2.1 mm i.d., 1.7 μm particle size), it can be shown that three masses are the main differentiators between the two samples: 569.1863, 437.1425, and 355.1034. Based on a database search in Pubchem, mass 355.1034 can be identified as chlorogenic acid and 437.1425 as phloridzin. The mass 569.184 could not be identified; however, seeing a mass peak where the fragment $C_5H_8O_4$ is lost indicates the loss of a C_5 sugar, leading to a fragment with the nominal mass of phloridzin. No further information can be extracted from the LC-MS data and therefore it was decided to transfer the separation to the analytical scale for LC-SPE-NMR/MS analysis. Here 5 μL of extract were injected on to a Phenomenex Prodigy column, 250×4.6 mm i.d., 5 μm particle size. Post-column SPE was set to search for the mass of 569.1863 and guide the corresponding LC peak into a Hysphere GP SPE cartridge (10×2 mm i.d.). The mass of interest was identified and the LC peak was trapped automatically. Elution of the trapped material into a 1.7 mm tube and measurement using a 1.7 mm cryoprobe (Bruker BioSpin) was performed running ^{1}H, COSY,[15] HSQC,[16] and HMBC[17] NMR experiments and also some selective excitation experiments.[18–20]

Figure 5.9 shows the chromatogram of the ultra-performance liquid chromatographic (UPLC) separation (upper trace) in comparison with the analytical-scale separation. For both separations the MS response and the

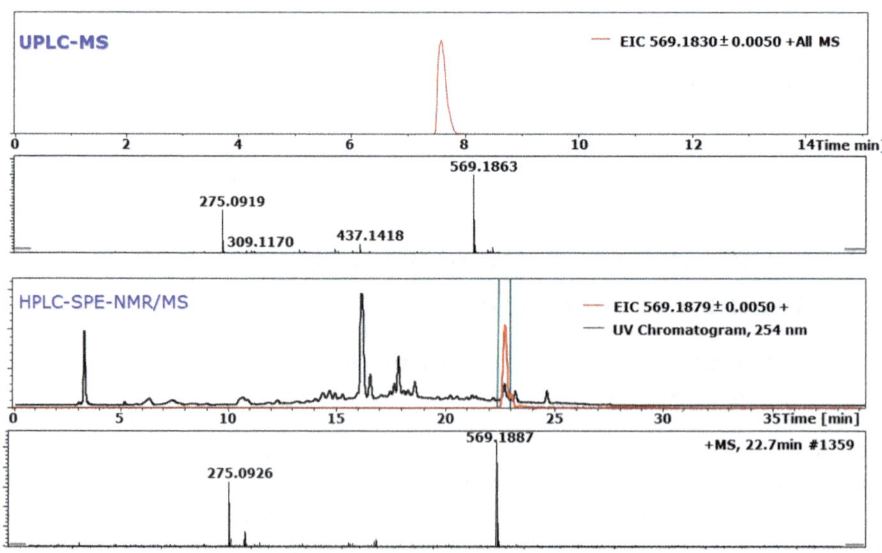

Figure 5.9 Transfer of a UPLC-MS method to analytical-scale HPLC for trapping of an unknown peak (phloridzin diglycoside) for NMR measurement.

UV trace are shown. The peak to be trapped is identified between the two blue bars.

The structure was assumed to be a diglycoside of phloretin and therefore the first experiments for structure elucidation performed were selective TOCSY experiments exciting the anomeric protons of the two expected sugar moieties that were assigned to the resonances in the 1D-NMR spectrum at 5.03 and 4.37 ppm, respectively. Figure 5.10 shows the results of three selective TOCSY experiments and the standard ^{1}H spectrum at the top. All sugar signals could be identified and associated with the 1′ and 1″ rings, and the final structure of the diglycoside is shown on the figure for reference.

The full proof for the correct structure, however, was obtained from ^{1}H/^{13}C inverse (HSQC) and inverse long-range (HMBC) correlated experiments. Figure 5.11 shows the overlay of the HSQC and the HMBC spectrum.

Starting from the anomeric proton on 1′ (see label), a long-range correlation to the closest carbon in the tetrasubstituted aromatic ring established the connectivity between the sugar ring with the ′-label and the aromatic skeleton. Also, the reverse connectivity is visible from the closest proton in the aromatic ring to the anomeric carbon 1′. The other important question of where the second sugar ring is connected is solved by observing the correlation from the proton on carbon C1″ to C6′. Carbon C6′ is easily identified as it shows two proton resonances for the two protons on C6′, the only CH_2 group in the sugar moiety.

This example nicely demonstrates the synergies between MS and NMR spectroscopy: MS allows the identification of the LC peaks of interest and the extraction of a molecular formula with high confidence. NMR allows the determination of the exact structure, and the performance of the NMR technique is vital in many cases where the sample amount per peak is limited. The use of post-column SPE with multiple trapping and of a cryogenic NMR probe can be keys to success in many cases for the elucidation of unknown natural products.

5.3.2 Example of the Total Analysis Concept SPE-LC-SPE-NMR/MS

The total analysis approach was created to facilitate the characterization of as many compounds in an extract as possible using NMR to perform a non-targeted screening independent of any LC peak information obtained using UV and MS detection. The procedure is as follows:

- SPE on a large scale using extraction columns with 250–500 mg of separation material.
- Elution with deuterated methanol.
- Partial removal of the solvent, then injection into an analytical-scale LC system with post-column small-scale SPE and DAD/MS detection.
- Trapping for 1 min per SPE cartridge then switching to the next cartridge. This is independent of any LC peak positions.

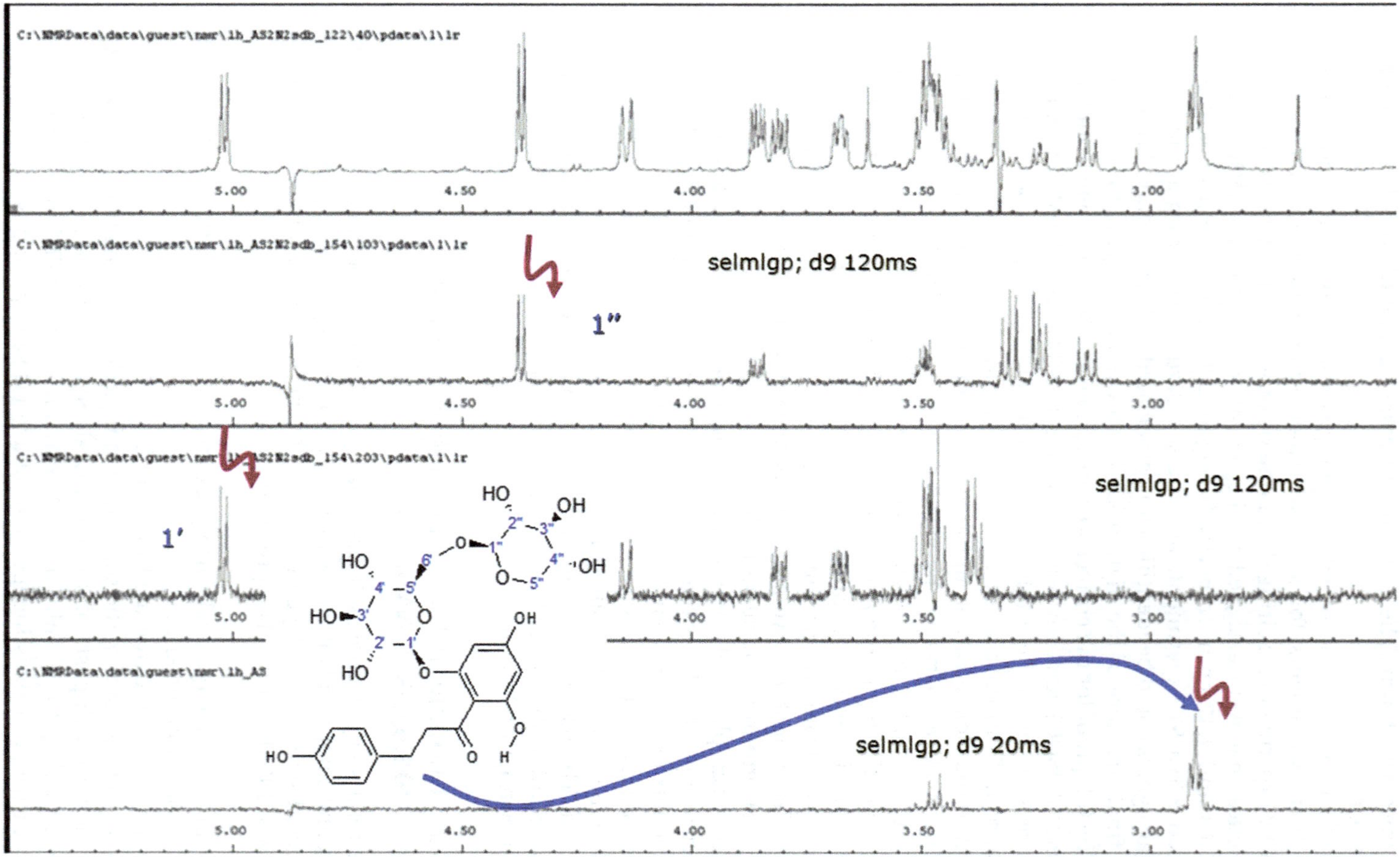

Figure 5.10 Selective TOCSY experiments on phloretin diglycoside, obtained through LC-SPE-NMR/MS, connecting the signals in the two sugar rings and the CH_2–CH_2 bridge between the two aromatic rings (600 MHz, 1.7 mm CryoProbe, mixing times as shown).

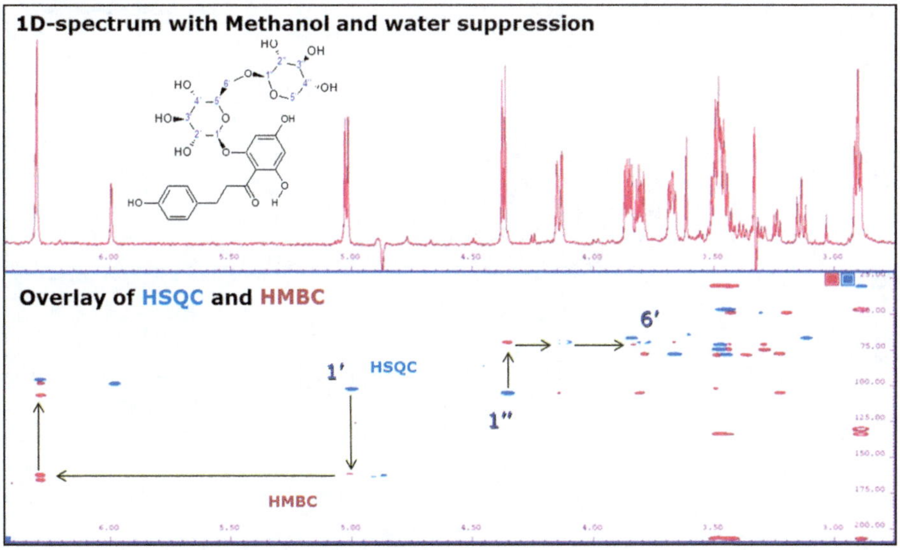

Figure 5.11 Superposition of the HSQC and HMBC spectra of the phloretin diglycoside identified in apple juice extracts.

- Elution of the trapped material in each cartridge into 1.7 mm NMR tubes and running the NMR spectra on the contents of each tube.
- Searching the spectra against a spectral database of pure compounds measured in the same solvent.
- If NMR peaks can be identified, then searching for the mass of the compounds verified for further confirmation.
- If unidentifiable NMR peaks are in the spectra, then running 2D-NMR spectra.
- If there is a mixture in the eluate of a cartridge, then reinjection into the LC-SPE-NMR/MS system.
- Running the peak detected by the trapping procedure. The chromatography can be optimized for a small retention time window trapped in the cartridges to obtain a clean spectrum of the isolated compound.
- Elution of the cartridges of the second trapping step into 1.7 mm NMR tubes and running the NMR spectra, preferably acquiring 2D-NMR spectra if the amount available in the sample allows a reasonable measurement time.
- Performing structure elucidation based on NMR and MS.

Figure 5.12 shows the chromatogram obtained for a 5 μL injection of the compound eluted from the large scale SPE on cranberry juice. High-resolution MS was performed in the negative ionization mode using base peak information over a mass range of 49–951 mass units with the UV data generated at 254 nm. Figure 5.13 shows the chromatogram after injection of 100 μL of the same eluate. In addition, the vertical lines show the switching

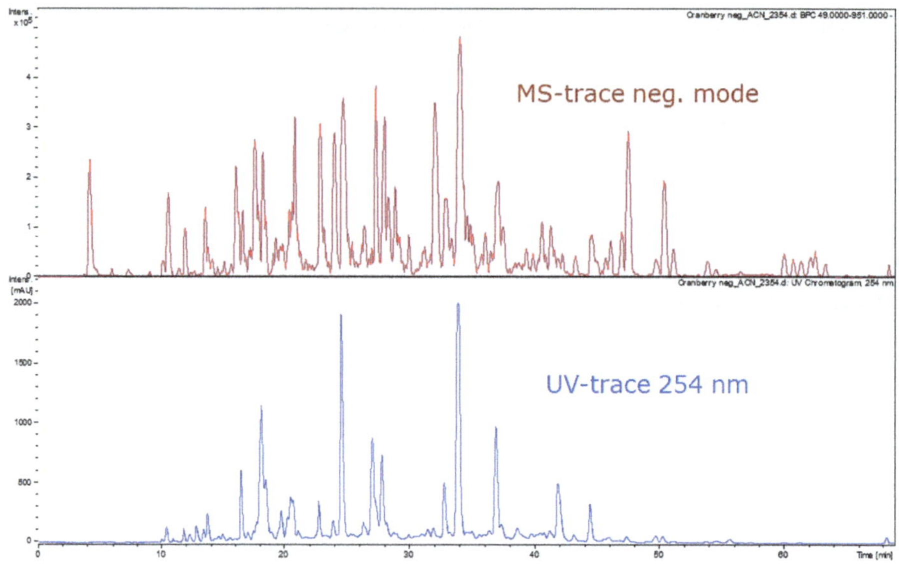

Figure 5.12 Chromatogram with UV (254-nm) and negative-mode MS base peak detection of a 5 μL injection from an SPE concentrate of cranberry juice.

of the cartridges into the flow line. As it takes a few seconds to change cartridges, there is a dead time of about 7 s where no trapping takes place. As can be seen, the trapping procedure starts at 5 min and ends after 75 min, meaning that 70 cartridges have been used for trapping. As the post-column SPE system used has a total of 192 cartridges, even finer gradations in time are possible, or longer runs can be executed. Even so, the chromatography looks totally overloaded with regard to UV, but the reduced sensitivity of the NMR technique moderates the picture and allows the generation of NMR spectra with usable purity in many cases, except where there are mixtures containing several peaks in, for example, a factor of 10–100 concentration scale. In the latter case, reinjection and LC peak-driven post-column SPE collection need to be conducted to purify the LC peaks.

Figure 5.14 shows the quality of NMR spectra obtained, where the spectra of each cartridge are placed into a pseudo on-flow spectrum. It is obvious that with this procedure many compounds can be made accessible to NMR detection.

If NMR signals are weak for some cartridges, then it is still possible to run the large-scale extraction in parallel on several cartridges and to combine the eluates. In order to increase the concentration for NMR further, partial evaporation of the elution solvent might be necessary. As this procedure is intended to deliver structure verification and elucidation of as many compounds as possible, it is not used quantitatively. After having resolved as many structures as possible and having pure spectra for input into a spectral database, then quantification can be performed on the SPE-NMR spectra of the large-scale extraction under precisely defined and quantitative conditions.

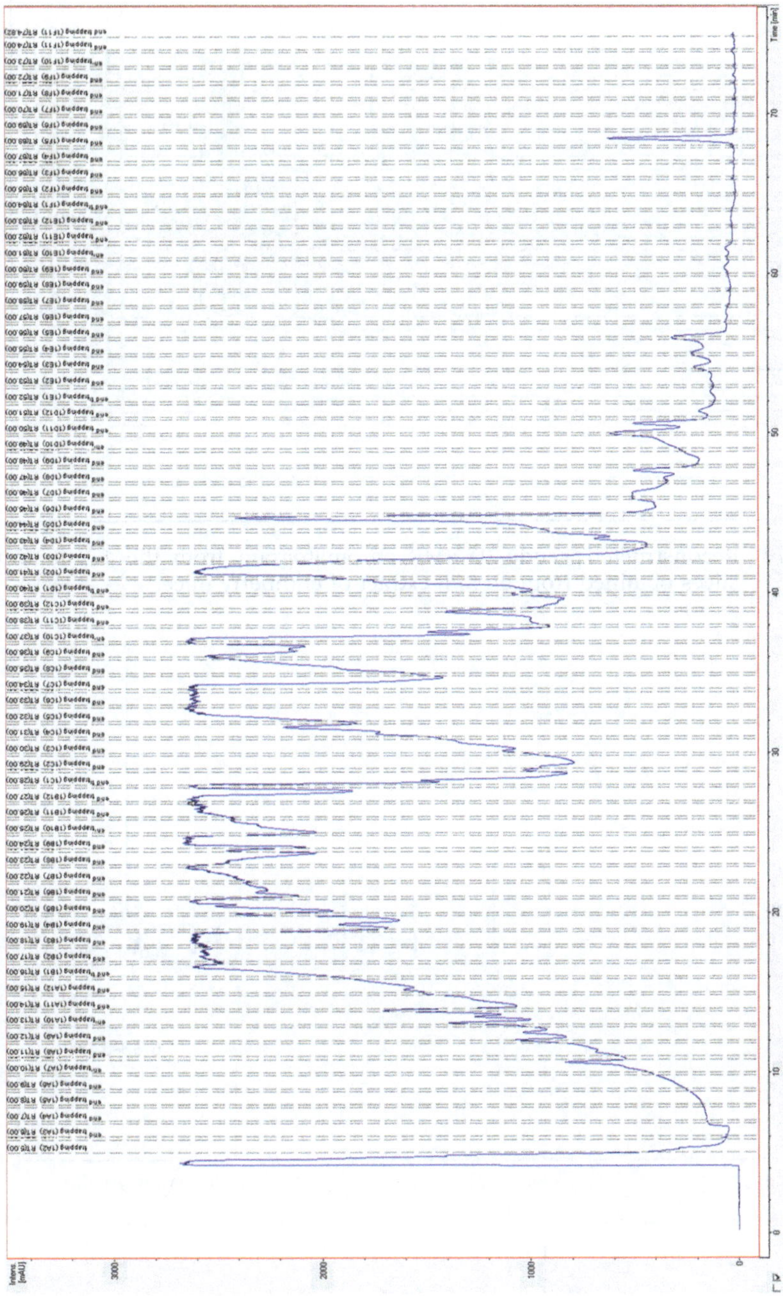

Figure 5.13 Visualization of the time slice SPE trapping process with 1 min slices applied to an SPE extract of cranberry juice with UV detection at 254 nm.

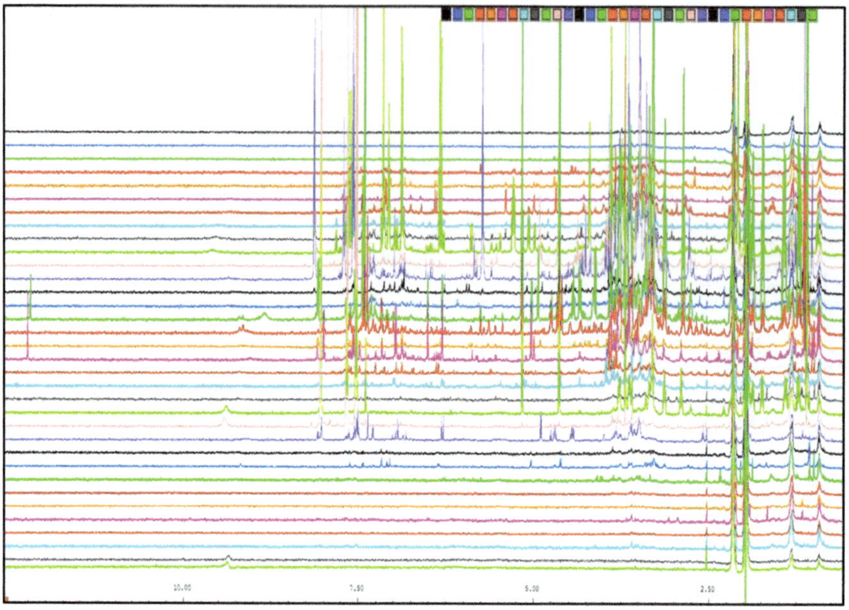

Figure 5.14 Reconstructed pseudo-2D-NMR chromatogram from the 1 min time slices obtained from an SPE extract of cranberry juice.

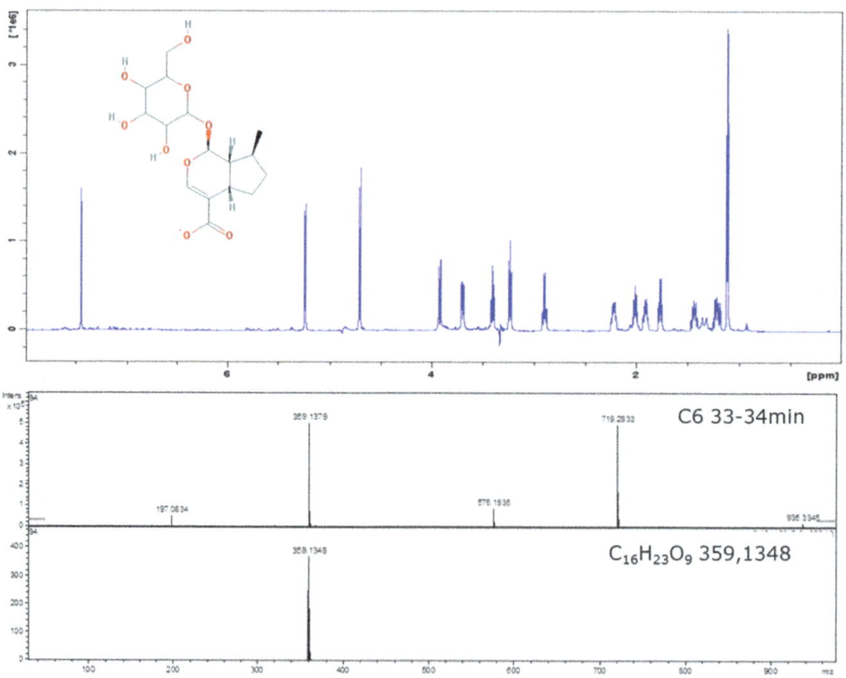

Figure 5.15 NMR and mass spectra obtained from a time slice of 34–35 min of the cranberry juice SPE extract injected into the LC-SPE-NMR/MS system and the structure of 7-deoxyloganic acid.

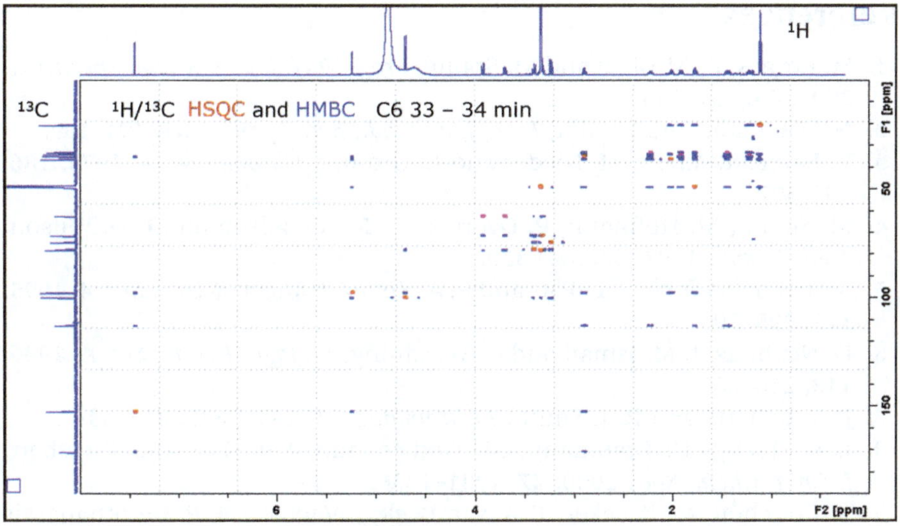

Figure 5.16 Overlay of HSQC and HMBC spectra of time slice 34–35 min of the cranberry juice SPE extract injected into the LC-SPE-NMR/MS system.

Figure 5.15 shows the 1D-NMR and mass spectra obtained for the cartridge containing the retention time window from 33 to 35 min. In this case, the NMR spectrum is pure enough to perform structure elucidation directly from an untargeted trapping procedure. Using the information from the $^{1}H/^{13}C$ inverse-detected HSQC and long-range HMBC correlation spectra shown in Figure 5.16, the compound is verified as 7-deoxyloganic acid, a compound not previously identified in cranberry juice. It should be obvious that the procedure described allows rapid dereplication and identification of unknown compounds using automation of the many steps described.

5.4 Conclusion

It has been demonstrated that LC-NMR can be integrated very efficiently into the structure verification and identification of natural product mixtures. The tools described allow us to increase NMR sensitivity in such a way that <1 μg components in the active volume can be accessed by NMR. The procedures described can be performed under full automation for most steps. Currently, the manual steps are the solvent evaporation and transfer of samples from large-scale SPE to the LC-SPE-NMR/MS setup. This is, however, something that may well be automated in the future. Software tools for the identification of compounds in a mixture, if the pure compounds exist in a spectral database, are already available under full automation. This means that, after NMR measurements of the small-scale SPE eluates for each cartridge, a listing of identified compounds can be generated automatically. Such approaches are discussed in Chapter 8 by Blunt *et al.*

References

1. H. Kovacs, D. Moskau and M. Spraul, *Prog. Nucl. Magn. Reson. Spectrosc.*, 2005, **46**, 131–155.
2. N. Watanabe and E. Niki, *Proc. Jpn. Acad., Ser. B*, 1978, **54**, 194–199.
3. E. Bayer, K. Albert, M. Nieder and E. Grom, *J. Chromatogr. A*, 1979, **186**, 497–507.
4. M. Spraul, M. Hofmann, P. Dvortsak, J. K. Nicholson and I. D. Wilson, *Anal. Chem.*, 1993, **65**, 327–330.
5. S. H. Smallcombe, S. L. Patt and P. A. Keifer, *J. Magn. Reson., Ser. A*, 1995, **117**, 295–303.
6. D. Neuhaus, I. M. Ismail and C.-W. Chung, *J. Magn. Reson., Ser. A*, 1996, **118**, 256–263.
7. J. K. Roberts and R. J. Smith, *J. Chromatogr. A*, 1994, **677**, 385–389.
8. L.-H. Tseng, U. Braumann, M. Godejohann, S.-S. Lee and K. Albert, *J. Chin. Chem. Soc.*, 2000, **47**, 1231–1236.
9. V. Exarchou, M. Krucker, T. A. van Beek, J. Vervoort, I. P. Gerothanassis and K. Albert, *Magn. Reson. Chem.*, 2005, **43**, 681–687.
10. O. Corcoran, P. S. Wilkinson, M. Godejohann, U. Braumann, M. Hofmann and M. Spraul, *Am. Lab. Perspect. Chromatogr.*, 2002, **34**, 18–21.
11. M. Godejohann, L.-H. Tseng, U. Braumann, J. Fuchser and M. Spraul, *J. Chromatogr. A*, 2004, **1058**, 191–196.
12. J. P. Shockcor, S. E. Unger, I. D. Wilson, P. J. Foxall, J. K. Nicholson and J. C. Lindon, *Anal. Chem.*, 1996, **68**, 4431–4435.
13. K. I. Burton, J. R. Everett, M. J. Newman, F. S. Pullen, D. S. Richards and A. G. Swanson, *J. Pharm. Biomed. Anal.*, 1997, **15**, 1903–1912.
14. I. F. Duarte, M. Godejohann, U. Braumann, M. Spraul and A. M. Gil, *J. Agric. Food Chem.*, 2003, **51**, 4847–4852.
15. W. P. Aue, E. Bartholdi and R. R. Ernst, *J. Chem. Phys.*, 1976, **64**, 2229–2246.
16. G. Bodenhausen and D. J. Ruben, *Chem. Phys. Lett.*, 1980, **69**, 185–189.
17. A. Bax and M. F. Summers, *J. Am. Chem. Soc.*, 1986, **108**, 2093–2094.
18. A. Bax and D. G. Davis, *J. Magn. Reson.*, 1985, **65**, 355–360.
19. H. Kessler, H. Oschkinat and C. Griesinger, *J. Magn. Reson.*, 1986, **70**, 106–133.
20. J. Stonehouse, P. Adell, J. Keeler and A. J. Shaka, *J. Am. Chem. Soc.*, 1994, **116**, 6037–6038.
21. M. Spraul, M. Hofmann, J. C. Lindon, D. Farrant, M. J. Seddon, J. K. Nicholson and I. D. Wilson, *NMR Biomed.*, 1994, 7, 295–303.

CHAPTER 6

Application of Non-uniform Sampling for Sensitivity Enhancement of Small-molecule Heteronuclear Correlation NMR Spectra

MELISSA R. PALMER,[a] RIJU A. GUPTA,[a] MARCI E. RICHARD,[a] CHRISTOPHER L. SUITER,[b] TATYANA POLENOVA,[b] JEFFREY C. HOCH[c] AND DAVID ROVNYAK*[a]

[a] Department of Chemistry, Bucknell University, Lewisburg, PA 17837, USA; [b] Department of Chemistry and Biochemistry, University of Delaware, Newark, DE 19716, USA; [c] University of Connecticut Health Center, 263 Farmington Avenue, Farmington, CT 06030, USA
*Email: drovnyak@bucknell.edu

6.1 Exponential Non-uniform Sampling and Sensitivity

Traditionally, time-domain NMR signals are acquired in equally spaced time increments (uniform sampling) and undergo a discrete Fourier transform (DFT) to recast these data as a frequency spectrum.[1,2] There has been rapid growth in the development and adoption of alternative approaches to acquire and process NMR data obtained for indirect evolution periods; such methods generally do not uniformly sample the NMR signal and also employ

Modern NMR Approaches To The Structure Elucidation of Natural Products,
Volume 1: Instrumentation and Software
Edited by Antony J. Williams, Gary E. Martin and David Rovnyak

Published by the Royal Society of Chemistry, www.rsc.org

methods other than the Fourier transform to obtain an NMR spectrum.[3–6] A large family of methods is based on non-uniform sampling (NUS),[7,8] defined here as the acquisition of a subset of samples selected from the uniformly sampled Nyquist grid. Non-uniform sampling is often implemented by retaining about 25–33% of the number of samples that would be acquired uniformly, but can be more sparse. The non-uniformly distributed samples often follow an exponential density (more samples at early times) and can be spread over evolution periods significantly longer than those if they were constrained to a uniform grid. Thus NUS has been widely used and investigated as a means to save total experiment time while simultaneously enhancing resolution,[7–15] and it has been observed empirically that NUS can be applied in any situation in which uniform sampling has sufficient sensitivity to yield interpretable spectra.[16–20]

The potential for enhancing signals by exponential NUS was recognized in seminal papers by Barna and co-workers in 1987,[7,8] and is based on the established principle that sensitivity in multi-dimensional NMR is favored when sampling is tailored to the signal envelope.[21] In 1991, Kumar *et al.* reported that signal enhancements in non-uniformly sampled time-domain data can be realized by taking more transients where the signal envelope is strongest.[22] The work of Kumar *et al.* perhaps has been underappreciated since they did estimate approximate signal enhancements, supported with example spectra in which the DFT was applied to data that had been acquired with transients distributed non-uniformly over the uniform Nyquist grid, a practice often dubbed non-uniformly weighted sampling (NUWS). Although NUWS incurs a line broadening of the detected signals when the DFT is used, the ability to test NUWS enhancements with the DFT, a power-conserving transform, clearly demonstrated that exponential sampling yielded signal enhancements. It is important to recognize that NUS and NUWS have the identical theoretical density of samples, so that the ability to obtain enhancements generalizes to either implementation of exponential sampling (NUS or NUWS).[22] Recently, the exact solution was reported for the enhancement of the intrinsic signal-to-noise ratio (S/N) of a signal in the time domain when applying non-uniform sampling to decaying signals, revealing that signal enhancements up to about twofold are possible for a given indirect evolution period.[23] The improvements can be compounded in multiple indirect dimensions to generate enhancements in excess of threefold.[24]

We review the sensitivity enhancement resulting from the use of NUS in an indirect evolution period (dimension) of a decaying signal and then present a number of example applications. Note that sensitivity is the S/N achieved per unit measurement time (strictly, per the square root of measurement time).[2] Since we will compare exclusively uniform and non-uniform acquisitions that consume identical total measurement times, we may use S/N and sensitivity interchangeably. Further, we have found it useful to distinguish the definitions of the intrinsic and apparent S/N values.[25] The intrinsic S/N refers to the raw acquired data, prior to any and all post-acquisition treatments, and is usually conveniently measured in the time domain, although it can be measured equivalently in the frequency domain if a DFT is applied as the sole operation. The apparent S/N refers to the appearance of spectra following digital signal processing such as apodization and linear prediction. In this chapter, we describe only the time evolution of the intrinsic S/N by NUS, as it is axiomatic that the intrinsic S/N is the sole criterion for judging if NUS is improving the actual levels of signal relative to noise in the acquired data. In summary, in any indirect dimension in which three conditions are met:

(i) the signal decays (the common case of monoexponentially decaying signals is explicitly considered),
(ii) the evolution time spanned by the non-uniform samples is in the range (2–3)T_2, and
(iii) the density of non-uniformly chosen samples resembles the signal envelope, such that more samples are distributed in regions that have large signal amplitude,

then the NUS approach will have greater intrinsic S/N than a uniformly incremented experiment spanning the same evolution time and consuming the same total experiment time. Depending on a number of acquisition parameters, the S/N improvement may be just 10–20%, but can often realistically achieve up to twofold improvement.[23,24] Criterion (i) can be generalized to state that any signal with a non-constant, time-domain envelope is a candidate for NUS-based enhancement, but this review will focus on exponentially decaying signals.

These three criteria immediately help to identify which types of experimentation will most benefit from NUS-based sensitivity enhancements. We briefly consider four cases: biological NMR in liquids, biological NMR in solids (biosolids NMR), small-molecule NMR in solids, and small-molecule NMR in liquids.

Biological NMR in liquids. In general, in protein NMR in liquids, there are modest opportunities to obtain NUS-based enhancements. For example, there is no possibility to enhance the S/N by performing NUS in a non-decaying period such as the constant-time periods commonly employed in biological *n*D-NMR experiments in liquids (criterion i).[23] Furthermore, the signal decay in liquid-state protein samples can be very long compared with accessible evolution times, such that even with NUS it may be difficult to reach times of (2–3)T_2 (criterion ii).[16] However, two-dimensional biomolecular liquids experiments such as 2D-HSQC spectra that are used to monitor chemical shift titrations could be enhanced by NUS.

Biological NMR in solids. In contrast, in biological solid-state NMR, there are a number of factors that are ideal for obtaining NUS-based signal enhancements.[24] For example, constant-time periods are not common in biosolids NMR experiments (criterion i). Further, T_2s are relatively short in solid state NMR of proteins, making it easy to reach (2–3)T_2 in indirect carbon and nitrogen dimensions of biosolids nD-NMR experiments (criterion ii). Thus biosolids NMR is well suited for achieving enhancements on the order of twofold in any one indirect dimension, and thus also allowing for the ability to compound such enhancements in multiple dimensions.[24] Recently, NUS-based signal enhancements in protein homonuclear correlation solid-state NMR experiments have also been demonstrated.[26]

Small-molecule NMR in solids. Similar considerations apply for small molecules in solid-state NMR, such as possessing fairly short T_2 relaxation times. NUS-based signal enhancements of up to twofold are realistic in this case also, as demonstrated for example on the MLF tripeptide.[24]

Small-molecule NMR in liquids. Finally, the study of complex small molecules in liquids with 2D-NMR often involves conditions that are favorable to achieving NUS-based enhancements, which can permit studies of very dilute samples. Constant-time evolution periods are not common in the acquisition of two-dimensional spectra such as HSQC, HMQC, *etc.*, for example (criterion i). Spectral crowding exhibited by natural products often requires operating at the limit of resolution by taking evolution times out to $(2–3)T_2$ (criterion ii). Indeed, in the NMR of complex small molecules it is necessary simultaneously to optimize sensitivity and resolution, and NUS provides a route to achieving long evolution times and improving sensitivity by up to twofold in a given dimension.

An aim of this chapter is to elaborate on a common approach to discussing the S/N and sensitivity of spectra acquired using NUS methods that is based solely upon the acquired time domain signal intensity prior to any form of post-acquisition signal manipulation, *i.e.* the intrinsic S/N. As noted earlier, a study of the NUS signal enhancement[22] reported estimates of the time-domain S/N by NUS that were later refined by the exact solution.[23] The analysis of S/N in the frequency domain may depend strongly upon computationally interpolated samples that were not part of the original data set,[27] and shows dependence on the chosen post-acquisition processing method.[18] Recent work also analyzed the sensitivity of the closely related practice of non-uniform weighted sampling (NUWS), in which all samples are acquired uniformly but transients are distributed non-uniformly;[28] however when processed by the DFT, then NUWS places an intrinsic windowing on the raw time-domain data that does not occur with NUS, requiring the authors to make a comparison of uniform sampling including post-acquisition apodization for noise suppression to an NUWS data set that received no post-acquisition noise suppression. Analysis of the raw time-domain data yields the only accurate determination of the sensitivity improvements afforded by NUS, regardless of the subsequent method used to manipulate the data and estimate spectra.[19,23,24] Any post-acquisition manipulation of the data, be it apodization prior to the FFT, artificial extension of the data by linear prediction, or the use of maximum entropy reconstruction, cannot change the intrinsic S/N of the raw data.[29] It is certainly true that an array of computational manipulations have been designed to distinguish authentic signals from Gaussian distributed noise, including some new approaches,[30,31] but in principle these are all reducible to computational models of the raw time-domain data.

6.2 Signal Enhancement by Non-uniform *Versus* Uniform Sampling

We wish to compare non-uniformly (NUS) and uniformly collected data in an indirect evolution period. Three broad criteria to assure that equitable comparisons are made will be discussed next: both acquisitions will consume the same total experimental time; both acquisitions will span the same total evolution time; and data will receive equitable treatment in processing. Next, the derivation of the enhancement is reviewed in Section 6.2.1, and non-uniform sampling densities are reviewed in Section 6.2.2, with a brief discussion of experimental validations in Section 6.2.3. We will use the definition that a "sample" corresponds to one precise indirect evolution time and represents both the real and imaginary free induction decays that are acquired in a typical States fashion for sign discrimination, but the generalization of NUS to partial component sampling, *i.e.* a sample may be only the real or imaginary part, has now been described.[32] Each sample is accumulated with some fixed number of transients.

Total experimental time. We need to specify *how* the uniform and NUS experiments will be adjusted to use identical experimental times. Clearly, it would invalidate any comparison if the relaxation delay, the FID length, or any other timing parameter in a given pulse sequence was varied. Pulse sequence parameters for delay times, pulse powers and durations, and receiver acquisition variables such as the gain must be strictly conserved in S/N comparisons. Only the number of transients that are acquired per sample in either uniform or non-uniform sampling can be varied. Specifically, since NUS acquires fewer samples in the indirect dimension compared with uniform sampling, the time saved by omitting samples *via* NUS can be used to increase the number of transients acquired for the remaining samples. For example, suppose one collects four transients per sample for 128 uniformly distributed samples, then one could collect 16 transients per sample for 32 non-uniformly distributed samples. Of course, this procedure is not feasible for the directly acquired FID.

Consistent evolution times. The last evolution time sampled in a given NUS schedule will be equal to that of uniform incrementation – but further consideration is needed. Several example NUS schedules are depicted in Figure 6.1, where each schedule retains the sample at time πT_2. A number of additional decisions are required on the nature of the NUS approach, as follows. (i) How should samples be distributed non-uniformly over the same evolution time that is spanned by uniform sampling? In analogy with the use of a matched filter in signal apodization, it is reasonable to propose sampling in a fashion that mirrors the intensity of the signal, where it is common to choose exponentially weighted sampling densities for performing NUS of decaying sinusoidal signals.[7,8] That is, the probability of choosing the non-uniform samples is weighted in proportion to the signal intensity, which has important implications for improving sensitivity by NUS. Continuing the analogy of a matched filter, we could use an exponentially weighted NUS sampling density that has the same time constant as the T_2 for the signal decay, a case which can be termed "matched NUS" and is depicted for the example of selecting 32 samples from a 128-sample Nyquist grid in Figure 6.1. (ii) What range of exponential sampling functions is feasible? Specifically, we may wish to bias the sampling to earlier times, where signal intensity is higher, by choosing a sampling density which decays more quickly than T_2, allowing one to allocate more samples and thus more transients to early times where the signal is stronger. Several cases of biased NUS are depicted in Figure 6.1, where it is observed that, when the exponential sampling density is biased to greater than about twofold *versus*

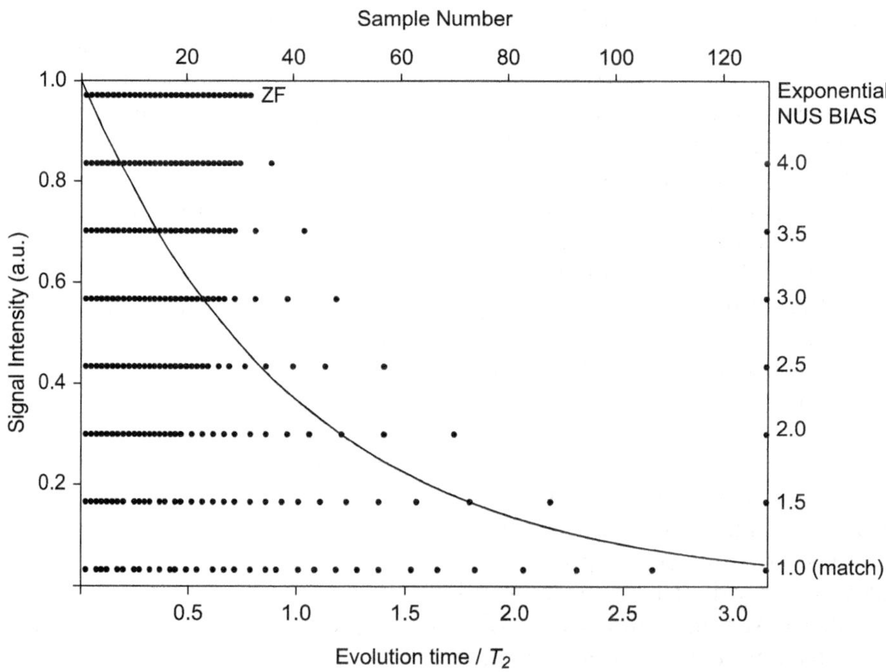

Figure 6.1 Examples of non-uniform sampling schedules selected according to exponentially weighted probabilities are superimposed on the signal decay (T_2). In the case of matched exponentially weighted NUS (bottom schedule) the probability density matches the signal decay, that is, the decay constant of the exponential probability density is $T_{smp} = T_2$. The non-uniform sampling schedules can then be biased to favor acquiring more samples at earlier times by choosing $T_{smp} = T_{2.}$, and examples of biased NUS schedules up to $T_2/T_{smp} = 4$ are given to facilitate comparison with the limiting case of signal truncation (top). These examples are based on selecting 32 samples from a uniform Nyquist grid of 128 samples.

the natural signal decay, the sampling approaches the trivial case of signal truncation, which risks forfeiting any resolution benefits of NUS. That is, although signal truncation can be very favorable to improving sensitivity, the value of the samples at long evolution times to improving spectral resolution is lost. (iii) What degree of sample reduction should be employed by NUS? Strictly, the context for our comments here is the subsequent use of maximum entropy reconstruction to process the non-uniform data.[1,33] A number of factors must be considered, and recommendations presented here are empirical in nature and emphasize conservative choices. As demonstrated previously, the greater the degree of sample reduction, the more closely the retained samples conform to the desired weighting.[23] That is, if one selects 64 samples exponentially from a 128-sample uniform grid, then there will be large tracts of uniformly spaced samples that do not conform to the desired exponential bias. In the other extreme, selecting eight samples out of 128 would appear to obviate this problem as there could be no uniform tracts, but such extreme reduction would have severe and complicated biases for the frequencies that are detectable (a.k.a. the point spread function[34,35]). In this and previous work, we consistently found that sample reduction of three- to fourfold is appropriate and somewhat conservative in a large range of cases, particularly when the uniform grid is on the order of 128–2048 samples. If the uniform grid is small (<128 samples), then threefold reduction is the practical limit in our experience, whereas if the uniform grid is large (>1024), then higher reductions such as a fivefold reduction can be performed.[10] (iv) Finally, how can samples be chosen to optimize the performance of subsequent spectral estimation? A great deal of progress has been made in identifying desirable factors for optimizing the distribution of non-uniformly chosen samples.[4,25,27,36–38] In brief, the size of the largest gap should be minimized, and the retained samples should have a random character to avoid introducing systematic frequency biases.

Equality of processing schemes. Spectra obtained from uniformly sampled data will be processed *via* the FFT algorithm, whereas spectra will be estimated from non-uniformly sampled data *via* maximum entropy reconstruction (MaxEnt). Identical digital resolution is specified in the frequency domain in each case. If apodization such as line broadening is applied to uniform data prior to the FFT, then the same result can be achieved in MaxEnt processing by specifying a convolution of the raw data with an exponential apodization function. MaxEnt is a non-linear reconstruction method, meaning that peak intensities in MaxEnt reconstructed spectra are not exact representations of their intensities in the time-domain samples.[1,39] Specifically, when there are multiple signals of different intensities in the raw time-domain non-uniform samples, then the resulting MaxEnt spectrum biases the strong peaks relative to the weak peaks. All peaks are still accurately detected at their correct frequencies, but their integrated areas cannot be directly interpreted, although it is possible to calibrate the non-linearity with injected signals, and then the peak areas may be evaluated precisely.[39] Recently, a straightforward extension of

MaxEnt has been described (MINT: maximum entropy interpolation) that imparts a high degree of linearity to the resulting frequency domain spectra and has been used to quantitatively evaluate NUS-based enhancements.[24,26] Forward maximum entropy (FM) also shows a linear response.[40] In this review, we will generally make qualitative comparisons that reflect routine user experiences with NUS regardless of which reconstruction algorithm is used, and which show unambiguously that NUS changes the detection limit of 2D heteronuclear correlation spectra.

6.2.1 Signal Enhancement of an Exponentially Decaying Signal by NUS

The most objective characterization of the signal and noise in NMR data must be of the raw time-domain data. In examining the time-domain data, the measured intensity for the *n*th sample is assumed to be separable as a sum of pure signal and pure noise contributions:

$$\text{intensity}\ (t_n) = \text{signal}\ (t_n) + \text{noise}\ (t_n) \tag{6.1}$$

The signal must add linearly for n samples, while the noise adds as the square root of the evolution time, *i.e.* the number of samples. It is convenient to work in the limit of continuous sampling, where the discrete sum of the pure signal is replaced by an integral, and the noise depends on the square root of the total acquisition time, t_{max}.[16,41] Then it can be shown that the S/N of a given FID develops in time as[16]

$$\text{S/N}(t) \propto \frac{T_2\left(1 - e^{-t_{\text{max}}/T_2}\right)}{\sqrt{t_{\text{max}}}} \tag{6.2}$$

where the proportionality reflects that there is a spectrometer-specific constant scaling factor that reflects variables dictating signal strength (*e.g.* preamplification of the signal) and the noise (*e.g.* cryogenic probes). A min–max computation shows that, regardless of the noise level, the maximum in S/N occurs at about $1.26T_2$.[16] Any samples after this time decrease the total S/N in the time domain whereas any samples before this time improve the S/N.

Enhancing the S/N in the time domain by non-uniform sampling is based on a strategy of "double dipping." First, one eliminates samples primarily from the region after $1.26T_2$. Discarding any sample after $1.26T_2$ will improve the total S/N of the remaining samples. Next, the time saved by omitting the late samples is used to acquire more transients for all of the remaining samples, further improving sensitivity. There are losses in signal intensity resulting from the reality that a small number of samples are discarded from times prior to $1.26T_2$. Since the time saved by discarding *any* sample is devoted to acquiring additional transients, the penalty for discarding a sample prior to $1.26T_2$ is partially mitigated.

The qualitative arguments above are reflected in the detailed mathematical treatment of the intrinsic S/N enhancement by non-uniform sampling. Specifically, we consider the case that time saved by omitting samples is used to acquire more transients spread evenly over the remaining samples. So, if 512 samples are selected non-uniformly from a uniform grid of 2048, and if the uniform acquisition employs four transients per increment, then the non-uniform acquisition will use 16 transients per increment. The uniform and non-uniform acquisitions to be compared must consume identical total measurement times. Pragmatically, this is most easily assured by setting the total number of acquired transients to be identical (as in the above example in which $16\times512=4\times2048=8192$ transients). We express this constraint in the continuous limit as requiring the areas of uniform and non-uniform sampling densities to be equal. We arbitrarily set the uniform sampling density to unity so that the area is just $1\times t_{\mathrm{max}}$ or simply t_{max}. Then we need only find a normalizing factor χ such that the area for the non-uniform sampling density equals t_{max}:

$$t_{\mathrm{max}}=\chi\int_0^{t_{\mathrm{max}}} h(t)\mathrm{d}t \tag{6.3}$$

where $h(t)$ is the non-uniform sampling density. Examples of several different normalized sampling densities $\chi h(t)$ are shown in Figure 6.2, where it may be appreciated that the areas are all identical. We focus this report on exponential sampling densities [*e.g.* $\exp(-t/T_{\mathrm{smp}})$, where T_{smp} is the decay constant of the exponential sampling density and we refer to the case of $T_{\mathrm{smp}}=T_2$ as matched NUS]. However, others report favorable experiences with Gaussian distributed NUS,[36] and we illustrate a Gaussian sampling density in the time domain in Figure 6.2 that is matched to the linewidth that would be expected from exponential T_2 decay, which is explained in more detail shortly.

Recognizing that the total number of transients is equivalent for the NUS and uniform acquisitions that we wish to compare, then both cases must have the same amount of noise, so it is not necessary to know or describe the noise further in the derivation. We then have only to describe the amount of signal obtained in the NUS and uniform approaches. The signal intensity for non-uniform sampling is the product of the normalized sampling density and the signal, which we take to be $\exp(-t/T_2)$:

$$S(\mathrm{NUS})=\int_0^{t_{\mathrm{max}}} \chi h(t)\mathrm{e}^{-t/T_2} \tag{6.4}$$

which is illustrated graphically for several exponentially weighted non-uniform sampling schemes in Figure 6.3. The enhancement by non-uniform *versus* uniform sampling is then the ratio

$$\eta=\frac{\int_0^{t_{\mathrm{max}}} \chi h(t)\mathrm{e}^{-t/T_2}}{\int_0^{t_{\mathrm{max}}} \mathrm{e}^{-t/T_2}}=\frac{\chi\int_0^{t_{\mathrm{max}}} h(t)\mathrm{e}^{-t/T_2}}{T_2(1-\mathrm{e}^{-t_{\mathrm{max}}/T_2})}. \tag{6.5}$$

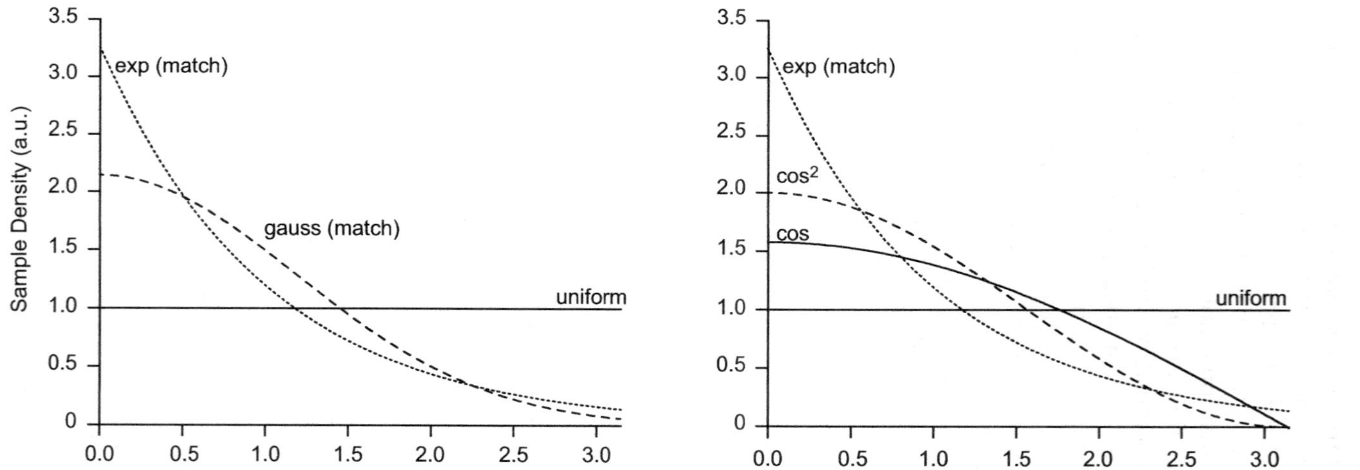

Figure 6.2 A survey of NUS sampling densities satisfying the criterion of having areas equal to that of uniform sampling, meaning that all depicted NUS schedules consume the identical experimental times.

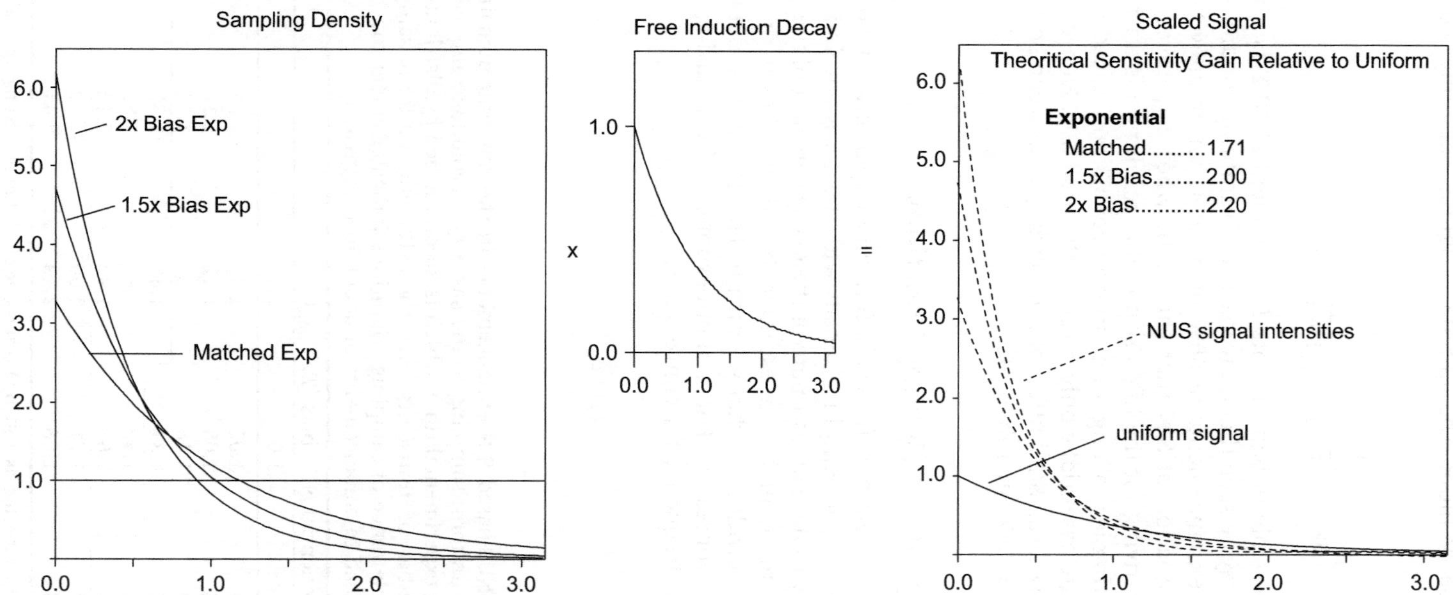

Figure 6.3 The origin of the signal enhancement in the time domain of NUS data is depicted graphically by recognizing that NUS delivers a scaled raw signal intensity. Conversely, this figure helps to understand that NUS cannot improve the sensitivity of a constant time signal since each sampling density would then be multiplied by unity; since the sampling densities all have equivalent areas in order to consume the same experimental time, they would then all result in the same signal intensity when applied to a constant-time signal.

For the specific case of monoexponentially weighted sampling, eqn (6.5) reduces to

$$\eta_{\text{exp}} = \frac{\chi\left\{1 - e^{-t_{\max}\left[(T_2/T_{\text{smp}})+1\right]/T_2}\right\}}{(T_2/T_{\text{smp}})(1 - e^{-t_{\max}/T_2})} \tag{6.6}$$

The range of accessible enhancements for exponential NUS is given in Table 6.1, where it may be seen that a diverse set of sampling conditions can be used to achieve enhancements of 50% or greater. An often-overlooked concern is the degradation of S/N that can occur when $t_{\max}$ exceeds $3T_2$. Indeed, a pitfall of striving for the highest possible resolution is mistakenly to set $t_{\max}$ to a time that is too long. In a uniform acquisition, those samples beyond $3T_2$ contribute essentially only noise to the total S/N. Hence the use of NUS is seen to provide an S/N buffer for mis-setting $t_{\max}$, as can be seen for the example of $t_{\max} = 3.5T_2$ in Table 6.1.

6.2.2 Evaluating NUS Weighting Functions

Maximal resolution is obtained for evolution times spanning up to about πT_2. In order to compare the sampling density with the decay envelope of the signal, it is necessary to have an estimate of the decay constant of the signal. It is often convenient to make an estimate based on known or typical frequency domain linewidths so that we turn to Fourier pairs to obtain the time domain decay constants. For a monoexponentially decaying signal which occurs at zero frequency for convenience:

$$S(t) = e^{-t/T_2} \tag{6.7}$$

Table 6.1 Survey of NUS-based S/N enhancements in the raw time-domain data, relative to uniform sampling to the same $t_{\max}$ evolution time using the same total experimental time, which is accomplished by distributing the same number of transients over the NUS and uniform samples; a considerable range of sampling conditions indicated under the stepped lines can lead to enhancements of about 50% or greater.

	Exponential NUS bias (T_2/T_{smp})				
$t_{\max}/T_2$	0.5	1.0	1.5	2.0	2.5
0.5	1.01	1.02	1.03	1.04	1.05
1.0	1.04	1.08	1.12	1.16	1.20
1.5	1.09	1.18	1.27	1.34	1.41
2.0	1.16	1.31	1.45	1.57	1.67
2.5	1.24	1.48	1.67	1.83	1.95
3.0	1.34	1.66	1.92	2.12	2.26
3.14	1.37	1.71	2.00[a]	2.20	2.35
3.5	1.45	1.86	2.18	2.41	2.58

[a]This value was incorrectly reported as 1.91 by Rovnyak *et al.*,[23] due to a typographical error.

We obtain its Fourier transform as

$$f(\nu) = \frac{1}{\pi} \frac{T_2}{\nu^2 + T_2^2} \tag{6.8}$$

where the full width at half-maximum (FWHM) is the linewidth:

$$\text{FWHM} = \frac{1}{\pi T_2} \tag{6.9}$$

If we anticipate a carbon linewidth of 4 Hz in the indirect dimension of an HSQC experiment, then we have a signal decay constant of 1/(3.14×4) or about 80 ms.

Gaussian distributed NUS has also attracted attention.[36,42] The Fourier transform of a Gaussian function returns another Gaussian function:

$$\mathrm{e}^{-t^2/2\sigma^2} \overset{\text{Fourier transform}}{\Rightarrow} \frac{1}{\alpha\sqrt{2\pi}} \mathrm{e}^{-\nu^2/2\alpha^2} \tag{6.10}$$

where $\alpha \equiv 1/2\pi\sigma$ and the frequency domain linewidth is $2\alpha\sqrt{2\ln 2}$. The algebra is then straightforward to find a given σ in the time-domain Gaussian such that $2\alpha\sqrt{2\ln 2}$ has the desired linewidth in the frequency domain. Specifically, we refer to a Gaussian sampling density as matched (*e.g.* Figures 6.2 and 6.4) if the corresponding Gaussian linewidth in the frequency domain matches the anticipated linewidth resulting from the exponential T_2 decay, that is, if $2\alpha\sqrt{2\ln 2} = 1/\pi T_2$. The enhancement in the time domain of Gaussian distributed NUS is illustrated in Figure 6.4 along with some representative values. Other densities are also showing high promise, notably one based on a portion of a sinusoid has the same sensitivity as a matched exponential schedule and leads to slightly improved lineshapes by maximum entropy reconstruction.[25]

6.2.3 Validation Using Linear Transforms

Testing the predictions of eqn (6.5) is not necessarily straightforward. Traditional applications of maximum entropy reconstruction are non-linear so that the S/N in the resulting spectra cannot be directly interpreted. Therefore, one approach to demonstrate that sensitivity enhancement of the time domain by NUS occurred is to show that the detection limit of a given experiment was altered by NUS on a scale comparable to the predicted enhancements.[23] Importantly, the predictions of eqn (6.5) have been validated by analysis of NUS data using an extended maximum entropy algorithm termed maximum entropy interpolation (MINT), which provides highly linear spectra from NUS data to permit quantitative comparisons of S/N in spectra obtained from uniform and non-uniform samples.[24] The enhancement of 2D NCACX spectra by NUS is demonstrated in Figure 6.5, in which MINT was employed to perform the spectral reconstructions so that S/N comparisons between uniform sampling (US) and non-uniform sampling

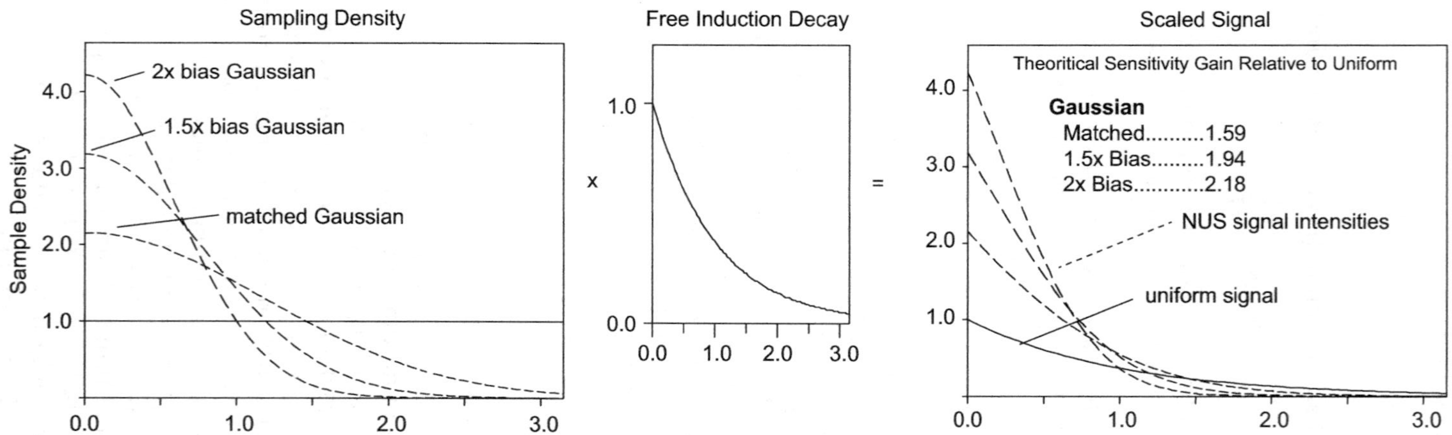

Figure 6.4 Analysis of several schemes for Gaussian distributed NUS, which can deliver compelling sensitivity improvements; however, Gaussian sampling densities decay more rapidly than their exponential counterparts.

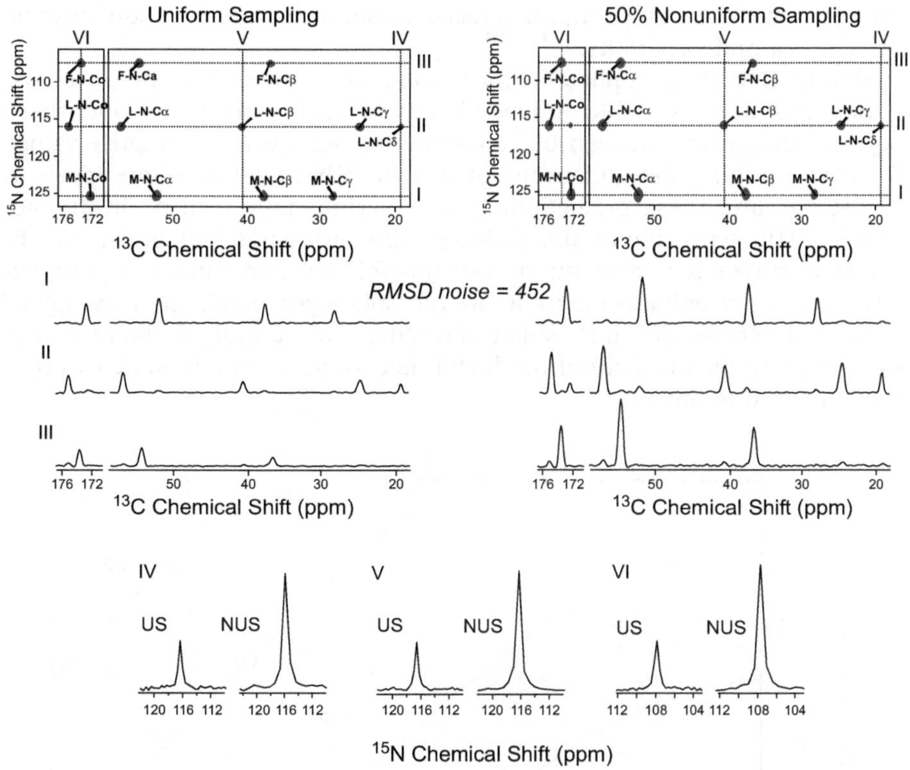

Figure 6.5 Comparison of 2D NCACX spectra of MLF. Top left, sampled uniformly (as a 1000×64 complex matrix, 4 transients per increment), and top right, non-uniformly in the indirect dimension (as a 1000×32 complex matrix, 8 transients per increment, using NUS1 sampling schedule). The RMS noise levels in the 2D NUS and US spectra are the same. Roman numerals I, II, and III correspond to 1D slices extracted along the direct dimensions for residues M, L, and F, respectively. Roman numerals IV, V, and VI correspond to 1D slices extracted along the indirect dimension for L–C^{δ}, L–C^{β}, and F–C°, respectively. It can be seen that in each case the NUS dataset yields an increase in peak intensity compared with the uniformly sampled slices displayed at the same noise level.
Reprinted with permission from Paramasivam *et al.*, *J. Phys. Chem. B*, 2012, **116**(25), 7416–7427.[24] Copyright 2012 American Chemical Society.

(NUS) could be rigorously made. The twofold improvement is clearly evident from the cross-sections of the data (see Paramasivam *et al.*[24] for additional details). The MINT approach is more computationally intensive than routine MaxEnt; therefore, since the enhancements of eqn (6.5) have been carefully validated with linearized tests, then the NUS enhancement of the raw time-domain data can be exploited by any of a number of spectral estimation techniques available to researchers.[43–59] Here we do not use MINT further, but continue to use MaxEnt, which has been shown in extensive studies to be robust, fast and easy to use.[1,4,9,12,33]

Finally, if NUS is applied in more than one dimension, then a signal enhancement is available independently in each dimension, and these separate enhancements can be compounded. Several representative values for compounding the enhancement in two NUS dimensions are given in Figure 6.6, and these predictions were experimentally realized in 3D biosolids NMR experimentation, yielding MINT-validated enhancements in excess of threefold,[24] a saving of over ninefold in time. Such compounded NUS sensitivity enhancements allow the solid-state NMR spectroscopy of difficult biomolecules and small molecules at natural abundance (*e.g.* pharmaceuticals and natural products) that would not be feasible by other methods at this time.

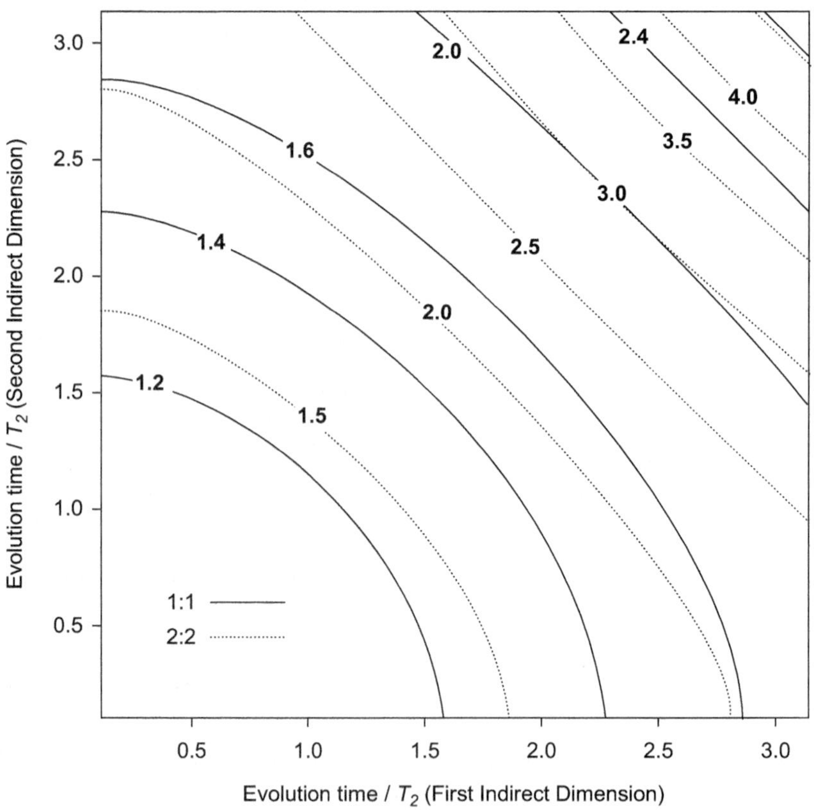

Figure 6.6 Compounded NUS-based S/N enhancements are depicted for two indirect evolution periods for cases when the exponential NUS densities are matched in both dimensions (solid lines) and are twofold biased in both dimensions (dashed lines). These predictions have been experimentally realized in the 3D solid-state NMR of protein assemblies.[24]

6.3 Application of NUS Enhancement to 2D Heteronuclear Correlations

The NMR study of complex small molecules (metabolites, steroidal compounds, natural products, *etc.*) has a number of challenges that are unique in comparison with other targets of NMR spectroscopy. In contrast to protein NMR in liquids, one cannot count on the predictability of chemical shift ranges or J couplings, for example. Generally no routes for isotopic enrichment are available, so it is not possible to resort to 3D multinuclear NMR spectroscopy to attain sufficient signal dispersion to perform assignments. Rather, it is often the case that the only option to resolve nearly degenerate lines is to obtain 2D spectra at ultra-high resolution in the indirect dimensions, where one approach that has been used to achieve this limit is intentionally to alias signals in the indirect dimension.[60] In general, with the long T_2^s that can often be encountered in small molecules (approximately <1000 kDa), indirect evolution times may be prohibitively long for uniform sampling without aliasing. Further, researchers must work with very small sample quantities on a milligram scale or much less, particularly in natural products work, further hindering attempts to acquire 2D heteronuclear correlation spectroscopy since only natural abundance spins are available. Importantly, innovations in hardware have led to dramatic improvements in mass sensitivity; for example, HSQC and HMBC spectra were acquired on an amount of strychnine sample as low as 5 μg employing a 1.7 mm microcryoprobe.[61]

When sensitivity is low, it can be difficult to justify evolutions beyond $t_{\max} \approx 0.7T_2$, as samples in the range $(0.7–1.26)T_2$ add negligibly to the total signal, whereas samples beyond 1.26 T_2 add more noise than signal [see eqn (6.2)].[16] In sum, small-molecule NMR faces the challenge of having insufficient signal to obtain the desired resolution, so NUS-based sensitivity enhancement can be used to help obtain spectra with sufficient resolution and sensitivity that would not be accessible by conventional uniform sampling.

To begin, we show an example in which NUS alters the detection limit of the experiment. A series of spectra were obtained on a 1 mM sample of deoxycholate, a dihydroxy bile acid present in the liver, and which is known to form micellar aggregates under many conditions, and also hydrogels in the presence of phosphate buffer at very low concentrations (<10 mM). In order to study micelle and hydrogel formation, sufficient sensitivity and resolution are needed to acquire 2D-HSQC spectra of pre-aggregate deoxycholate monomer. In this series, each 2D-HSQC consumed 7 h of measurement time, and it is important first to compare Figure 6.7a and b, which are processed by FFT and MaxEnt, respectively, to appreciate that MaxEnt processing is not being used to suppress noise artificially in these data sets.

A challenging area in small-molecule NMR spectroscopy is to resolve the signal from multiple aromatic groups. A window from an aromatic-only

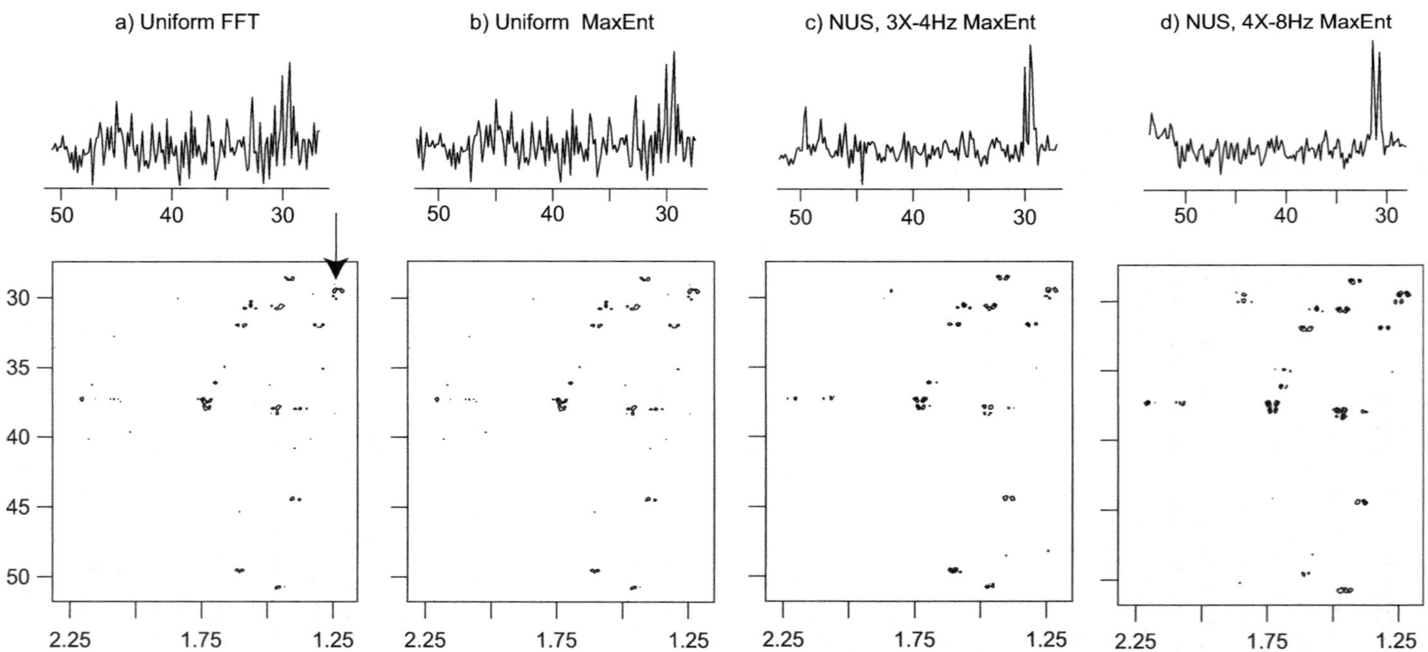

Figure 6.7 A series of GHSQC spectra are shown (600 MHz, inverse RT $^{1}H/^{13}C/^{15}N$ probe, 25 °C: see Section 6.5 for further details) of a 1 mM deoxycholate solution, each acquired in 7 h. The complex spectrum can be fully resolved by using evolution times on the order of $3T_2$. It is recognized in comparing panels (a) and (b) that MaxEnt cannot be used to improve the sensitivity of the uniform data. When the NUS density is approximately matched to the expected 4 Hz linewidths, significant improvements are recognized in comparing (a) and (c). Biasing the NUS density by about twofold results in further improvement as seen in (d). The theoretical enhancements in (c) and (d) are about 1.7- and 2.1-fold, respectively. It was shown previously that uniform acquisitions that are extended to match the predicted enhancements of NUS acquisitions show close agreement in their sensitivities.[23] That is, to obtain a uniformly sampled data set with comparable sensitivity to (d), one would require $(2.1)^2 \times 7 = 31$ h.

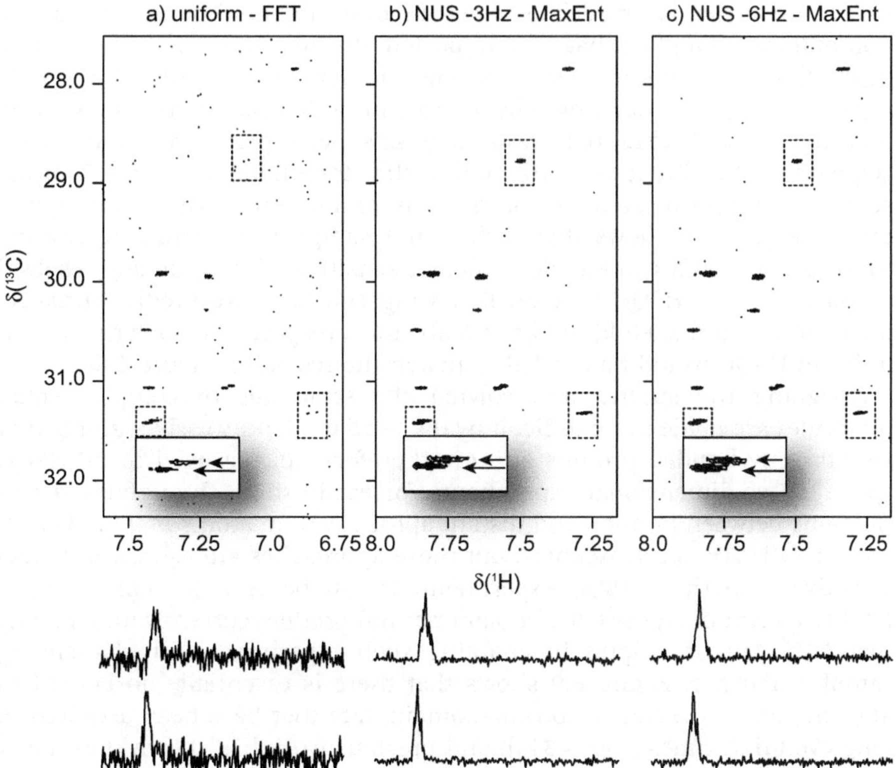

Figure 6.8 NUS-based signal enhancements change the detection limit of 2D-HSQC spectroscopy. A series of aromatic GHSQC spectra are shown (600 MHz, inverse 5 mm RT $^{1}H/^{13}C/^{15}N$ probe, 25 °C: see Section 6.5 for further details) of a 3 mM solution containing a polyaryl ligand, acquired in 12 h in each. Several peaks are detected only with the aid of NUS enhancement, while the enhancement also improves the ability to observe the lineshapes. The chosen cross-sections illustrate that resolution in the non-uniformly sampled dimension is not compromised in either NUS scheme: pure lineshapes are detected for a doublet of just a few hertz in the ^{13}C dimension in both (b) and (c).

HSQC spectrum is shown in Figure 6.8 for a polyaromatic diphosphine moiety (1 or 5 mM in $CDCl_3$, courtesy Prof. R. Stockland) acquired by uniform, exponential NUS assuming a 3 Hz linewidth, and exponential NUS assuming a 6 Hz linewidth sampling methods and processed by MaxEnt. Each experiment required 12 h. A number of features may be highlighted. First, it is evident that the uniform data set is missing peaks and very poorly detects others. Missing peaks are restored by NUS sensitivity enhancement in either the conservative 3 Hz (approximately matched) or 6 Hz biased exponential sampling strategies, confirming that NUS authentically transforms the detection limit of 2D-NMR.[23] In addition, representative cross-sections through the data support a significant improvement in the sensitivity of the observed lineshapes. Second, it should be noted that the non-uniform sampling has not impacted the resolution negatively. Two peaks that are extremely close to one another are magnified (only 6 Hz separation in ^{13}C); they yield clean slices in both NUS spectra, supporting that the needed ultra-high resolution has been preserved in the NUS acquisition. Finally, it is worth noting that for dilute samples exhibiting severely overlapped aromatic spectra, it is common to neglect their assignments. Figure 6.8 shows that with 1 mM samples on room-temperature, 5 mm probes, NUS can enable sufficient sensitivity for maximally resolved aromatic ^{13}C–^{1}H HSQC spectra. Following eqn (6.5), we predict enhancements of 1.7- and 2.0-fold in Figure 6.8b and c, respectively, such that a 48 h uniform HSQC would be needed to match the results in Figure 6.8c.

Assigning the spectra and solving the structures of complex small molecules are made more difficult by the need to identify carbon atoms that lack directly bonded protons and are therefore not observed in 2D-HSQC spectra. Two-dimensional experiments for establishing through-bond correlations between protons and distant aprotic carbon atoms include HMBC and ADEQUATE spectroscopies, but these approaches are significantly less sensitive than the GHSQC experiment. It can be seen in representative HMBC spectra in Figure 6.9 of a plant natural product currently under study that NUS can be helpful in enabling such experiments for challenging samples. Further, Figure 6.9 shows that there is essentially no benefit to applying linear prediction to time-domain data that have been acquired to long evolution times (*e.g.* $\sim 3T_2$), and previous work has shown that linear prediction cannot distinguish peaks if evolution times are such that the digital resolution is larger than the peak resolution.[17]

Finally, we look at an example that provides a perspective on the question of whether NUS should be employed in all situations. Suppose, as in Figure 6.10a, that a good-quality HSQC spectrum can be obtained on a moderately challenging sample (5 mM strychnine). What criteria might one consider to decide whether the use of NUS would offer sufficient advantages? Spectra obtained by MaxEnt reconstruction of non-uniform data are shown in Figure 6.10b and c. One difference is that certainly there is an improvement in spectral quality, as demonstrated by a representative ^{1}H slice that is attributable principally to the sensitivity enhancement and not to the use of MaxEnt. Although it is always desirable to work with stronger signals, the case could be made that the signals from uniform sampling in Figure 6.10a are strong enough. As also discussed in relation to the data in Figure 6.8, the resolution is certainly not compromised in the NUS data, where a magnified region in Figure 6.10 shows two peaks that are essentially equally resolved in the ^{13}C dimension in the uniform and NUS cases. However, an often overlooked point might be appreciated from inspection of Figure 6.10a in which contours have been chosen such that some weak artifacts can be seen in the spectrum obtained by Fourier transformation of uniform data. In order that all spectra in Figure 6.10 consume the identical measurement time, just two transients per increment were employed in the uniform acquisition, whereas

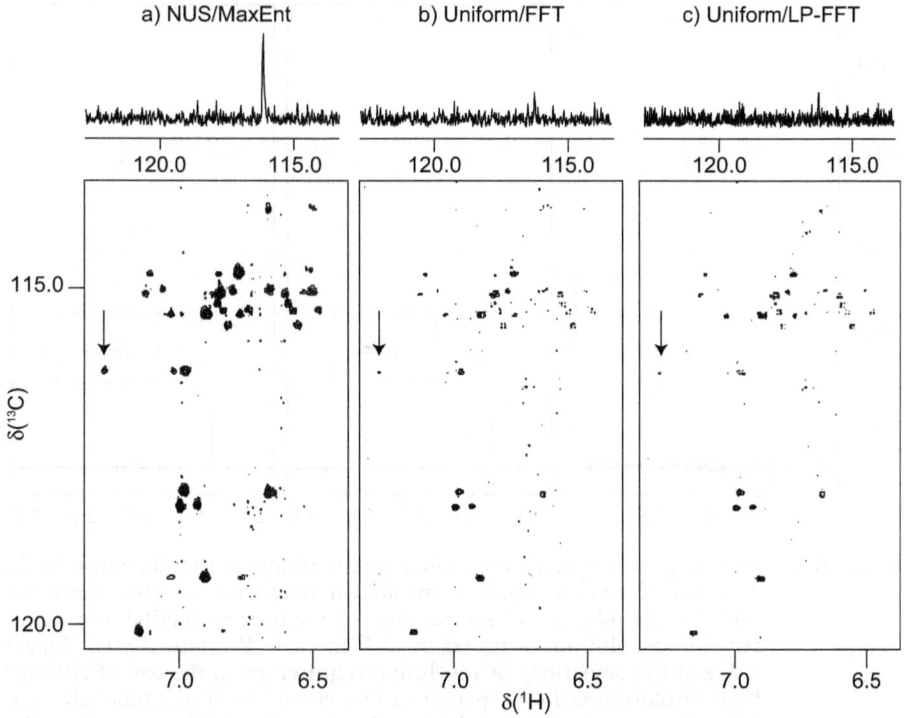

Figure 6.9 An example of a 2D-HMBC of a natural product (courtesy Prof. G. Henry, Susquehanna University) acquired by NUS (a) and uniform sampling (b, c) in 5.6 h for each case. A large number of peaks in this aromatic region are easily and correctly detected in the NUS/MaxEnt spectrum that are indistinguishable from the noise in the uniform acquisition, regardless of whether linear prediction is applied or not. A representative one-dimensional cross-section is illustrated for the peak indicated with an arrow.

the non-uniform acquisitions used eight transients per increment. Hence the NUS data are able to benefit from more extensive phase cycling and artifact reduction than the uniformly acquired data. Although gradient coherence selection and modern quadrature detection have certainly greatly diminished the role of artifacts in data, phase cycling has not been obviated and still provides additional suppression of artifacts and preservation of coherence pathways.

6.4 Critique and Outlook

The use of exponentially weighted NUS schedules to improve resolution and/or save total experiment time when recording an indirect dimension containing decaying signals is now well established. In general, matched NUS can be taken to be a conservative and trustworthy choice provided that one has a reasonable estimate of the expected linewidths. The ability to achieve

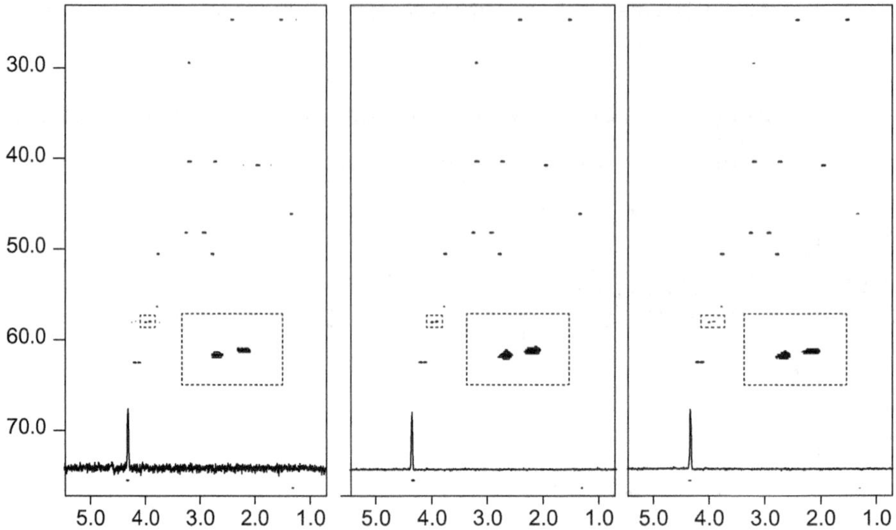

Figure 6.10 NUS improves spectra even when not working at the detection limit. A series of GHSQC spectra are shown (600 MHz, inverse 5 mm RT $^{1}H/^{13}C/^{15}N$ probe, 25 °C: see Section 6.5 for further details) of a 5 mM strychnine solution, acquired in 12 h in each. Without *a priori* knowledge of the sensitivity or resolution requirements, the use of NUS for high-resolution GHSQC spectra can be viewed as simultaneously optimizing resolution and sensitivity. The use of NUS often results in the ability to use more transients per sample, which can aid in artifact reduction and preservation of coherence transfer pathways.

signal enhancement of the raw time-domain data by exponential NUS in the same experimental time as uniform sampling is now well established by multiple experimental investigations.[7,8,22–26,30] Few or negligible effects on lineshape occur for conservative exponential NUS, while larger S/N enhancements exceeding twofold are available for more aggressive NUS that will incur some line broadening. Compounding the enhancement in multiple NUS dimensions results in improvements in excess of threefold in biosolids and small-molecule solid-state NMR research. Since NUS enhances the raw, unprocessed time-domain data, any subsequent spectral estimation will benefit; however, we find MaxEnt to be robust, easy, and fast to apply to data sets such as those shown in this chapter. Small-molecule liquid-state NMR and solid-state NMR of macromolecules and small molecules are all especially well suited to achieve optimal NUS enhancements in natural abundance 2D-HSQC spectra that require ultra-high resolution and the highest available sensitivity.

6.5 Methods and Materials

All spectra were acquired on a Varian (Agilent) 600 MHz VNMRS four-channel spectrometer at 25 °C using an indirect detection triple resonance

($^1H/^{13}C/^{15}N$) 5 mm probe for maximum sensitivity on ^1H. The vendor-supplied pulse sequence GHSQC was used without any modification except for arraying the indirect evolution period according to the specified non-uniform sampling schedules by using a custom-built macro. Sampling schedules were generated with the Sampsched program, which is a part of the Rowland NMR Toolkit (RNMRTK).[1] At this time, Agilent and Bruker have implemented integrated support for non-uniform sampling and spectral reconstruction. All spectra were processed using the RNMRTK, and using a high-performance workstation (4 × 6-core Xeon X5670 2.93 GHz CPUs, 48 GB RAM, Red Hat Enterprise 5). The 2D-HSQC spectra obtained by non-uniform sampling were processed with maximum entropy reconstruction as implemented in RNMRTK as the program msa.

Deoxycholate and strychnine were obtained commercially and used without any further purification. The polyaryl ligand studied for Figure 6.8 was provided courtesy Prof. R. Stockland (Bucknell University). The plant natural product (9 mg in 0.5 mL of $CDCl_3$) was provided courtesy Prof. G. Henry (Susquehanna University). Specific parameter choices for the experiments were as follows (swH = ^1H spectral width, swC = ^{13}C spectral width, ni = number of increments, acq = receiver gating time): deoxycholate: swH = 5434.8, swC = 6485.1, ni = 256, recycle = 1.2 s, acq = 0.20 s, 7 h each [nt = 32 (uni), 96 (nus3×-6 Hz), 128 (nus4×-12 Hz)]; polyaryl ligand: swH = 5605.4, swC = 6032.7, ni = 2048, recycle = 2 s, acq = 0.50, 12 h each [nt = 4 (uni), 16 (nus4×-3 Hz), 16 (nus4×-6 Hz)]; plant natural product: swH = 5531, swC = 5280, ni = 1400, recycle = 1.4 s, acq = 0.40 s, 5.6 h each [nt = 4 (uni), 16 (nus4×-6 Hz)]; strychnine: swH = 5896, swC = 9048.75, ni = 2048, recycle = 1.5 s, acq = 0.50 s, 5 h each [nt = 2 (uni), 8 (nus4×-4 Hz), 8 (nus4×-8Hz)].

Acknowledgements

We are grateful to Prof. R. Stockland (Bucknell University) for access to the ligands shown in Figure 6.8 and to Prof. G. Henry (Susquehanna University) for access to the plant natural product shown in Figure 6.9. We thank Brian Breczinski for assistance with the NMR spectrometers and Jeremy Dreese for computing support. T.P. acknowledges the support of the National Institutes of Health (NIH Grant R01GM085396).

References

1. J. C. Hoch and A. S. Stern, *NMR Data Processing*, Wiley, New York, 1996.
2. R. R. Ernst, G. Bodenhausen and A. Wokaun, *Principles of Nuclear Magnetic Resonance in One and Two Dimensions*, Oxford University Press, Oxford, 1987.
3. K. Kazimierczuk, J. Stanek, A. Zawadzka-Kazimierczuk and W. Kozminski, *Prog. Nucl. Magn. Reson. Spectrosc.*, 2010, **57**, 420.

4. M. W. Maciejewski, M. Mobli, A. D. Schuyler, A. S. Stern and J. C. Hoch, *Top Curr. Chem.*, 2012, **316**, 49.
5. M. W. Maciejewski, A. S. Stern, G. F. King, J. C. Hoch, in *Modern Magnetic Resonance*, ed. G. A. Webb, Springer, Dordrecht, 2006, p. 1305.
6. M. Mobli and J. C. Hoch, *Concepts Magn. Reson., Part A*, 2008, **32A**, 436.
7. J. C. J. Barna and E. D. Laue, *J. Magn Reson.*, 1987, **75**, 384.
8. J. C. J. Barna, E. D. Laue, M. R. S. Mayger, J. Skilling and S. J. P. Worrall, *J. Magn. Reson.*, 1987, **73**, 69.
9. D. Rovnyak, D. P. Frueh, M. Sastry, Z. Y. J. Sun, A. S. Stern, J. C. Hoch and G. Wagner, *J. Magn. Reson.*, 2004, **170**, 15.
10. J. A. Kubat, J. J. Chou and D. Rovnyak, *J. Magn. Reson.*, 2007, **186**, 201.
11. P. Schmieder, A. S. Stern, G. Wagner and J. C. Hoch, *J. Biomol. NMR*, 1993, **3**, 569.
12. P. Schmieder, A. S. Stern, G. Wagner and J. C. Hoch, *J. Biomol. NMR*, 1994, **4**, 483.
13. A. D. Schuyler, M. W. Maciejewski, H. Arthanari and J. C. Hoch, *J. Biomol. NMR*, 2011, **50**, 247.
14. E. Kupce and R. Freeman, *J. Biomol. NMR*, 2003, **25**, 349.
15. V. Y. Orekhov, I. Ibraghimov and M. Billeter, *J. Biomol. NMR*, 2003, **27**, 165.
16. D. Rovnyak, J. C. Hoch, A. S. Stern and G. Wagner, *J. Biomol. NMR*, 2004, **30**, 1.
17. D. Rovnyak, C. Filip, B. Itin, A. S. Stern, G. Wagner, R. G. Griffin and J. C. Hoch, *J. Magn. Reson.*, 2003, **161**, 43.
18. Y. Matsuki, M. T. Eddy, R. G. Griffin and J. Herzfeld, *Angew. Chem.*, 2010, **49**, 9215.
19. H. Heise, K. Seidel, M. Etzkorn, S. Becker and M. Baldus, *J. Magn. Reson.*, 2005, **173**, 64.
20. W. T. Franks, H. S. Atreya, T. Szyperski and C. M. Rienstra, *J. Biomol. NMR*, 2010, **48**, 213.
21. M. H. Levitt, G. Bodenhausen and R. R. Ernst, *J. Magn. Reson.*, 1984, **58**, 462.
22. A. Kumar, C. B. Brown, M. E. Donlan, B. U. Meier and P. W. Jeffs, *J. Magn. Reson.*, 1991, **95**, 1.
23. D. Rovnyak, M. Sarcone and Z. Jiang, *Magn. Reson. Chem.*, 2011, **49**(8), 483.
24. S. Paramasivam, C. L. Suiter, G. J. Hou, S. J Sun, M. Palmer, J. C. Hoch, D. Rovnyak and T. Polenova, *J. Phys. Chem. B*, 2012, **116**, 7416.
25. M. R. Palmer, B. R. Wenrich, P. Stahlfeld and D. Rovnyak, *J. Biomol. NMR*, 2014, **58**, 303.
26. C. L. Suiter, S. Paramasivam, G. Hou, S. Sun, D. Rice, J. C. Hoch, D. Rovnyak and T. Polenova, *J. Biomol. NMR*, 2014, **59**, 57.
27. S. G. Hyberts, K. Takeuchi and G. Wagner, *J. Am. Chem. Soc.*, 2010, **132**, 2145.
28. C. A. Waudby and J. Christodoulou, *J. Magn. Reson.*, 2012, **219**, 46.
29. D. L. Donoho, I. M. Johnstone, A. S. Stern and J. C. Hoch, *Proc. Natl. Acad. Sci. U. S. A.*, 1990, **87**, 5066.

30. S. G. Hyberts, S. A. Robson and G. Wagner, *J. Biomol. NMR*, 2013, **55**, 167.
31. H. S. Taylor, R. Haiges and A. Kershaw, *J. Phys. Chem. A*, 2013, **117**, 3319.
32. A. D. Schuyler, M. W. Maciejewski, A. S. Stern and J. C. Hoch, *J. Magn. Reson.*, 2013, **227**, 20.
33. J. C. Hoch and A. S. Stern, *Methods Enzymol.*, 2001, **338**, 159.
34. M. Mobli, A. S. Stern and J. C. Hoch, *J. Magn. Reson.*, 2006, **182**, 96.
35. M. R. Gryk, J. Vyas and M. W. Maciejewski, *Prog. Nucl. Magn. Reson. Spectrosc.*, 2010, **56**, 329.
36. M. T. Eddy, D. Ruben, R. G. Griffin and J. Herzfeld, *J. Magn. Reson.*, 2012, **214**, 296.
37. J. C. Hoch, M. W. Maciejewski and B. Filipovic, *J. Magn. Reson.*, 2008, **193**, 317.
38. M. Mobli, M. W. Maciejewski, A. D. Schuyler, A. S. Stern and J. C. Hoch, *Phys. Chem. Chem. Phys.*, 2012, **14**, 10835.
39. P. Schmieder, A. S. Stern, G. Wagner and J. C. Hoch, *J. Magn. Reson.*, 1997, **125**, 332.
40. S. G. Hyberts, G. J. Heffron, N. G. Tarragona, K. Solanky, K. A. Edmonds, H. Luithardt, J. Fejzo, M. Chorev, H. Aktas, K. Colson, K. H. Falchuk, J. A. Halperin and G. Wagner, *J. Am. Chem. Soc.*, 2007, **129**, 5108.
41. R. G. Spencer, *Concepts Magn. Reson., Part A*, 2010, **36A**, 255.
42. Y. Matsuki, T. Konuma, T. Fujiwara and K. Sugase, *J. Phys. Chem. B*, 2011, **115**, 13740.
43. V. Y. Orekhov, I. V. Ibraghimov and M. Billeter, *J. Biomol. NMR*, 2001, **20**, 49.
44. Y. Matsuki, M. T. Eddy and J. Herzfeld, *J. Am. Chem. Soc.*, 2009, **131**, 4648.
45. R. Bruschweiler, *J. Chem. Phys.*, 2004, **121**, 409.
46. H. S. Atreya and T. Szyperski, *Proc. Natl. Acad. Sci. U. S. A.*, 2004, **101**, 9642.
47. A. Gutmanas, P. Jarvoll, V. Y. Orekhov and M. Billeter, *J. Biomol. NMR*, 2002, **24**, 191.
48. E. Kupce and R. Freeman, *J. Am. Chem. Soc.*, 2004, **126**, 6429.
49. E. Kupce and R. Freeman, *J. Magn. Reson.*, 2003, **162**, 158.
50. R. Freeman and E. Kupce, *J. Biomol. NMR*, 2003, **27**, 101.
51. B. E. Coggins, R. A. Venters and P. Zhou, *J. Am. Chem. Soc.*, 2005, **127**, 11562.
52. T. Szyperski and H. S. Atreya, *Magn. Reson. Chem*, 2006, **44 Spec. No.**, S51.
53. D. Malmodin and M. Billeter, *J. Am. Chem. Soc.*, 2005, **127**, 13486.
54. K. Kazimierczuk, M. Misiak, J. Stanek, A. Zawadzka-Kazimierczuk and W. Kozminski, *Top. Curr. Chem.*, 2012, **316**, 79.
55. K. Kazimierczuk and V. Y. Orekhov, *Angew. Chem.*, 2011, **50**, 5556.
56. M. Gal, P. Schanda, B. Brutscher and L. Frydman, *J. Am. Chem. Soc.*, 2007, **129**, 1372.

57. J. W. Werner-Allen, B. E. Coggins and P. Zhou, *J. Magn. Reson.*, 2010, **204**, 173.
58. S. G. Hyberts, A. G. Milbradt, A. B. Wagner, H. Arthanari and G. Wagner, *J. Biomol. NMR*, 2012, **52**, 315.
59. A. S. Stern, D. L. Donoho and J. C. Hoch, *J. Magn. Reson.*, 2007, **188**, 295.
60. D. Jeannerat, *J. Magn. Reson.*, 2007, **186**, 112.
61. G. E. Martin, B. D. Hilton, D. Moskau, N. Freytag, K. Kessler and K. Colson, *Magn. Reson. Chem.*, 2010, **48**, 935.

CHAPTER 7

NMR Spectroscopy Using Several Parallel Receivers

RAY FREEMAN*[a] AND ĒRIKS KUPČE*[b]

[a] Jesus College, Cambridge University, Cambridge CB5 8BL, UK;
[b] Agilent Technologies, Yarnton, Oxford, OX5 1QU, UK
*Email: rf110@hermes.cam.ac.uk; eriks.kupce@bruker.com

7.1 Introduction

Speed has supplanted sensitivity as the key parameter for modern multidimensional NMR experiments. Today, the overall duration of a typical measurement is more likely to be determined by the total number of evolution increments than by the required signal-to-noise ratio, particularly when cryogenically cooled probes are employed. Complex biomolecules, often with global isotopic enrichment, demand higher dimensional experiments, and the sampling procedure must satisfy the Nyquist condition and resolution requirements in all evolution dimensions. These investigations are said to be "sampling limited" rather than "sensitivity limited." There have been many innovations aimed at speeding up such measurements – selective excitation with Hadamard encoding,[1–4] covariance spectroscopy,[5–7] spatially selective single-scan methods,[8–15] G-matrix Fourier transform NMR,[16–20] sparse sampling of evolution space,[21–29] projection-reconstruction,[30–42] and several schemes designed to reduce the delays for spin–lattice relaxation.[43–45] This chapter demonstrates how these existing fast schemes can be complemented by a new approach designed to increase the amount of information extracted from a single measurement.

Modern NMR Approaches To The Structure Elucidation of Natural Products, Volume 1: Instrumentation and Software
Edited by Antony J. Williams, Gary E. Martin and David Rovnyak

Published by the Royal Society of Chemistry, www.rsc.org

7.2 Multiple Receivers

The recent introduction of spectrometers with several receivers operating in parallel[46,47] addresses the speed problem by increasing the information content of each experiment with little or no increase in the duration of the measurement. Whereas the accepted custom is to run several standard experiments one after the other, this innovation combines these in a single entity and records signals from several nuclear species in a single pass. The basic requirement is a standard "triple resonance" or "broadband" radiofrequency probe with a separate receiver for each nuclear species to be investigated. Each receiver channel comprises a separate preamplifier, local oscillator, mixer, amplifier, digitizer and controller. The control software allows the acquisition of the different signals either simultaneously or staggered at different stages of a composite sequence. The detection is optimized to favour the low-sensitivity nuclei, for example, ^{13}C and ^{15}N. Where necessary, cryogenically cooled receiver coils and preamplifiers are used. This technique has been called PANSY (Parallel Acquisition NMR Spectroscopy). Figure 7.1 shows simultaneous ^{1}H, ^{13}C, ^{15}N and ^{31}P spectra of 5′-guanosine triphosphate enriched in ^{13}C and ^{15}N, recorded on a 600 MHz spectrometer using four receivers operating in parallel.

However, the concept is not limited to this "bread-and-butter" application. There are more exciting extensions that combine two or more different standard pulse sequences into a single unit, allowing several different items of structural information to be recorded in a single pass. Candidates for incorporation into a parallel acquisition scheme include HSQC (heteronuclear single-quantum correlation), HMBC (heteronuclear multiple-bond correlation), INADEQUATE[48,49] (incredible natural abundance double-quantum transfer experiment), COSY (homonuclear correlation spectroscopy), INEPT (insensitive nuclei enhanced by polarization transfer) and TOCSY (total correlation spectroscopy). For example, the 600 MHz ^{1}H–^{1}H TOCSY and ^{1}H–^{13}C correlation spectra of brucine (3% in $CDCl_3$) have been recorded in parallel in a measurement of 20 min duration.[46]

Since many molecules of interest to the pharmaceutical industry incorporate fluorine atoms, it can be particularly useful to measure two-dimensional correlations between heteronuclei (such as ^{13}C or ^{15}N) with *both* ^{1}H and ^{19}F simultaneously. The basic requirement is a radiofrequency probe double-tuned to ^{1}H and ^{19}F, and equipped with a separate coil that can be tuned to ^{13}C or ^{15}N. This PANSY–HSQC sequence is initiated with ^{1}H–^{13}C and ^{1}H–^{19}F INEPT segments with timing adjusted to take into account the different coupling constant magnitudes. As an illustrative example, Figure 7.2 shows the 600 MHz aromatic region of the dual-HSQC spectrum of 2-bromophenyl-3-trifluoromethyl-5-methylpyrazole (Scheme 7.1) with the natural ^{13}C isotopic abundance.[47] The combined spectrum comprises one direct ^{19}F–^{13}C correlation peak (red) together with five single-bond ^{1}H–^{13}C correlation peaks (black), recorded at the same time in a single PANSY run of 22 min duration. Long-range correlations to ^{13}C can be measured in a

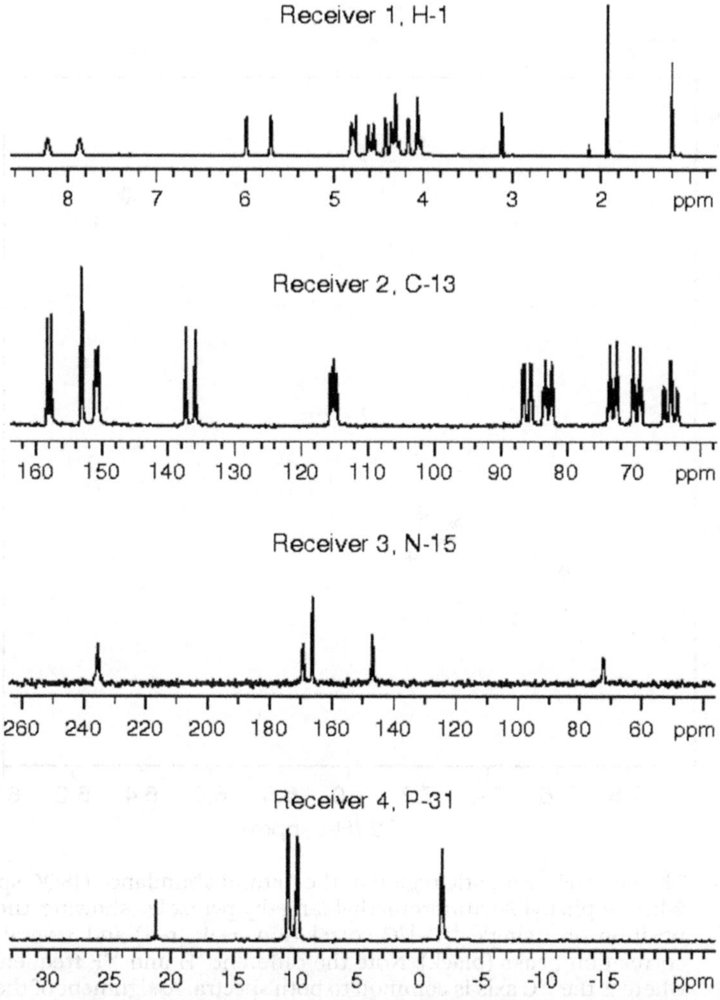

Figure 7.1 The use of a 600 MHz NMR spectrometer equipped with four independent parallel receivers to record ^{1}H, ^{13}C, ^{15}N and ^{31}P spectra simultaneously. The sample is 5′-guanosine triphosphate enriched in ^{13}C and ^{15}N.

similar manner, and the method can be extended to detect the corresponding long-range correlations to ^{15}N, albeit in an experiment of appreciably longer duration.[47]

7.3 PANACEA

Further possibilities are offered by a more general application of multiple receivers. The combination of several carefully chosen standard NMR pulse sequences into a single entity can deliver the complete structure of a small organic molecule. In many cases, the INADEQUATE technique[48,49] is the key

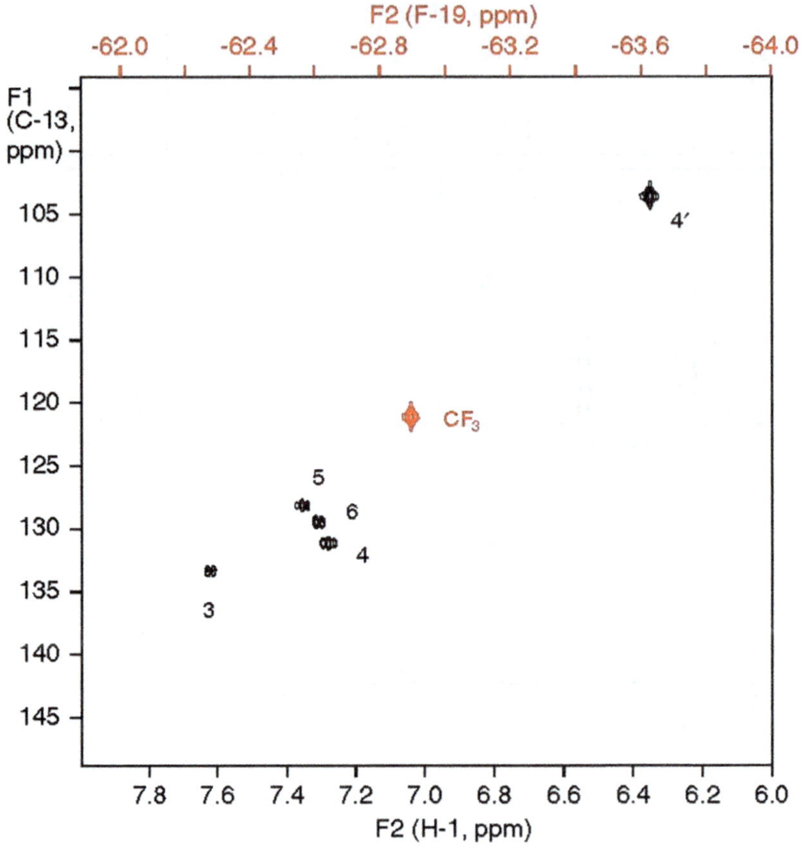

Figure 7.2 The 600 MHz aromatic region of the natural abundance HSQC spectra of 2-bromophenyl-3-trifluoromethyl-5-methylpyrazole, showing the superposition of a single ^{19}F–^{13}C correlation peak (red) and several ^{1}H–^{13}C correlation peaks (black). Note the different ^{1}H and ^{19}F frequency axes, whereas the ^{13}C axis is common to both spectra. Assignment of the phenyl carbons is based primarily on the proton multiplicities. The two measurements were made in parallel with an experimental duration of 22 min. Reproduced from Kupče *et al.*[47] with permission of John Wiley & Sons, Ltd.

ingredient because it establishes the basic carbon framework of the molecule in an unambiguous manner. This general method[50–53] has been named "PANACEA" (Protons And Nitrogen And Carbon *Et Alia*). There is an inherent synergy in such a multifunctional "spin choreography" scheme. With careful design, these experiments can make the best use of all the available components of nuclear magnetization. For example, spin coherence used to generate one observable NMR spectrum can be refocused and exploited to obtain further information. This idea can be taken even further. Signals detected at an early stage of a sequence can be used "on the fly" to set the operating parameters for a subsequent stage without operator intervention. As an example, ^{13}C chemical shifts measured at the beginning

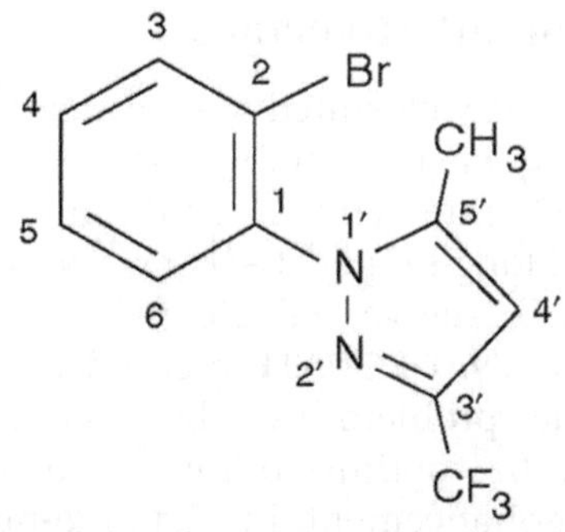

Scheme 7.1 2-Bromophenyl-3-trifluoromethyl-5-methylpyrazole.

of a sequence may be subsequently exploited to set up an INADEQUATE experiment using multiple selective excitation encoded by a Hadamard matrix, thereby speeding up the measurement (see Section 7.3.3.1). One might think of this as enlisting artificial intelligence to control complex spin manipulations that would normally have to be designed and programmed before the experiment is started.

The incorporation of all the requisite pulse schemes in a single entity offers a second important feature. Whereas the conventional protocol would involve a suite of different pulse sequences carried out at different times (or even on different days), PANACEA condenses everything into a single shot. Consequently, all the different kinds of NMR data are recorded under essentially identical environmental conditions; any drifts in spectrometer parameters, or any slow deterioration of the sample, can be confidently dismissed. Because PANACEA delivers a *comprehensive* result, there is no need for some later reinvestigation of the sample to check back for supplementary structural information. For the first time, a chemist may say, "run the NMR" in the same manner as he might say, "get the mass spectrum" or "record the infrared." Indeed this "all-in-one" protocol lends itself to the investigation of an entire run of different samples delivered by an automatic sample changer without further operator intervention.

The family of PANACEA experiments can be divided into three main categories, depending on the type of chemical application. The first[50] comprises experiments on small organic compounds with the goal of deriving the molecular structure from just a single measurement, usually incorporating the INADEQUATE technique to establish the carbon framework of the molecule. The second category[51] addresses experiments that require dispersal of the information into a third frequency dimension. One such application is the precise measurement of long-range carbon–proton couplings, which span a wide range of magnitudes. A third scheme[52] focuses on the all-important factor of speed – obtaining the requisite PANACEA information in the shortest possible time. Finally, the general concept of multifunctional pulse sequences can be extended[53] to deal with the added complexities of biochemical molecules, often globally enriched in the isotopes ^{13}C and ^{15}N.

7.3.1 Structure of Small Molecules

Small organic molecules are well suited to investigation by a combination of standard NMR pulse sequences, notably the INADEQUATE technique to monitor direct ^{13}C–^{13}C interactions, multiplicity-edited heteronuclear correlations (HSQC) and long-range ^{13}C–^{1}H or ^{15}N–^{1}H correlations (HMBC). The key initial step is to derive reliable information about the carbon framework. Although the INADEQUATE method is handicapped by its low intrinsic sensitivity, the problem can be moderated by employing a cryogenically cooled probe, optimized for ^{13}C detection. This offers an approximately 10-fold enhancement in signal-to-noise ratio, further improved by a processing scheme[50,54] based on the symmetry properties of these spectra in the acquisition dimension. Even so, the INADEQUATE feature can often be the rate-determining step for natural-abundance samples.

Figure 7.3 shows a schematic PANACEA representation that comprises INADEQUATE, HSQC and HMBC elements. Phase-sensitive INADEQUATE spectra are recorded by incrementing the phase of the first three pulses in 45° steps, thus shifting the double-quantum coherences by 90°, giving two signal components in quadrature. Very little is wasted. The strong signals from isolated ^{13}C sites (suppressed during the INADEQUATE element) are refocused and employed for other purposes – for recording the conventional ^{13}C spectrum, and for the HSQC feature that distinguishes CH, CH_2 and CH_3 groups. Because of the intervening ^{16}O or ^{14}N sites, the preliminary carbon

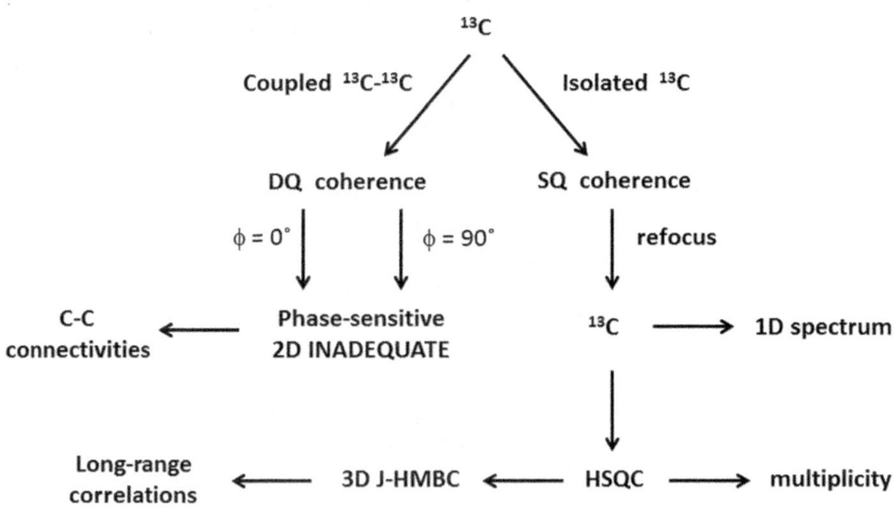

Figure 7.3 Schematic representation of the PANACEA experiment combining INADEQUATE, HSQC and HMBC features. It provides a one-dimensional decoupled ^{13}C spectrum, a two-dimensional ^{13}C–^{13}C correlation spectrum (INADEQUATE), multiplicity-edited ^{1}H–^{13}C single-bond correlation spectra and three-dimensional multiple-bond ^{1}H–^{13}C correlation spectra. The sequence is readily extended to include direct and long-range ^{1}H–^{15}N correlation spectra when a parallel ^{15}N receiver is available.

framework is often incomplete, but the isolated fragments can be linked together by information from the long-range correlations recorded in the HMBC measurements, and from evidence based on the ^{13}C chemical shifts.

7.3.1.1 The First Practical Test

Consider a representative test case applied to a supposedly "unknown" sample. The molecular mass is found to be 232.28. The decoupled one-dimensional ^{13}C spectrum establishes that there are x chemically distinct carbon sites (Figure 7.4). These ^{13}C shifts are used in a symmetrization routine that enhances the mean signal-to-noise ratio in the two-dimensional INADEQUATE spectrum (Figure 7.5) by a factor of two. This spectrum identifies a six-membered ring with a branched "isopropyl" side-chain, together with a fragment of two attached carbon atoms, and a single isolated carbon. As part of the same measurement, multiplicity-edited ^{13}C-HSQC spectra are recorded (Figure 7.6 shows one example) indicating how these sites are protonated. Two NH moieties are identified in the corresponding ^{15}N-HSQC measurements.

The next stage is to link the various fragments (Figure 7.7a) into a full structure using the corresponding HMBC measurements of the long-range ^{13}C–^{1}H interactions. Two- and three-bond NH interactions in the H–N–C–C

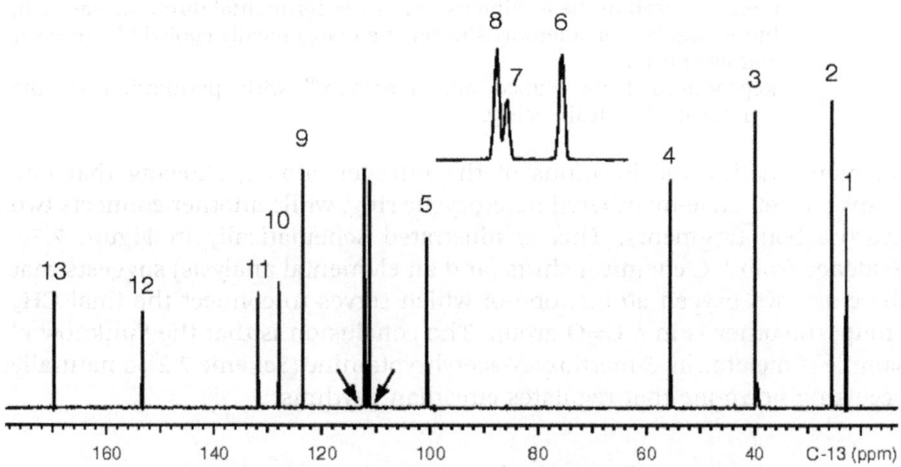

Figure 7.4 The decoupled ^{13}C spectrum of the first "unknown" test sample recorded as part of the PANACEA experiment on a 600 MHz spectrometer equipped with three parallel receivers. The sample was made up of 260 mg dissolved in 500 μL of DMSO-d_6. The narrow frequency range indicated by the two arrows has been expanded (inset) to show three close resonances (C6 and C8 are later shown to be directly coupled). Reproduced from Kupče and Freeman[50] with permission of the American Chemical Society.

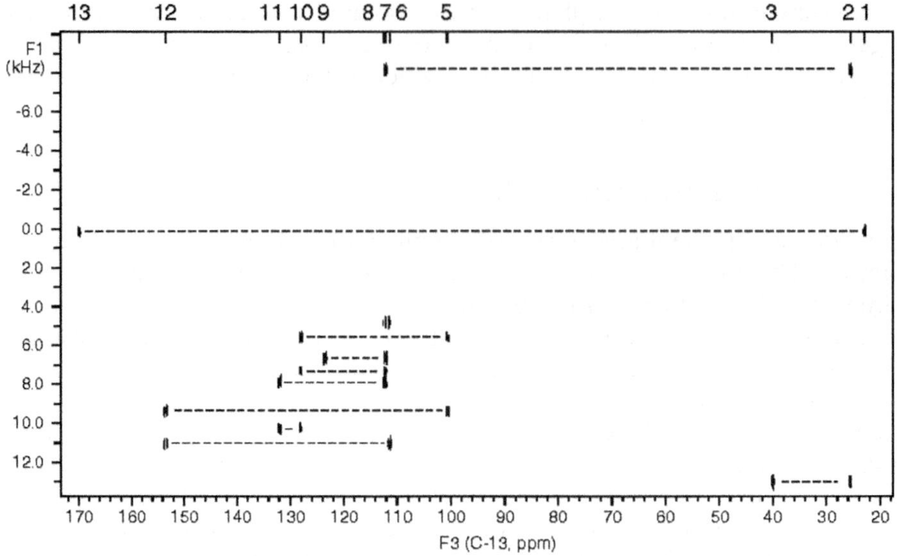

Figure 7.5 The two-dimensional INADEQUATE spectrum of the first "unknown" test sample recorded as part of the PANACEA experiment on a 600 MHz spectrometer with a standard room-temperature ^{1}H, ^{13}C, ^{15}N "triple-resonance" probe. The numbering scheme is that of Figure 7.4. Horizontal dashed lines indicate the correlations. Direct coupling between C6 and C8 forms a strong-coupling AB pattern near 111 ppm on the ^{13}C axis. The correlation at the lower right corner has been deliberately aliased in the F_3 dimension. The experimental duration was 12 h, but would be considerably shorter if a cryogenically cooled ^{13}C receiver coil were used.
Reproduced from Kupče and Freeman[50] with permission of the American Chemical Society.

fragments define the locations of the nitrogen atoms, showing that one forms part of a five-membered heterocyclic ring, while another connects two hydrocarbon fragments. This is illustrated schematically in Figure 7.7c. Evidence from ^{13}C chemical shifts (and an elemental analysis) suggests that there are two oxygen atoms, one of which serves to connect the final CH_3 group; the other is in a C=O group. The conclusion is that the "unknown" sample is melatonin, 5-methoxy-*N*-acetyltryptamine (Scheme 7.2), a naturally occurring hormone that regulates circadian rhythms.

7.3.1.2 The Second Practical Test

A second, more challenging, test molecule has a molecular mass of 324.44 and a carbon skeleton made up of 20 atoms, as indicated by the high-resolution ^{13}C spectrum (Figure 7.8). This spectrum was processed at the same time as the INADEQUATE spectrum (Figure 7.9), generating the connectivity pattern illustrated in Figure 7.10a. The multiplicity-edited CH correlation measurements identify the attached protons (Figure 7.10b).

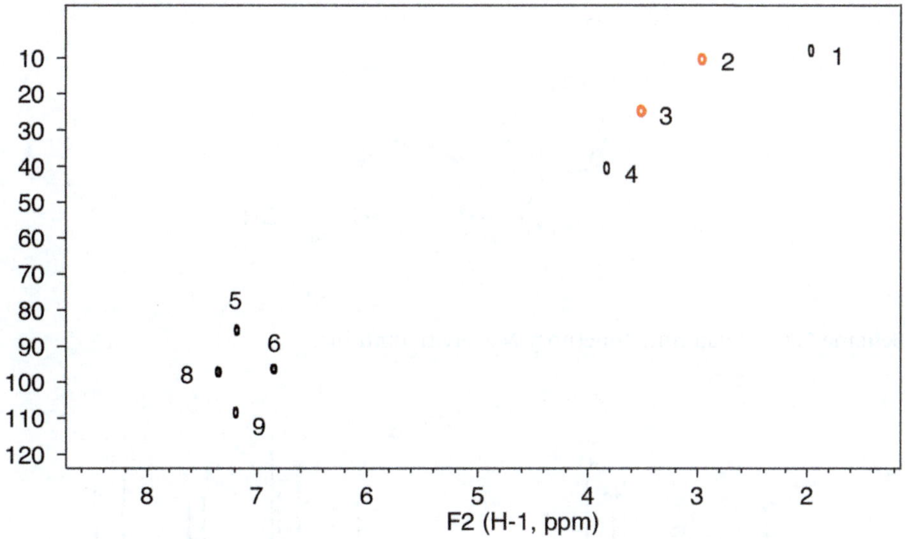

Figure 7.6 Multiplicity-edited HSQC spectrum of the first "unknown" test sample, showing responses from CH and CH_3 (black) and inverted signals from CH_2 (red). The vertical dimension is the ^{13}C axis in ppm. These results were recorded in parallel with the HMBC and INADEQUATE measurements.
Reproduced from Kupče and Freeman[50] with permission of the American Chemical Society.

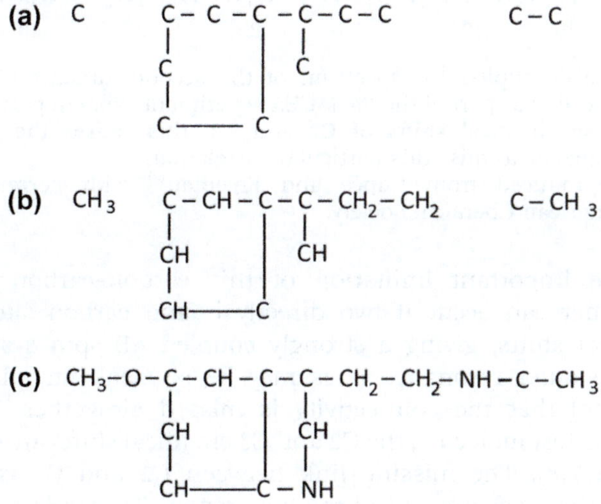

Figure 7.7 (a) The carbon–carbon connectivity pattern derived from the INADEQUATE data shown in Figure 7.5. (b) The effects of the multiplicity-edited single-bond HSQC experiments. (c) Inclusion of the HMBC long-range CH and NH correlation measurements, which serve to link the three fragments together and also close a five-membered heterocyclic ring. The sample is in fact melatonin (Scheme 7.2).

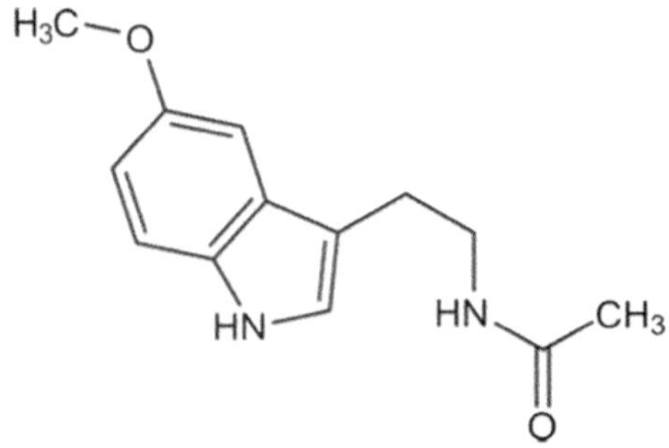

Scheme 7.2 Melatonin, 5-methoxy-*N*-acetyltryptamine.

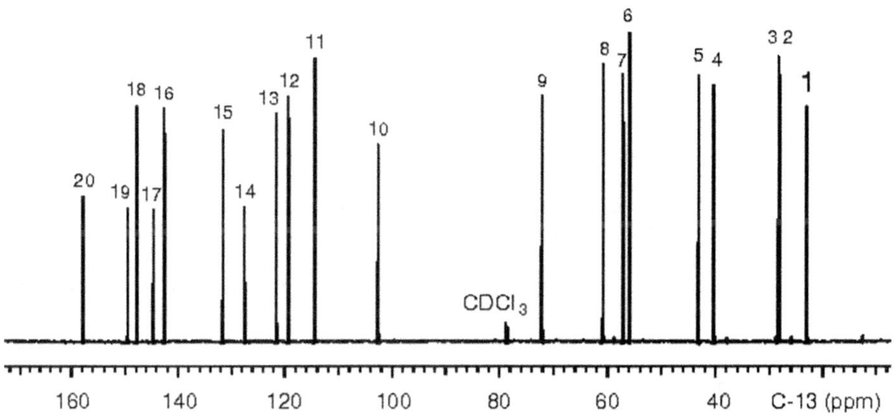

Figure 7.8 The decoupled ^{13}C spectrum of the second "unknown" test sample recorded as part of the PANACEA experiment. Note in particular the very close chemical shifts of C2 and C3; this causes the INADEQUATE sequence to miss this particular correlation.
Reproduced from Kupče and Freeman[50] with permission of the American Chemical Society.

There is one important limitation of this carbon–carbon connectivity experiment that can occur if two directly bound carbon sites have very close chemical shifts, giving a strongly coupled AB spin system. In this situation, the outer resonance lines may be so weak (in relation to the baseline noise) that the connectivity is missed altogether. This indeed happens in the test molecule; the C2 and C3 chemical shifts are only 0.14 pm apart (Figure 7.8). The missing link between C2 and C3 is only made evident from the measurement of the long-range CH correlations, recorded in parallel. Long-range NH correlations define the locations of the nitrogen atoms in the heterocyclic rings (Figure 7.10c). Carbon chemical shift evidence indicates that the methyl group is attached to the rest by means of an oxygen atom, and that there is an attached hydroxyl group. The molecule is in fact quinine (Scheme 7.3).

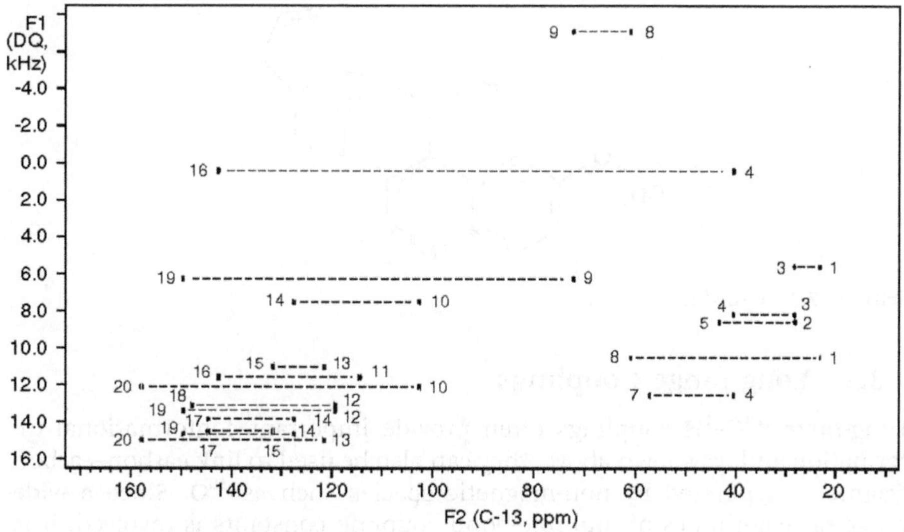

Figure 7.9 The two-dimensional ^{13}C–^{13}C INADEQUATE spectrum of the second "unknown" test sample, recorded with some controlled aliasing in the F_2 dimension. Horizontal dashed lines indicate the correlations. The measurement fails to find a correlation between C2 and C3 owing to the very small chemical shift difference (0.14 ppm).
Reproduced from Kupče and Freeman[50] with permission of the American Chemical Society.

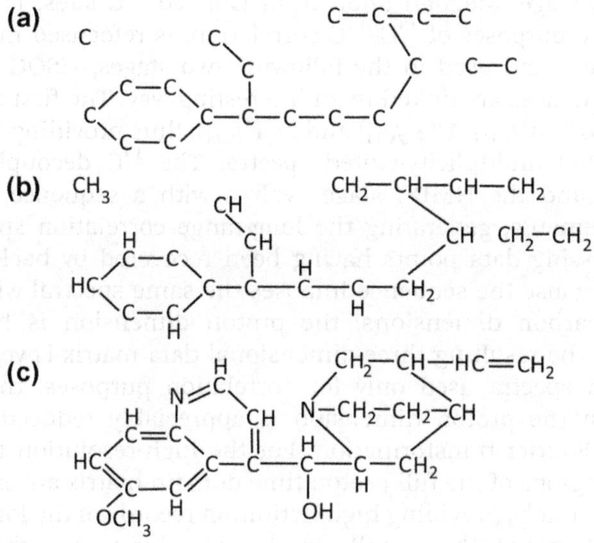

Figure 7.10 (a) Carbon connectivity pattern derived for the second "unknown" test sample from the INADEQUATE feature of the PANACEA experiment. (b) Result of incorporating the multiplicity-edited single-bond CH correlation measurement (HSQC). (c) Inclusion of the long-range CH and NH correlation results, establishing ring closures and confirming that the C2 and C3 sites are indeed directly bonded.

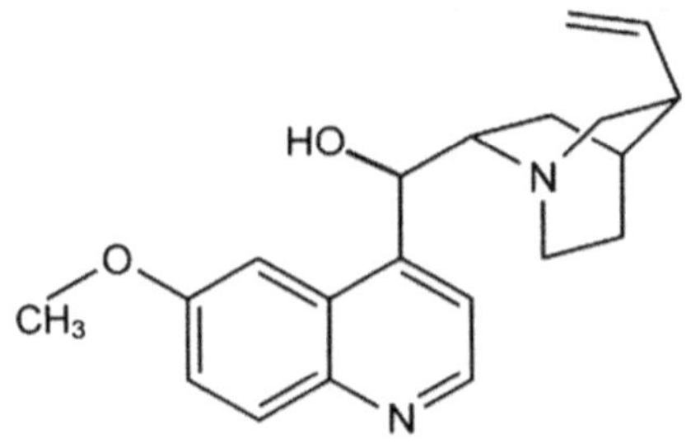

Scheme 7.3 Quinine.

7.3.2 Long-range Couplings

Long-range ^{13}C–^{1}H couplings often provide important conformational information and, as shown above, they can also be used to link carbon–carbon fragments separated by non-magnetic species such as ^{16}O. Since a wide range of magnitudes of multiple-bond coupling constants is involved, it is important to ensure that none are overlooked by an unfortunate choice of timing parameters. The remedy is to introduce a third frequency dimension where the long-range couplings are allowed to evolve, and where they can be measured with high precision. High resolution can be achieved in the direct (proton) dimension with only a negligible increase in overall measurement duration.

This three-dimensional PANACEA sequence[51] starts with the INADEQUATE stage. Magnetization from isolated ^{13}C sites, normally suppressed for the purposes of ^{13}C–^{13}C correlation, is refocused in a constant-time experiment and used in the following two stages, HSQC and HMBC. These two sequences are linked in an interesting way. The first three timing steps are set to $1/(4\,^{1}J_{CH})$, $1/(2\,^{1}J_{CH})$ and $3/(4\,^{1}J_{CH})$, thus providing the required two-dimensional multiplicity-edited spectra. The ^{13}C decoupling is then switched off, and the HMBC stage evolves with a sequence of repeated $1/(4\,^{1}J_{CH})$ increments, generating the long-range correlation spectrum, the first three missing data points having been recovered by backward linear prediction. Because the sequence imposes the same spectral widths on the proton and carbon dimensions, the proton dimension is heavily oversampled, and the resulting three-dimensional data matrix is very large. For low-resolution spectra used only for correlation purposes, the degree of digitization in the proton dimension is appreciably reduced to avoid a cumbersome Fourier transformation. For the high-resolution HMBC spectra, selected regions of the full proton time-domain matrix are extracted and processed separately, providing high-definition records of the long-range CH interactions. Although the overall experimental duration of this variant of PANACEA can be fairly long, it is determined predominantly by the stringent digitization demands of the HMBC stage, rather than by the sensitivity requirements of INADEQUATE. A cold probe would not speed up this particular application appreciably.

7.3.2.1 A Practical Example: Methyl Salicylate

This high-resolution version of PANACEA is illustrated by reference to 600 MHz spectra of methyl salicylate ($\sim$50% by volume in $CDCl_3$). The three-dimensional feature allows free evolution of the long-range CH interactions, and ensures that none are overlooked; Figure 7.11 shows the appropriate projections onto the carbon–proton plane. Couplings between ring carbons and the hydroxyl proton (on C1) offer useful information about the conformation of the C–O–H link. These particular long-range splittings are illustrated in Figure 7.12, and their values are measured with high accuracy ($\pm$0.05 Hz) by adopting the method of *J*-doubling.[55] In practice, threefold doubling is employed. Appropriate narrow regions of the proton spectrum are extracted, back-transformed into the time domain and multiplied by the function $\cos(\pi J^*t)\cos(2\pi J^*t)\cos(4\pi J^*t)$, where J^* is a computer-generated variable frequency. When J^* reaches J, there is mutual cancellation of 14 antiphase signals, and the integral of the absolute magnitude of the corresponding frequency-domain spectrum passes through a well-defined minimum. The resulting couplings for methyl salicylate suggest that the OH group is oriented towards the C=O group in such a way as to form a short hydrogen bond.

7.3.3 Fast Measurements

The acronym INADEQUATE was always intended as a gentle reminder that there might be some problems with its inherent sensitivity for samples with the natural ^{13}C abundance. It is not advisable to criticize Mother Nature, but only one useful molecule in 857 might seem a little parsimonious. Nevertheless the technique has been made a key component of PANACEA because it provides unambiguous evidence about the basic carbon skeleton before the full molecular structure is fleshed out. The INADEQUATE stage consequently acts as a serious brake in the speed of the measurement – a kineticist would call it the rate-determining step. There is therefore much to be gained by speeding up the acquisition or by improving the sensitivity of this particular feature.

Two-dimensional INADEQUATE traces in the F_2 dimension possess important symmetry properties. These four-line spectra possess *global* symmetry with respect to the point of intersection with the double-quantum diagonal, and a *local* symmetry with respect to the chemical shift of each coupled site. These features may be exploited to improve the signal-to-noise ratio, making use of the fact that random noise is not identical at the four sites.[50] However, small corrections need to be made to the positions of these centres of symmetry. The location of the global centre is slightly affected by the coarse digitization in the double quantum (F_1) dimension. The position of the local centre of symmetry with respect to the usual ^{13}C chemical shift is slightly shifted by the secondary isotope shift because each ^{13}C atom now has a ^{13}C neighbour. Furthermore, when there is strong coupling, the local

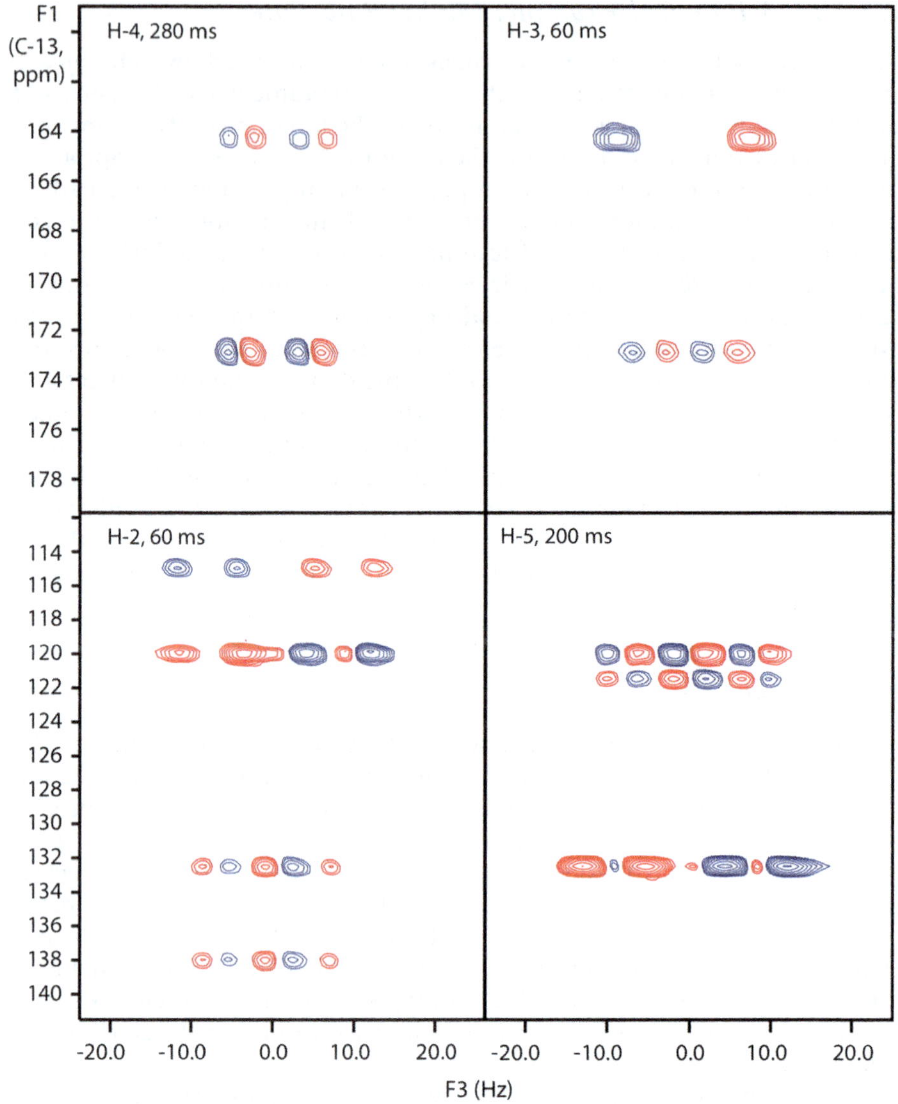

Figure 7.11 Long-range ^{13}CH couplings (antiphase patterns) extracted from the three-dimensional HMBC spectrum of methyl salicylate recorded as part of a PANACEA experiment on a 600 MHz spectrometer. Red and blue signals have opposite phases. The timings (top left) were chosen to display the best F_1F_3 planes of the three-dimensional matrix. The duration of this experiment was principally determined by the high definition required for the three-dimensional HMBC feature, rather than the intrinsically low sensitivity of the INADEQUATE sequence. Reproduced from Kupče and Freeman[51] with permission of John Wiley & Sons.

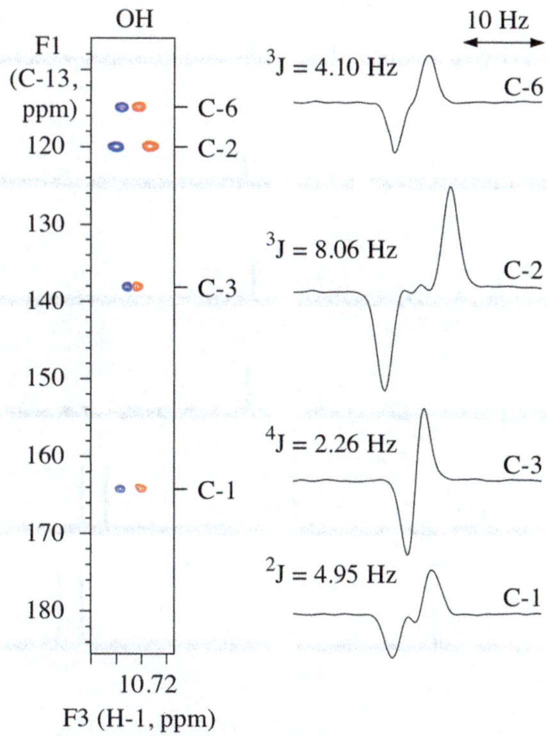

Figure 7.12 Long-range splittings between four ring carbons of methyl salicylate and the hydroxyl proton (on C1), measured by the HMBC element of the PANACEA sequence, recorded in parallel with the INADEQUATE and multiplicity-edited HSQC measurements. Red and blue signals have opposite phases. The *J*-doubling method was employed to measure these splittings, giving an accuracy estimated to be ±0.05 Hz.
Reproduced from Kupče and Freeman[51] with permission of John Wiley & Sons.

symmetry centre in an AB system is slightly displaced (by a calculable amount) from the actual chemical shift frequency.

Accurate values for the ^{13}C chemical shifts are obtained from the one-dimensional measurement and can be used to set up the symmetrization routine. The corrections to these frequencies are fairly small, and the range of possible ^{13}C–^{13}C coupling constants is well known, so a limited-range search algorithm can quickly locate the four resonance lines that make up a typical F_2 trace. Allowance can be made for low-sensitivity situations where one of the expected four components is not detected; in this case, the remaining three intensities are combined.[54] The symmetry algorithm discriminates against thermal noise, accidental overlap of extraneous signals, and spectral artefacts. Figure 7.13 illustrates a mean twofold sensitivity enhancement achieved for F_2 traces taken from the INADEQUATE spectrum of a sugar derivative.

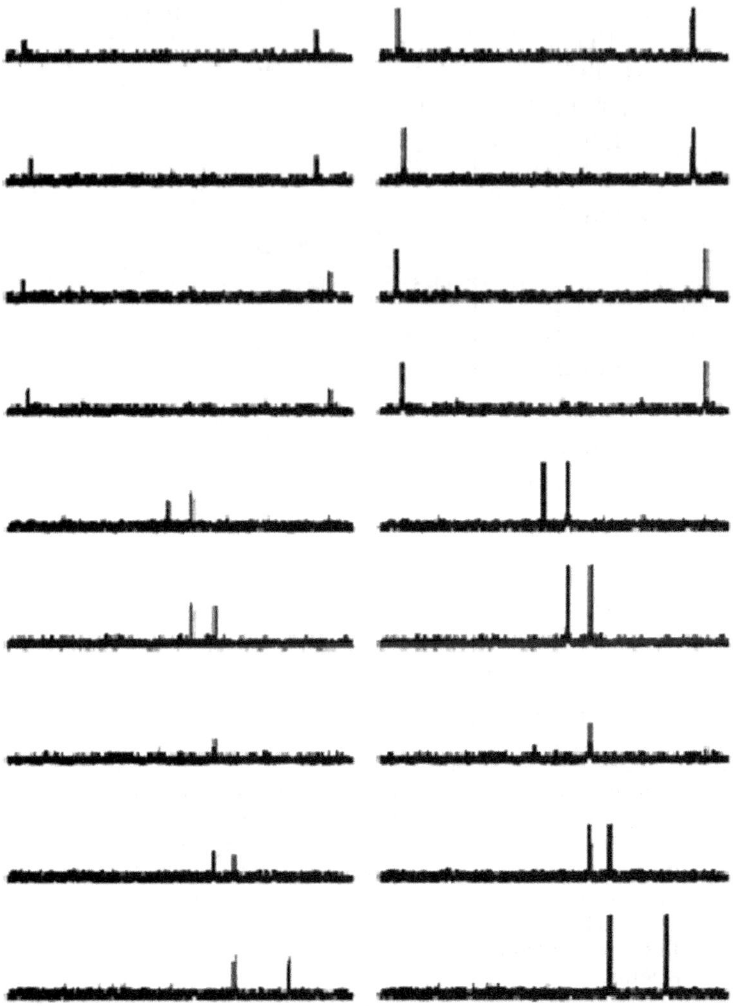

Figure 7.13 Sensitivity enhancement for nine F_2 traces extracted from the ^{13}C INADEQUATE spectrum of a sugar derivative. A symmetrization program based on global and local symmetry properties has been applied to the raw data (left) to generate the enhanced data (right). The mean improvement factor is two.
Reproduced from Kupče and Freeman[50] with permission of the American Chemical Society.

The enhancement of sensitivity by symmetrization only works effectively if the combined responses have comparable intensities; the inclusion of lines with much lower intensities (and hence poorer signal-to-noise ratios) actually degrades sensitivity. An important example arises in the relatively uncommon case of a strongly coupled pair of ^{13}C spins. If the intensities of the inner lines are more than $\sqrt{2}$ times the intensity of the outer lines, the latter would contribute more than their "fair share" of noise, and should therefore be excluded from the local symmetrization procedure.

7.3.3.1 Hadamard-encoded Spectra

As mentioned above, a very useful feature of PANACEA is that it provides several different types of NMR information "on the fly" during the sequence. In particular, a one-dimensional decoupled ^{13}C spectrum is retrieved and can be stored separately, giving accurate values for the ^{13}C chemical shifts. This prior knowledge provides the starting point for a Hadamard-encoding scheme designed to speed up the measurement. First, homonuclear double-quantum coherence (DQC) is excited for all the directly bonded ^{13}C–^{13}C pairs in the molecule by the standard pulse sequence:

$$90°(X) - \tau - 180°(X) - \tau - 90°(X) \rightarrow \mathrm{DQC}$$

where $\tau = 1/(4J_{CC})$. Normally, the next step would be free evolution of the double-quantum coherence; instead, this stage is replaced by a set of simultaneous selective radiofrequency pulses (I_X), each tuned to the ^{13}C chemical shift of a specific carbon site. Suppose that there are n distinct chemical sites in the molecule under investigation. The n excitation frequencies are encoded (+ or −) with a Hadamard matrix[1] of order N, where $N \geq n$. Altogether N scans are made, each with a new encoding pattern defined by the rows of this matrix. Since (N - n) columns serve no useful purpose, it is advantageous to match N with n as closely as possible. Because these matrices are known for N = 4, 8, 12, 16, 20, 24, 28, 32, *etc.*, it is easy to find the nearest efficient encoding scheme. The speed gain arises because only N scans are made, whereas the conventional scheme involves K scans, where K is the required number of evolution increments, set by the Nyquist condition and the resolution requirements in the double-quantum dimension. The ratio K/N can easily reach an order of magnitude.

In the product operator formalism,[56] a selective radiofrequency pulse I_X applied to a "source site" converts part of the double-quantum coherence into observable (antiphase) magnetization at the "target site" (the S spins):

$$2I_XS_Y + 2I_YS_X \rightarrow 2I_XS_Y + 2I_ZS_X \quad (7.1)$$

In practice, evolution *during* the selective pulse under the $2I_ZS_Z$ operator allows an in-phase signal to be generated:

$$2I_ZS_X \rightarrow S_Y \quad (7.2)$$

Thus one particular column of the Hadamard matrix (the source site, defined by I_X) is correlated with the target site (defined by the response S_Y). This single coherence transfer establishes that I and S are directly coupled. In principle, this would be all the information needed for correlation, but irradiation at another column of the matrix by the selective pulse S_X generates the reverse transfer:

$$2I_XS_Y + 2I_YS_X \rightarrow 2I_XS_Z + 2I_YS_X \quad (7.3)$$

which evolves under the $2I_ZS_Z$ operator:

$$2I_XS_Z \rightarrow I_Y \tag{7.4}$$

This has the effect of confirming the first result.

Figure 7.14 shows schematically how the Hadamard processing works for a simple illustrative case of an 8 × 8 matrix. Eight successive scans are performed with the eight selective radiofrequency pulses modulated (plus or minus) according to the rows of this matrix. Consider, for example, the case of selective irradiation of site 3 (highlighted in red). In each new scan the sense of this particular radiofrequency pulse is alternated according to the signs in column 3. As a result, only NMR signals modulated in this particular pattern (+ − + − + − + −) are retained; signals derived from the other seven columns are modulated by different patterns, and vanish. Note that success depends on completion of all eight scans, although less than eight sites may be irradiated.

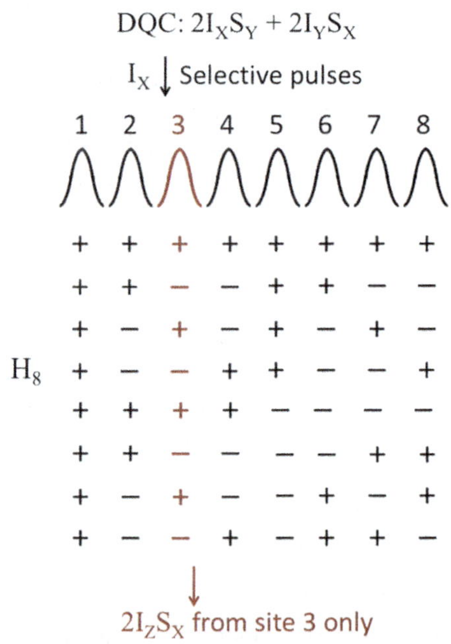

Figure 7.14 Schematic diagram representing the Hadamard-encoded ^{13}C–^{13}C correlation experiment. The 8×8 Hadamard matrix is shown for illustrative purposes, although a 16×16 matrix was actually used in practice. Eight simultaneous selective radiofrequency pulses (I_X) are employed. Suppose one representative pulse (highlighted in red) is applied at the frequency of a chosen site (C3). In each successive scan, this pulse is modulated according to column 3 of the matrix. Consequently, the corresponding NMR signals are similarly modulated. After decoding according to the same pattern, the coherence transfer response S_Y from the coupled site is detected and establishes the correlation; signals from the other seven columns vanish identically. Selective irradiation of the remaining columns determines all eight ^{13}C–^{13}C correlations.

The Hadamard spectroscopy results are recorded as a set of one-dimensional traces where the detected NMR response appears at the chemical shift of the target site, while the source site is identified by the corresponding irradiation frequency (set by the appropriate column of the Hadamard matrix). As shown above, there is a twofold redundancy, corresponding to the forward and backward coherence transfers. A straightforward reformatting of this set of one-dimensional traces allows the two-dimensional INADEQUATE spectrum to be reconstructed in its more familiar form. Figure 7.15 shows how the genuine correlations are distinguished from uncorrelated ^{13}C–^{13}C pairs by comparing all possible pairs of experimental traces and testing them for reflection symmetry, using a well-known algorithm that compares ordinates and retains only the lower absolute value. The accepted traces are then reassembled as a function of the known double-quantum frequencies along the F_1 axis (Figure 7.15d).

7.3.3.2 A Practical Example: Menthol

A practical example is provided by the Hadamard-encoded INADEQUATE spectrum of menthol (30% in $CDCl_3$) recorded on a 500 MHz spectrometer with a cold probe optimized for ^{13}C detection.[52] There are 10 carbon sites in this molecule, and in principle the H_{12} matrix would have sufficed for encoding; in practice, the H_{16} matrix is employed because it has been found that the "power-of-two" Hadamard matrices give cleaner responses than the "4x" matrices, being less sensitive to spectrometer imperfections. The spectrum in Figure 7.16 establishes all the expected experimental correlations with a good signal-to-noise ratio in a measurement that lasts only 56 s, including the corresponding HSQC and HMBC measurements. At long last, the old sensitivity handicap of INADEQUATE appears to have been overcome, suggesting that many small molecules can be attacked by the PANACEA protocol.

7.4 Biochemical Samples

PANACEA was conceived for measurements on small molecules. However, the concept of combining two or more standard pulse sequences into a single unit can be extended to biomolecules such as proteins, where much of present-day NMR research is focused. This presents important new challenges – consolidating complex pulse sequences without sacrificing sensitivity or resolution. In addition to the problem of molecular complexity, the new sequences need to be adapted to take account of global isotopic enrichment in ^{13}C and ^{15}N, a standard procedure for spreading the NMR information into new frequency dimensions. The INADEQUATE method, a key feature of the small-molecule experiments, is unsuitable for large biomolecules.

There exists a whole family of "biomolecular" pulse sequences that might be considered candidates for parallel acquisition. Just one illustrative

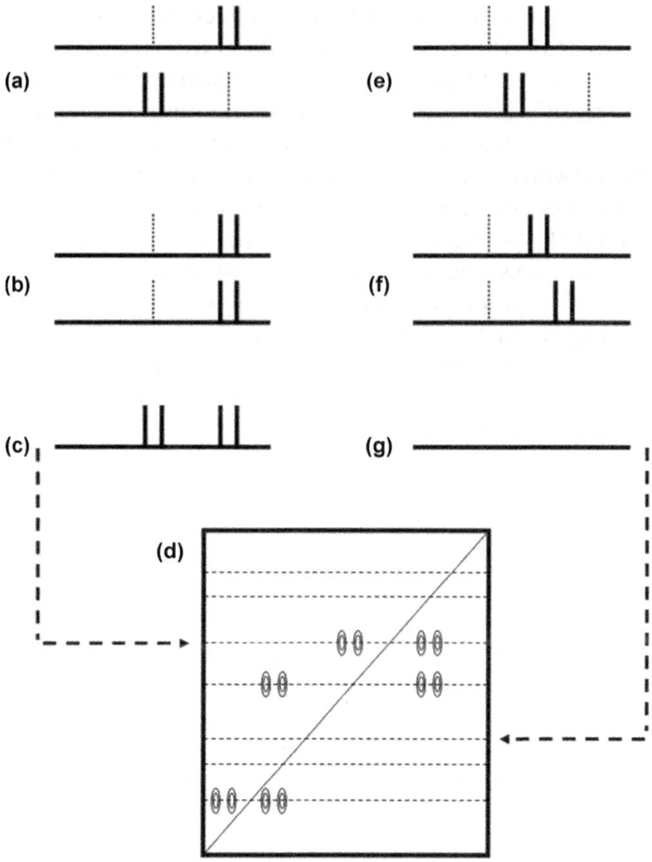

Figure 7.15 The processing scheme for reconstructing an INADEQUATE spectrum in its familiar format from a set of one-dimensional traces obtained by Hadamard-encoded selective irradiation experiments. All possible combinations of traces are compared in pairs. True correlations (left column) are defined by the reflection symmetry of the forward and backward coherence-transfer traces. Uncorrelated traces (right column) are recognized by their lack of this reflection symmetry. In this test, a lower value algorithm is applied to each pair of traces; this retains the signals (c) for genuine correlations but cancels the signals (g) for uncorrelated pairs. The reconstruction of the familiar two-dimensional contour display makes use of the known double-quantum frequencies (vertical axis) so that all four-line patterns are centred on the skew diagonal.
Reproduced from Kupče and Freeman,[52] copyright 2010, with permission of Elsevier.

example is considered here. For sensitivity reasons, many NMR investigations of proteins have focused on proton detection, but recently interest has been rekindled in the idea of *direct detection* of the low-gamma nuclei ^{13}C or ^{15}N, partly because they are less susceptible than protons to broadening by paramagnetic species. Consider the case of a protein of

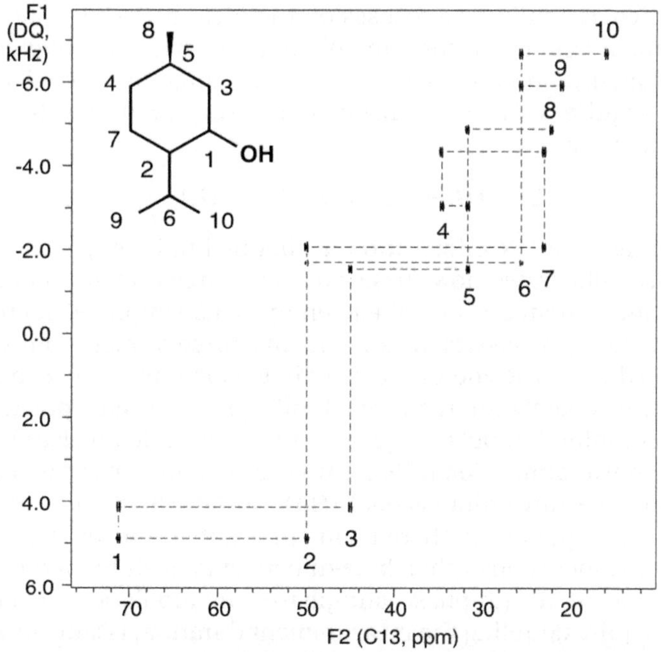

Figure 7.16 The INADEQUATE spectrum of menthol (30% in $CDCl_3$) recorded with the Hadamard-encoded selective irradiation scheme. The 500 MHz spectrometer was equipped with a cold probe optimized for ^{13}C detection. A 16×16 Hadamard matrix was used to encode the signals from the 10 carbon sites. This required 16 scans (rows of the matrix) but only 10 columns were used. The reconstruction of the two-dimensional spectrum followed the scheme shown in Figure 7.15. This measurement formed part of a PANACEA sequence that also provided HSQC and HMBC information; it was completed in only 56 s.
Reproduced from Kupče and Freeman,[52] copyright 2010, with permission from Elsevier.

moderate size, nuclease A inhibitor (143 amino acid residues) studied as a 1 mM aqueous solution (10% D_2O) on a 600 MHz spectrometer. Suppose that the principal interest is in the direct detection of the ^{13}CO resonances from a two-dimensional (HA)CACO sequence. It would be interesting to derive further structural information by recording the evolution of the ^{15}N spins, if this could be obtained without significant prejudice to the ^{13}C measurement. It can be shown that the Fourier transform of an exponentially decaying free induction signal delivers the optimum signal-to-noise ratio when the truncation occurs a point 1.26 times the time constant of the decay. (Surprisingly, this optimum is *independent* of the general level of the noise in comparison with the NMR signal.) Once the required ^{13}C signal has been detected, there remains a very weak response (the "afterglow") from the unused tail of the truncated free induction decay.[53]

This particular application focuses on the idea that the weak "afterglow" may be transferred to protons for observation with the higher intrinsic proton sensitivity. After refocusing, this tiny afterglow signal can be exploited to acquire a three-dimensional spectrum according to the overall magnetization flow:

$$\mathrm{HA} \rightarrow \mathrm{CA} \rightarrow \mathrm{CO} \rightarrow {}^{15}\mathrm{N} \rightarrow \mathrm{NH}$$

In this manner, the ^{15}N information is obtained indirectly.

Figure 7.17 illustrates how these two- and three-dimensional sequences have been incorporated into a single entity. This simplified representation should not be construed to mean that the three-dimensional element is merely tacked on to the end of the two-dimensional part: the two sequences are, in fact, intimately interconnected, offering important practical advantages (the combination pulse sequence is set out in detail elsewhere).[53] For example, the two-dimensional (HA)CACO results are augmented by signals derived from the three-dimensional (HA)CA(CO)NNH feature by summing over all ^{15}N data points, so there is no appreciable sensitivity penalty associated with the inclusion of this three-dimensional feature. In the (HA)CACO sequence, the IPAP (in-phase–anti-phase) manipulation, which would otherwise require doubling the measurement duration, is subsumed into the two data sets used for ^{15}N quadrature detection.

Naturally, the combination of two different sequences involves a certain amount of compromise; there is some trade-off between the resolution in

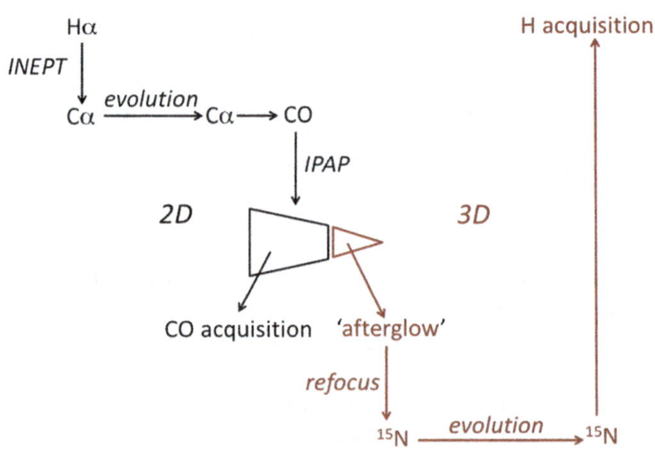

Figure 7.17 Schematic representation showing how a two-dimensional (HA)CACO pulse sequence (black) is combined with a three-dimensional (HA)CA(CO)NNH pulse sequence (red) without any appreciable sacrifice in sensitivity or resolution. The directly-detected CO signal is manipulated with an IPAP (in-phase–anti-phase) routine to give a "decoupled" response. The relative size of the much weaker CO "afterglow" is exaggerated for the purpose of illustration. This is refocused and transferred to ^{15}N to obtain the nitrogen spectrum before back-transfer to protons for acquisition with increased sensitivity.

the two-dimensional ^{13}C stage and the proton sensitivity achieved in the three-dimensional part. The ^{13}C resolution could be improved by increasing the ^{13}C acquisition time (giving a weaker afterglow), but with the danger that proton signals associated with broader CO sites might be lost. For applications where spin–spin relaxation effects are more severe, it might be advisable to shorten the ^{13}C acquisition time further in order to leave enough afterglow for the subsequent three-dimensional feature.

Figure 7.18 shows the ^{13}C direct detection Cα–CO correlation spectrum[53] of nuclease A inhibitor recorded at 25 °C. To provide some indication that larger proteins could be considered, these measurements were then repeated at 2 °C. Earlier ^{15}N relaxation studies[57] have shown that this molecule tumbles with effective correlation times of 8.8 ns at 25 °C and 17.5 ns at 2 °C. Despite the expected reduction in signal-to-noise ratio due to line

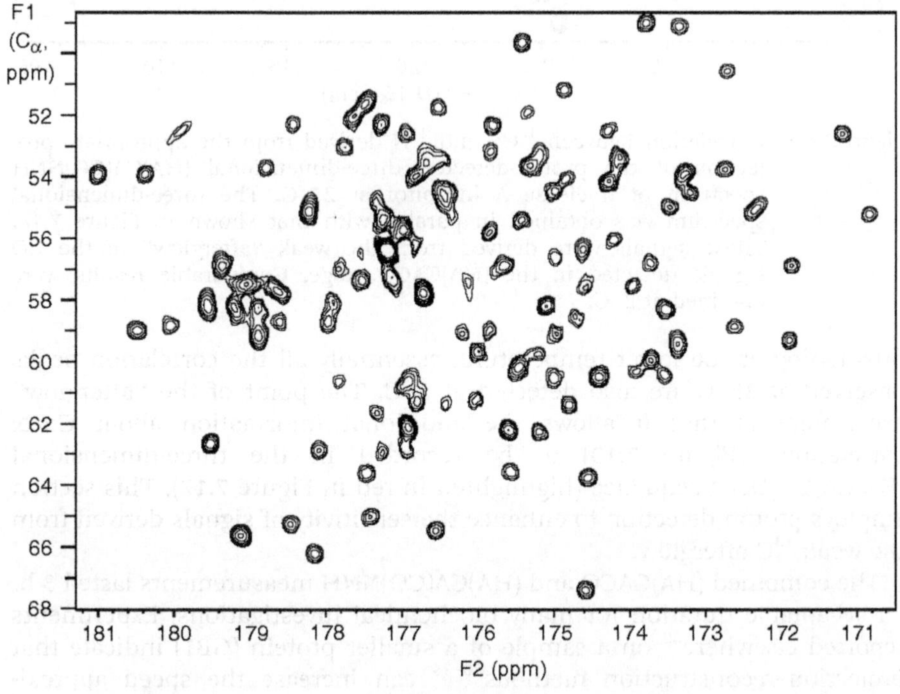

Figure 7.18 The two-dimensional directly-detected (HA)CACO correlation spectrum of a 1 mM aqueous solution (10% D_2O) of the 143-residue, globally enriched nuclease A inhibitor recorded at 25 °C. The horizontal axis shows the CO shifts. The 600 MHz spectrometer was equipped with a proton–carbon–nitrogen probe with cryogenically cooled ^{13}C and ^{1}H receiver coils and preamplifiers. Comparable results were obtained at 2 °C where molecular tumbling is twice as slow, suggesting that even larger protein molecules could be investigated.
Reproduced from Ē. Kupče, L. E. Kay, R. Freeman, *J. Am. Chem. Soc.*, 2010, **132**, 18008–18011, with permission of the American Chemical Society.

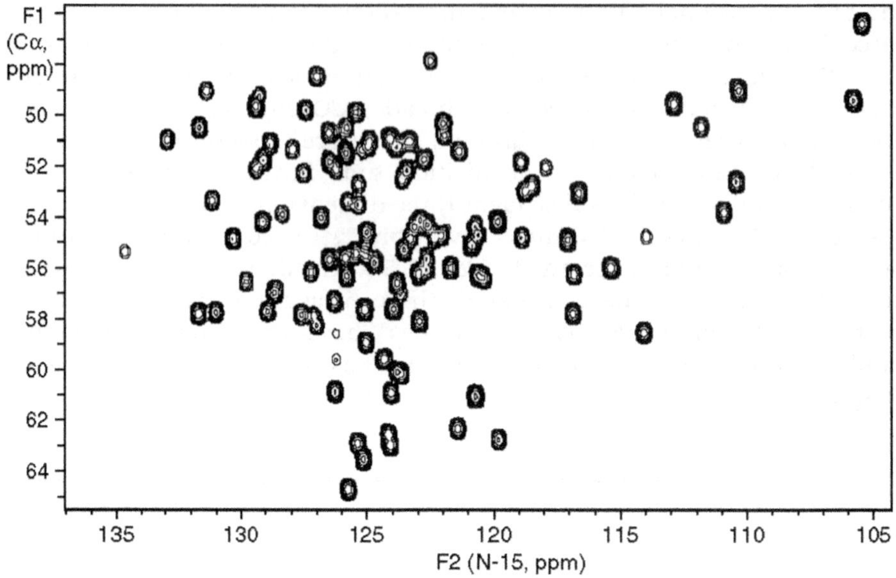

Figure 7.19 Correlation between ^{13}Cα and ^{15}N derived from the appropriate projection of the proton-detected three-dimensional (HA)CA(CO)NNH spectrum of nuclease A inhibitor at 25 °C. The three-dimensional spectrum was obtained in parallel with that shown in Figure 7.18. These signals were derived from the weak "afterglow" of the CO signals detected in the (HA)CACO stage. Comparable results were obtained at 2 °C.

broadening at the lower temperature, essentially all the correlation peaks observed at 25 °C are also detected at 2 °C. The point of the "afterglow" experiment is that it allows the additional information about Cα–N correlations (Figure 7.19) to be recorded in the three-dimensional (HA)CA(CO)NNH sequence (highlighted in red in Figure 7.17). This section employs proton detection to enhance the sensitivity of signals derived from the weak ^{13}C afterglow.

The combined (HA)CACO and (HA)CA(CO)NNH measurements lasted 3 h, an acceptable duration for many biochemical investigations. Experiments reported elsewhere[53] on a sample of a smaller protein (GB1) indicate that projection–reconstruction methods[30–35] can increase the speed approximately 12-fold compared with the conventional stepwise acquisition of evolution data on a Cartesian grid. These results suggest that the parallel acquisition protocol can be extended to other combinations of standard "biochemical" pulse sequences.

7.5 Conclusion

The introduction of multiple NMR receivers operating in parallel has made possible important new NMR procedures, for example, simultaneous ^{13}C–^{1}H and ^{13}C–^{19}F (or ^{15}N–^{1}H and ^{15}N–^{19}F) correlation measurements in a single experiment. It is also the basis for a family of PANACEA experiments – molecular structure determination, long-range coupling measurements and a fast version. Two worked examples of structural determination on small organic molecules are presented. Long-range ^{13}C–^{1}H interactions have been studied by extending PANACEA into three frequency dimensions. A fast version of PANACEA has been implemented by employing Hadamard-encoded multiply selective experiments. The extension of the general concept to a biomolecular sample, nuclease A inhibitor, has been carried out with an acceptable experimental duration of a few hours. It seems clear that spectrometers equipped with two or more receivers operating in parallel have a promising future in NMR.

Acknowledgements

The authors acknowledge extensive technical support for parallel acquisition experiments by Boban K. John. The sample of nuclease A inhibitor was kindly provided by Robert E. London. The PANACEA acronym "Protons And Nitrogen And Carbon *Et Alia*" was suggested by Malcolm Levitt; it replaces our earlier formulation "Parallel Acquisition NMR an All-in-one Combination of Experimental Applications."

References

1. J. Hadamard, *Bull. Sci. Math.*, 1893, **17**, 240.
2. Ē. Kupče and R. Freeman, *J. Magn. Reson.*, 2003, **162**, 300.
3. Ē. Kupče and R. Freeman, *J. Magn. Reson.*, 2003, **163**, 56.
4. Ē. Kupče, T. Nishida and R. Freeman, *Progr. NMR Spectr.*, 2003, **42**, 95.
5. R. Brüschweiler and F. Zhang, *J. Chem. Phys.*, 2004, **120**, 5253.
6. N. Trbovic, S. Smirnov, F. Zhang and R. Brüschweiler, *J. Magn. Reson.*, 2004, **171**, 277.
7. F. Zhang and R. Brüschweiler, *J. Am. Chem. Soc.*, 2004, **126**, 13180.
8. L. Frydman, T. Scherf and A. Lupulescu, *Proc. Natl. Acad. Sci. U.S.A.*, 2002, **99**, 15858.
9. L. Frydman, T. Scherf and A. Lupulescu, *J. Am. Chem. Soc.*, 2003, **125**, 9204.
10. Y. Shrot and L. Frydman, *J. Am. Chem. Soc.*, 2003, **125**, 11385.
11. B. Shapira, A. Lupulescu, Y. Shrot and L. Frydman, *J. Magn. Reson.*, 2004, **166**, 152.
12. Y. Shrot, B. Shapira and L. Frydman, *J. Magn. Reson.*, 2004, **171**, 163.
13. Y. Shrot and L. Frydman, *J. Chem. Phys.*, 2006, **125**, 204507.
14. Y. Shrot and L. Frydman, *J. Chem. Phys.*, 2008, **128**, 52209.
15. M. Gal, L. Frydman, in *Multidimensional NMR Methods for the Solution State*, ed. G. A. Morris and J. W. Emsley, Wiley, Chichester, 2010, ch. 3.
16. K. Ding and A. Gronenborn, *J. Magn. Reson.*, 2002, **156**, 262.

17. T. Szyperski, D. C. Yeh, D. K. Sukumaran, H. N. Moseley and G. T. Montelione, *Proc. Natl. Acad. Sci. U.S.A.*, 2002, **99**, 8009.
18. W. Kozminski and I. Zhukov, *J. Biomol. NMR*, 2003, **26**, 157.
19. S. Kim and T. Szyperski, *J. Am. Chem. Soc.*, 2003, **125**, 1385.
20. T. Szyperski and H. S. Atreya, *Magn. Reson. Chem.*, 2006, **44**, 51.
21. J. C. J. Barna, E. D. Laue, M. R. Mayger, J. Skilling and S. J. P. Worrall, *J. Magn. Reson.*, 1987, **73**, 69.
22. J. Chen, V. A. Mandelshtam and A. J. Shaka, *J. Magn. Reson.*, 2000, **146**, 363.
23. P. Schmieder, A. S. Stern, G. Wagner and J. C. Hoch, *J. Biomol. NMR*, 1993, **3**, 569.
24. I. Ibraghimov and M. Billeter, *J. Biomol. NMR*, 2003, **27**, 165.
25. A. J. Dunn and P. J. Sidebottom, *Magn. Reson. Chem.*, 2005, **43**, 124.
26. K. Kazimierczuk, A. Zawadzka, W. Kozminski and I. Zhukov, *J. Biomol. NMR*, 2006, **36**, 157.
27. K. Kazimierczuk, W. Kozminski and I. Zhukov, *J. Magn. Reson.*, 2006, **179**, 323.
28. M. Misiak and W. Kozminski, *Magn. Reson. Chem.*, 2006, **45**, 171.
29. Ē. Kupče and R. Freeman, *J. Magn. Reson.*, 2008, **191**, 164.
30. R. Freeman and Ē. Kupče, *J. Biomol. NMR*, 2003, **27**, 101.
31. Ē. Kupče and R. Freeman, *J. Biomol. NMR*, 2003, **27**, 383.
32. Ē. Kupče and R. Freeman, *J. Am. Chem. Soc.*, 2003, **125**, 13958.
33. Ē. Kupče and R. Freeman, *J. Am. Chem. Soc.*, 2004, **126**, 6429.
34. Ē. Kupče and R. Freeman, *Concepts Magn. Reson.*, 2004, **22A**, 4.
35. Ē. Kupče and R. Freeman, *J. Biomol. NMR*, 2004, **28**, 391.
36. R. A. Venters, B. E. Coggins and P. Zhou, *J. Am. Chem. Soc.*, 2004, **126**, 1000.
37. B. E. Coggins, R. A. Venters and P. Zhou, *J. Am. Chem. Soc.*, 2005, **127**, 11562.
38. S. Hiller, F. Fiorito, K. Wüthrich and G. Wider, *Proc. Natl. Acad. Sci. U.S.A.*, 2005, **102**, 10876.
39. D. Malmodin and M. Billeter, *J. Am. Chem. Soc.*, 2005, **127**, 13486.
40. J. W. Yoon, S. Godsill, Ē. Kupče and R. Freeman, *Magn. Reson. Chem.*, 2006, **44**, 197.
41. F. Fiorito, S. Hiller, G. Wider and K. Wüthrich, *J. Biomol. NMR*, 2006, **35**, 27.
42. Ē. Kupče and R. Freeman, *J. Am. Chem. Soc.*, 2006, **128**, 6020.
43. P. Schanda and B. Brutscher, *J. Am. Chem. Soc.*, 2005, **127**, 8014.
44. P. Schanda, Ē. Kupče and B. Brutscher, *J. Biomol. NMR*, 2005, **33**, 199.
45. Ē. Kupče and R. Freeman, *Magn. Reson. Chem.*, 2007, **45**, 2.
46. Ē. Kupče, R. Freeman and B. K. John, *J. Am. Chem. Soc.*, 2006, **128**, 9606.
47. Ē. Kupče, S. Cheatham and R. Freeman, *Magn. Reson. Chem.*, 2007, **45**, 378.
48. A. Bax, R. Freeman and T. A. Frenkiel, *J. Am. Chem. Soc.*, 1981, **103**, 2102.
49. A. Bax, R. Freeman, T. A. Frenkiel and M. H. Levitt, *J. Magn. Reson.*, 1981, **43**, 478.

50. Ē. Kupče and R. Freeman, *J. Am. Chem. Soc.*, 2008, **130**, 10788.
51. Ē. Kupče and R. Freeman, *Magn. Reson. Chem.*, 2010, **48**, 333.
52. Ē. Kupče and R. Freeman, *J. Magn. Reson.*, 2010, **206**, 147.
53. Ē. Kupče, L. E. Kay and R. Freeman, *J. Am. Chem. Soc.*, 2010, **132**, 18008.
54. T. Nakazawa, H. Sengstschmid and R. Freeman, *J. Magn. Reson., Ser. A*, 1996, **120**, 269.
55. L. McIntyre and R. Freeman, *J. Magn. Reson.*, 1992, **96**, 425.
56. O. W. Sørensen, G. W. Eich, M. H. Levitt, G. Bodenhausen and R. R. Ernst, *Prog. NMR Spectrosc.*, 1983, **16**, 163.
57. N. A. Farrow, R. Muhandiram, A. U. Singer, S. M. Pascal, C. M. Kay, G. Gish, S. E. Shoelson, T. Pawson, J. D. Forman-Kay and L. E. Kay, *Biochemistry*, 1994, **33**, 5984.

Part 2
Data Processing and Informatics

CHAPTER 8

^{1}H-NMR Spectroscopy: The Method of Choice for the Dereplication of Natural Product Extracts

JOHN BLUNT,[a] MURRAY MUNRO*[a] AND ANTONY J. WILLIAMS[b]

[a] Department of Chemistry, University of Canterbury, Christchurch, New Zealand; [b] ChemConnector Inc., Wake Forest, NC 27587, USA
*Email: murray.munro@canterbury.ac.nz

8.1 Natural Product Chemistry

Natural product chemists are the hunter–gatherers of the chemical world who untiringly scour the biota in a never-ending search to locate new structures from new sources. Before the start of our recorded history, the various shaman, medicine men, or witch doctors associated with the tribes of early civilization garnered specialist knowledge of the herbs, trees, fungi, and animals of their regions that had medicinal, hallucinogenic, poisonous, or antidote properties and, *de facto*, were the first natural product chemists. The use of natural products in these ways is as old as mankind itself and the facts and fancies of those times have been revealed in the recorded histories of the Babylonian, Assyrian, Egyptian, Persian, Indian, and Chinese civilizations. Interestingly, as noted in der Marderosian's 1969

Modern NMR Approaches To The Structure Elucidation of Natural Products, Volume 1: Instrumentation and Software
Edited by Antony J. Williams, Gary E. Martin and David Rovnyak

Published by the Royal Society of Chemistry, www.rsc.org

review article "Marine Pharmaceuticals," which details much of this early history, it was "left to Grecian times to see the beginnings of the dissociation of medicine from magic and religion."[1] The development of a scientific approach stagnated through the Dark Ages and it took until the 19th century before chemical isolations and pharmacological studies were initiated, and with it, the start of modern natural products chemistry. A good example is the extraction and purification of morphine in 1804 from the opium poppy plant.[2,3] In the two centuries that have followed, very few areas on Earth have remained unexplored in the quest for new ecological niches that might harbor unexplored species and potentially new compounds. There appears to be consensus that the currently accepted total number of species recognized on Earth is 1.75–1.9 million.[4–7] In time this will increase to 3–5 million,[4] 10–14 million,[5,7] or even as high as perhaps 100 million species,[6,7] as microscopic or not economically important groupings are explored along with the abyssal depths and the amazing biodiversity associated with tree-tops in the tropical forests (both areas have distinct problems associated with collection strategy). Of the 1.9 million currently accepted species, only 310 000 belong to the Plantae,[5] with the majority (1.42 million) belonging to the Animalia, which is totally dominated by the terrestrial Arthropoda (>1 million). Because of the sheer abundance and diversity of the terrestrial Arthropods, for which there is no marine equivalent, the biodiversity of species between the terrestrial (1.67 million) and hydrospheres (0.23 million) is very much in favor of the terrestrial sphere.[8] In the main, the terrestrial natural product chemists have explored the Plantae (~302 000), while marine natural product chemists have examined both the marine-based Plantae (8750) and Animalia (~193 000),[8] but it is worth noting that studies on one relatively small Animalia phylum, the Porifera, have contributed about one-third of all the publications (9200) reporting new marine natural products.[9] Although a definitive figure for the total number of natural products isolated and characterized to date is not possible, a total of ~176 000 is accepted.[10] The majority are from terrestrial plants, with dicotyledons the most studied, followed by actinomycetes and fungi, algae, and a contribution of ~25 000 from marine origins.[9]

8.2 Dereplication

8.2.1 Concept and Definitions

Is it known? Is it new? When it comes to dereplication, that is the catch-cry. For natural product chemists, the answers to these questions are of paramount importance and it is within these questions that the whole concept of dereplication is defined. The origin of the term dereplication is not clear but is of relatively recent origin, appearing first in the 1970s.[11] It was in the Foreword to the 1980 edition of the *CRC Handbook of Antibiotic*

Compounds that Langlykke put the use of dereplication into historical perspective:[12]

> . . . early recognition of duplication in a new active agent was essential. . . . In the earliest procedures extensive use of biological activity and resistance patterns served to detect similarities. . . . Now however, the great convenience of chemical and physical instrumental methods lays the groundwork for more specific identification and dereplication.

From that historical account, the etymology of dereplication is obvious, since it means the "removal of replicas" and the term is certainly appropriate. Other definitions focus more on bioactive/pharmaceutical aspects such as that from Webster's Online Dictionary, where dereplication is the process of testing samples of mixtures that are active in a screening process, so as to recognize and eliminate from consideration those active substances already studied.[13] At the outset, dereplication was used for quickly identifying known chemotypes.[14] Examples include the use of a phorbol dibutyrate binding assay combined with HPLC-UV detection to dereplicate active compounds rapidly,[15] and a sterol-dependent antifungal assay in the search for new antifungals.[16] These early dereplication exercises did not necessarily lead to the isolation of specific compounds. Such "partial identifications" were useful for the identification and elimination from consideration of "nuisance" compounds such as tannins, polyphenols, and sulfated polysaccharides that all show general, but non-specific, biological activities.[17] Another aspect of dereplication that has arisen is that of chemical screening in which use is made of a range of solid-phase extraction (SPE) cartridges to obtain chromatographic profiles of bioactive components in a natural product extract to assist in the subsequent isolation phases.[18,19] Despite variations in definitions for dereplication, and the nuances of the various dereplication outcomes, the definition of dereplication used in this chapter is the outright *identification* of known metabolites and *recognition* of novel compounds in natural products extracts. Put simply: is it known? Is it new?

8.2.2 Why Dereplication is Necessary

Dereplication is the vital, initial, and often time-consuming exercise that all natural product chemists must practice if they are to find new compounds or known compounds with a new bioactivity, or simply to prioritize extracts for further study. Most natural product extracts are complex mixtures that may contain hundreds, or even thousands, of components made up from the products of primary metabolism mixed with secondary metabolites (the natural products). In this matrix, there are possibly new compounds, or bioactive compounds of interest. Until the matrix is simplified (dereplicated), it is not possible to answer these questions. The collection of ~176 000 known natural products is structurally diverse, and even though the number of species available to study is over 10 times greater than this

(~1.9 million), a realistic number of species available is very much less than this, probably of the order of 550 000 (400 000 terrestrial and 150 000 marine). For many of these ~550 000 species, the problems associated with accessibility, availability of sufficient mass, or culturability drive that number down to such an extent that the probability that a new crude extract will contain a new compound is not high.

The linkage between natural products and biomedical applications is strong, and natural products play a vital part in our pharmacopeia. In many parts of the world, pharmaceuticals are not available, but barks, leaves, seeds, extracts, skin, horn, *etc.*, are available for medicinal purposes. In the Western world, the opposite is true. In a 1962 survey,[20] it was estimated that over 47% of all new prescriptions filled contained a drug of natural origin as the sole ingredient, or as one of two or more ingredients, while the monumental 2012 survey by Newman and Cragg[21] examined all sources of new pharmaceuticals over the period 1981–2010. Of the 1073 new chemical entities introduced, 64% were a natural product, derived from a natural product, or a synthetic compound containing a pharmacophore derived from a natural product. Natural products continue to play a pivotal role in our well-being. As many dereplication exercises are driven by high-throughput screening assays, the other major outcome of any dereplication exercise is the discovery of a new use for a known compound.

8.3 Approaches to Dereplication

8.3.1 Time, Scale, Cost

When it comes to dereplication, the three factors of time, scale, and cost are closely interrelated and have implications for the identification techniques used. Ideally, dereplication should be rapidly accomplished using non-complex methodology. The time taken for each sample and the technology employed in the process have a very direct bearing on the cost of dereplication and have to be carefully evaluated against the outcomes. For example, HPLC-SPE-NMR, effective in some circumstances, requires expensive robotic equipment, is not robust, is not rapid, is relatively insensitive, is not cost-effective, and consequently is a poor choice for routine dereplication.

To consider the time, scale, and cost requirements, it is necessary to look at the typical steps involved in dereplication. First, there is a separation step on the crude extract, usually accompanied by the collection of fractions. Invariably, this is a chromatographic step, usually employing HPLC or SPE with collection into microtiter plates or some other array. The separation is usually accompanied by concomitant spectroscopic assessment of the HPLC effluent or collected fractions. If there were biological assays available, this would be carried out on each fraction at this point. The final step would be identification or recognition either of a new entity or bioactive component based on spectroscopic assessment.

Because of the mass-handling limitations of an analytical or microbore HPLC column, perhaps only 10–100 μg of crude extract are used for dereplication. The subsequent concentrations of the individual peaks eluting from the column are more than sufficient for the acquisition of excellent mass and UV spectra, but until recently were far short of the mass requirements for the acquisition of ^{1}H-NMR data. Typically, detection limits are about 10^3–10^4 times lower for mass spectrometry (MS) and 10^2 times lower for UV spectroscopy in comparison with ^{1}H-NMR spectroscopy. Potential identity or novelty can be determined from the UV and MS data, but structural confirmation usually requires acquisition of definitive NMR data. Until the last decade, this would have required repeating the chromatography on a larger scale to reisolate the compound of interest with sufficient mass to acquire the necessary 1D- and 2D-NMR data, extending the time and cost for dereplication and also the complexity of the process. Developments over the last decade have seen a steady drop in the mass requirements for the acquisition of ^{1}H-NMR data. With the use of capillary[22] or micro-cryoprobes[23] (see Chapter 4), the mass requirements have fallen to the 2–20 μg range for acquisition of a full dataset of 1D and 2D data, well within the mass range that can be obtained from a single injection onto an HPLC column. With the potential for the more or less simultaneous acquisition of UV, MS, and ^{1}H-NMR data, a full and definitive dereplication exercise can be launched immediately following data collection.[23]

8.3.2 Existing Methodologies

As the pressure to find new (bioactive) compounds increases, so does the importance of efficient dereplication practices. "Time is money," so the more rapidly dereplication can be accomplished the lower is the cost. Although many variations are possible, most assemblies would include an analytical HPLC system with diode-array detection (DAD) and evaporative light scattering (ELSD) detectors with a split effluent to an electrospray mass spectrometer (ESMS) operating in positive and negative ion modes and/or an ion trap mass spectrometer for the ready acquisition of MS/MS data. This configuration would generate the chromatographic profile (ELSD) for the extract and the UV profile (DAD), and molecular mass/molecular formula (ESMS) data for each component. With an ion trap, MS^n information would also be available.

Typically, a standard water–acetonitrile gradient is used for chromatography of 100–500 μg of extract on a reversed-phase column, usually octadecyl (C_{18}), and the effluent is collected. In one approach,[24] collection of the effluent into a master microtiter plate is started after the solvent front has passed through and 88 samples (250 μL each) are collected at 15 s intervals (2.5–24.5 min). The nature of the gradient used is such that all likely compounds of (biological) interest have eluted over this period. The acquired retention times (TR) for the components, acquired under standard conditions for one column type, are remarkably consistent and can be used for comparison purposes and compared against external standards.[25]

Daughter plates can be generated from the master plate for a range of biological assays as necessary. Using a centrifugal evaporator, the master microtiter plate can be taken to dryness and, using a capillary or micro-cryoprobe, a range of 1D- and 2D-^{1}H-NMR data are obtained. The UV and MS data for searching in libraries acquired during the HPLC run can be directly correlated with specific wells on the microtiter plate while the biological and NMR data are acquired directly from specific wells. This tight integration of the data is of great advantage when correlating results, as this can be done on a well-by-well basis.[24]

8.4 Databases

To dereplicate a crude natural product extract effectively, it is necessary to search the definitive information about each component of interest against appropriate databases. The efficiency of this process is very much a function of access to appropriate databases. At this point of the investigation, the probable taxonomy of the organism will be known, and which peaks in the chromatographic profile are bioactive and also the molecular mass/molecular formula, UV, and ^{1}H-NMR spectra of the components of interest will have been determined. With appropriate databases, this is usually sufficient to complete the dereplication of the sample. There are literally thousands of chemistry databases documenting the physical, spectroscopic, and chemical properties of compounds that can be accessed and manipulated, but only a fraction of this large body of data/databases is immediately relevant to natural product chemists. These databases can be segregated into three domains based on availability: public, commercial, and private (Table 8.1).[9,10,25–55]

Those in the public list are freely available for consultation without a fee. This is in contrast to the commercial databases such as Chemical Abstracts

Table 8.1 A selection of databases that deal with natural products.

Public	Commercial	Private
ChemSpider[26]	CAS Registry[34–37]	All Pharma
CSLS[27]	SpecInfo[38]	GVK Biosciences NPD[47]
PubChem[28]	Reaxys[39]	UC UV DB[48]
NMRShift DB[29]	ACD/Labs[40,41]	DTU UV/MS DBs[49–51]
Naproc-13[30]	NAPRALERT[42]	Marine NP DB[52]
SuperNatural[31]	NIST 11[43]	InterMed UV DB[25]
SDBS[32]	*Dictionary of Natural Products*[10]	InterMed NMR DB[25]
Binding DB[33]	*Dictionary of Marine Natural Products*[44]	Novartis IR DB[53]
	AntiBase[45]	National Centre for Plant and Microbial Metabolomics[54]
	MarinLit[9]	CH-NMR-NP[55]
	AntiMarin[46]	Merck & Co.[56]

DB, database.

Service (CAS) and the CAS Registry.[34–37] The last category, the private databases,[25,47–56] are privileged and are usually associated with large pharma or specialist collections and not generally accessible. However, there is little doubt that these private domain databases will likely contain the full ranges of spectral, taxonomic, and biogeographical data that can be utilized as indicated schematically in Figure 8.1.[56]

Databases of relevance for dereplication purposes, along with their essential features, are listed in Table 8.2.

8.4.1 Taxonomic Information

With taxonomy in hand, it is possible immediately to consult relevant databases and ascertain what chemistry has come before from the species collected and perhaps even prioritize the collection based on taxonomy even before bioassays have been completed. Within the selected databases in Table 8.2 the coverage of taxonomy is good, with the possibility of searching on given genera or species. However, three of the public domain databases do *not* include taxonomic capability. These are ChemSpider,[26] CSLS,[27] and PubChem.[28] For specialist taxonomic databases, refer to either the Catalogue of Life[57] or the World Register of Marine Species (WoRMS).[58]

8.4.2 Biological Data

Biological data for known compounds in the highlighted databases (Table 8.2) is not comprehensive. Probably the best coverage is in the *Dictionary of Natural Products* (DNP)[10] and the *Dictionary of Marine Natural Products* (DMNP).[44] However, a recent advance in the MarinLit database[9] is the linking of structural motifs in this marine literature database to BindingDB, an open-access Web resource with published binding data for ~442 000 small molecules against ~6800 protein targets.[33]

8.4.3 UV Spectral Data

UV data in combination with TR and/or molecular weight (MW) or molecular formula (MF) data can be a quick and effective method of dereplication, but UV matching alone cannot be a universal approach, as a UV spectrum is only indicative of a chromophore *within* a structure and is not definitive of the structure. Furthermore, not all compounds contain UV chromophores. Searchable UV databases have been developed but are not in the public or commercial domains. It is clear that, with these databases, the matching of UV spectra from HPLC analysis of extracts is a rapid and powerful dereplication tool with ready comparison between spectra and the added advantage of linkage to TR values. As analyses of crude extracts are generally run under standard conditions, this is a powerful additional point of comparison based on the similarity of UV spectra and TR and also identifies compounds with similar UV profiles but a different TR[25,49] (also see Figure 8.1).[56]

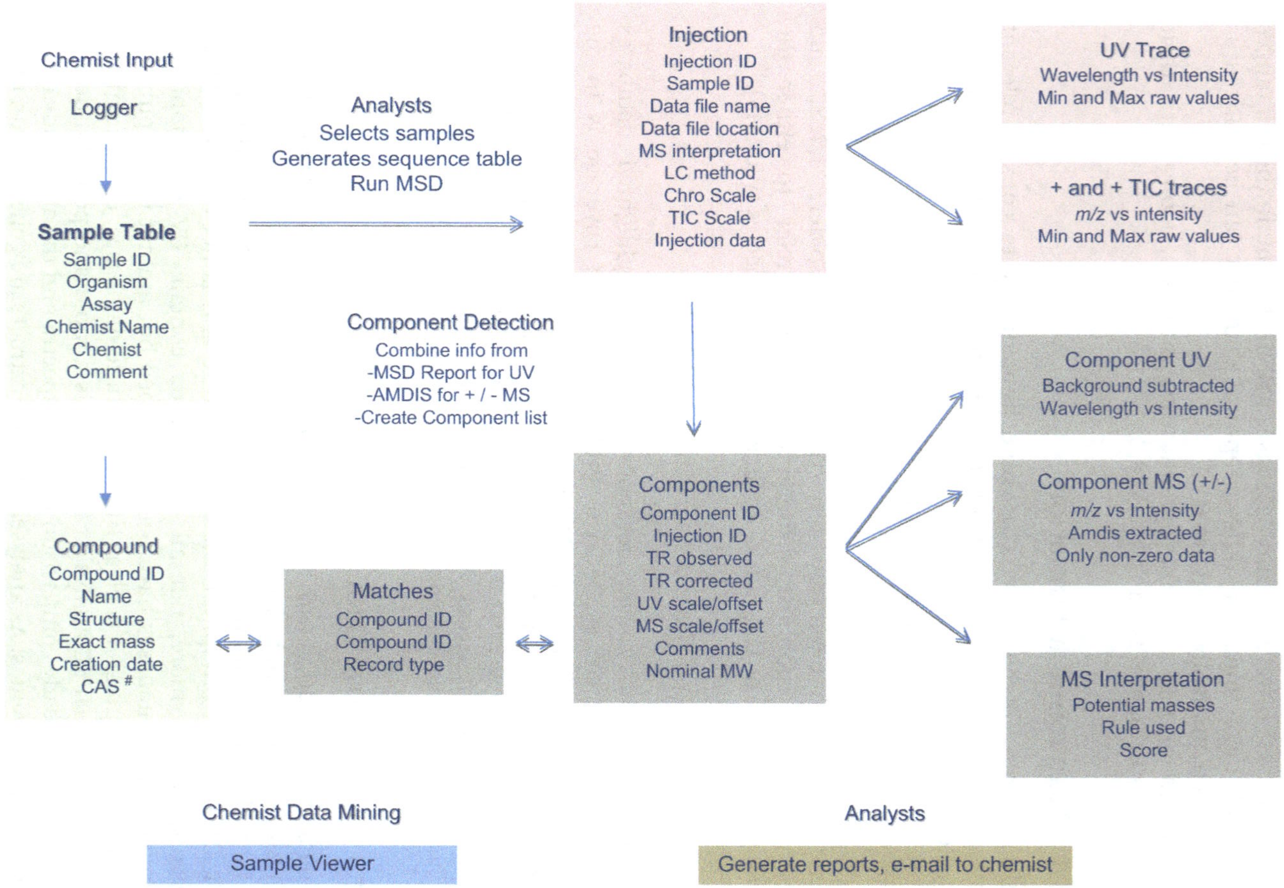

Figure 8.1 Dereplication protocol used at Merck, Rahway, NJ, USA. Adapted from Web presentation.[56]

Table 8.2 Databases that are of potential use for the dereplication of natural product extracts.

Database	No. of compounds[a] Total	Natural products	Current up to	MW	MF	UV[c] λ	SSS[d]	Tax.[e]	Biol.[f]	NMR data[b] δ	Spectra	^{1}H-SF	HSQC/DEPT
CAS Registry	$>8.5\times10^{7}$	~250 000	2012	+[g]	+	–	+	+	+	–	+	–	–
CSLS	4.7×10^{7}	Extracts	~2011	–	+	–	+	–	?	–	–	–	–
ChemSpider	3.0×10^{7}	?	2013	+	+	–	+	–	–	+[h]	+[h]	–	+[h] –
PubChem	3.0×10^{7}	?	2012	+	+	–	+	–	+	–	–	–	–
Reaxys	$>10^{7}$	170 000	2012	+	+	–	+	+	?	–	–	–	–
ACD/Labs DB	322 000	?	~2011	+	+	–	+	–	–	+	+	–	–
Dictionary of Natural Products	241 000	175 720	2013	+	+	+	+	+	+	–	–	+	–
Dictionary of Marine Natural Products	43 842	29 525	2011	+	+	+	+	+	+	–	–	–	–
AntiMarin		62 852	2012	+	+	+[i]	+	+	+[i]	–	–	+	–
AntiBase		38 666	2012	+	+	+	+	+	+	+	+[i]	–	–
MarinLit		24 663	2013	+	+	+	+	+	+[i]	+	+	+	+

[a]Where possible an estimate is given for the number of natural products in the database.
[b]Four options for NMR data: the δ values (calculated or actual), spectra, ^{1}H NMR structural features (^{1}H-SF), or calculated HSQC/DEPT spectra.
[c]Actual λ (ε) values for UV data as opposed to a reference to the data.
[d]Substructure searching capability.
[e]Taxonomic data.
[f]Biological activity data.
[g]In the current version of SciFinder the extraction of molecular mass data is not straightforward.
[h]ChemSpider contains NMR data for ~2500 compounds.
[i]Partial data only.

Several of the commercial and public domain databases listed in Table 8.2 contain UV data, but have only searchable λ_{max} values, not searchable spectra. These limitations diminish the current value of the UV approach to dereplication.

8.4.4 Mass Spectrometric Data

LC-ESMS is the approach mostly widely used for the dereplication of natural product extracts. This can be carried out under low-resolution (LR) or high-resolution (HR) conditions. The resultant MW or MF data can then be searched directly in most databases for possible matches. All of the databases listed in Table 8.2, except CSLS,[27] can be searched for both MW and MF data. The MW and MF are physical properties of a compound, but are not necessarily unique when considered against the >85 million compounds in the CAS Registry,[34–37] meaning that these data are seldom discriminatory when considered in isolation and will inevitably return many options. Little *et al.*[59] reported on searching for "known unknowns" in both commercial (CAS Scifinder) and freely accessible (ChemSpider) databases. The approach uses searching of monoisotopic (mI) masses and refining the search results by sorting the number of references associated with each compound in descending order. Such an approach has been shown to be very effective for the identification of various types of chemicals in databases containing millions of chemical structures. Although this approach could in theory be of value for the dereplication of natural products, the searches would need to be performed against only a slice of the overall data, as it is common for too many compounds to be returned using mI mass searches against such databases. As an example, the mI for suregadolide C (see Figure 8.2) is 348.1937 Da. A search for this mass $\pm$ 0.001 against the ChemSpider database gives 572 compounds, many of them natural products. Clearly, mass-based searches in such large resources are of less value than searching in specialized databases. However, if the MS data are taken in conjunction with UV and ^{1}H-NMR data, then the choices can be dramatically reduced.

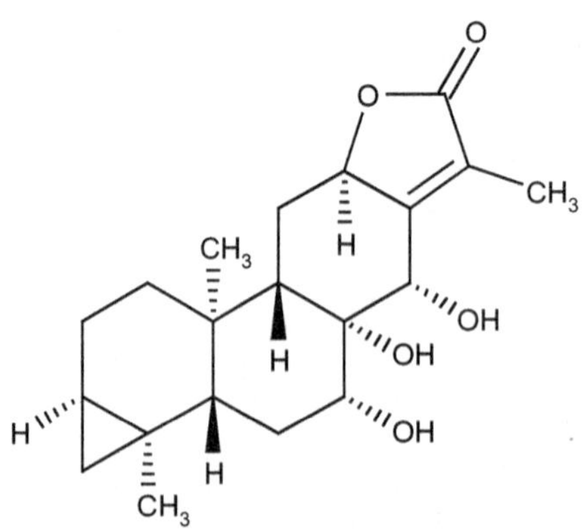

Figure 8.2 Structure of suregadolide C.

Apart from the difficulty associated with sensitivity arising from compounds unable to ionize under positive or negative conditions, other problems in the interpretation of LC-MS data from a crude extract include deciding which ion in the mass spectrum corresponds to the molecular ions $[MH]^+$or $[M - H]^-$, whether it is an adduct (MNa^+, $MNH_4{}^+$, $MHCOO^-$, *etc.*), dimer or trimer, or a fragment ion resulting from the ready loss of HCOOH, CH_3COOH, H_2O, or CO_2. Another common problem arises from the presence of minor component(s) that ionize more readily than the compound of interest and generate the major intensity peaks in the spectrum. Any of these can result in an erroneous assignment of molecular mass, complicating the decision-making process. Useful publications in this area have been from the Danish Technical University (DTU), which covered the ESMS behavior of 474 compounds, followed by a later publication covering a further 719 microbial natural products.[50,51] Using MS/MS approaches partially circumvents this problem but in a study of 1020 commercially available and in-house stocked compounds it was noted that in two-thirds of the examples different fragment ions were noted from $[MH]^+$ versus $[MNa]^+$, but with no comprehensive database available for consultation the obvious disadvantage of a library concept with a limited number of compounds is highlighted.[60] The NIST 11 database contains MS/MS ion trap spectra for 4628 compounds and collision cell spectra for another 3877 compounds, but is not natural product based.[43]

A powerful approach that can lead to *new* compounds is to search the MH^+ and MNa^+ results against specialist databases looking for no hits for the acquired MW or MF data.

8.4.5 ^{1}H-NMR Data

Like UV and mass spectra, ^{1}H-NMR spectra are information-rich, but in contrast allow for the ready recognition of substantial portions of the molecule on inspection. There is a wide variety of functional groups that are easily recognizable, such as methyl groups, acetal protons, α-protons in peptides, carbinol, and olefinic protons and aromatic substitution patterns, all of which occur at characteristic chemical shifts in a ^{1}H-NMR spectrum and give clues to the environment in which they exist. However, there are at least two factors that have stifled the use of ^{1}H-NMR data for dereplication purposes. First, there is the difficulty of acquiring high-quality ^{1}H-NMR data on the same scale that dereplication is typically carried out (10–100 μg). However, with the advent of capillary[22] and micro-cryoprobes,[23] that deficiency has now been addressed. Without ^{1}H-NMR data, the UV and MS data and perhaps taxonomic considerations were used to simplify the complexity to a few candidates only. This then required isolation efforts to obtain adequate material for the generation of NMR data before deciding whether the compound was new or known and for completion of the dereplication exercise. The lack of ready access to appropriate NMR-based databases was the second factor and often required a full structural assignment of a compound, only to discover that it had been previously identified. That too has now been addressed with NMR data being included in several specialist databases (ACD/HNMR DB,[40,41] MarinLit,[9] AntiBase,[45] AntiMarin,[46] and DNP[10]), but, with the exception of SciFinder,[37] ^{1}H-NMR data are generally not available in databases.

Accessing the ^{1}H-NMR databases can be carried through at two different levels that represent different approaches to the use of ^{1}H-NMR data for dereplication. One approach relies on chemical shift matching (ACD/HNMR, AntiBase, and MarinLit databases) and is an absolute approach that is widely used in the metabolomics field for the detection of metabolites,[61] but can also be applied to dereplication as discussed below. The alternative approach is that of pattern matching using databases that log the numbers and types of easily recognized NMR features in a molecule (MarinLit,[9] AntiMarin,[46] and DNP[10]). The pattern recognition database approach does not rely on the assignment of chemical shifts, running spectra under standard conditions, or analysis of correlations. No analysis of the data is required beyond the counting of recognizable groups and then consultation of a database. These alternative approaches to the use of ^{1}H-NMR data for dereplication are considered the following in more detail.

8.5 Pattern-matching Approach to Dereplication

An experienced eye cast over ^{13}C- or ^{1}H-NMR spectra quickly assesses the "pattern" of the resonances and conclusions can be drawn as to the likely structural class involved. Assessment of the "pattern" of resonances was part of an approach to the further development of the MarinLit database in collaboration with ACD/Labs, when consideration was given to the inclusion of predicted, not actual, NMR chemical shift information in the database. The conclusion at the time, 1999, was that calculated and not actual data would be sufficient as it was the "pattern" of resonances that was important, not the precise chemical shift. Besides, the accuracy of the calculated data even then was very high, quoted to be an average deviation of <0.3 ppm between experimental and predicted chemical shifts. Thus, the idea of using numbers and types (patterns) of NMR resonances for dereplication purposes was conceived and initial work was started. In 2001, Bradshaw *et al.* demonstrated that the counting of the numbers of methyls, methylenes, and methines in a molecule (as determined from ^{13}C-DEPT spectra), in combination with mass data, was sufficient to reduce the count of known compounds that fit the criteria to <10.[62] This confirmed the validity of a pattern-matching approach to dereplication.

The initial thinking of using numbers and types of NMR resonances had been based on ^{13}C spectral data, but the possibility of acquiring ^{13}C data from samples at the sample size that was being used even a decade ago was remote. It was clear that the sensitivity of the ^{1}H-NMR experiments was rapidly increasing and it would soon be possible to obtain at least 1D-^{1}H-NMR spectra on samples generated during a routine dereplication exercise (10–100 μg scale), and so consideration was given to extending the pattern recognition beyond methyls, methylenes, and methines to easily recognizable ^{1}H-NMR features. Within the same timeframe, this approach was also taken by the German pharmaceutical company InterMed Discovery for their own in-house NMR/MS/UV dereplication database.[25] The NMR component of this discrete database was based on the DNP library by arrangement with Chapman and Hall (see also Section 8.5.3).

8.5.1 Searchable ^{1}H-NMR Features and the MarinLit Database

The most obvious, easily recognizable ^{1}H-NMR features are methyl groups, but also included in the pattern recognition exercise were primary and secondary carbinols/ethers, acetals/ketals, formyl, all alkenes, aromatic substitution patterns, sp^3 methylenes and methines, and sp^2 hydrogens. These are groups that can be readily recognized in a ^{1}H-NMR spectrum by appearance (multiplicity) and/or chemical shift values. A specifically developed algorithm recognizes these features within a structure leading to each structure in the MarinLit database[9] being coded with the numbers present for each of these features observable in a ^{1}H-NMR spectrum. Each of these values is placed in a searchable field. It is the number of each of these features in a molecule that is important, not the exact chemical shift. This has the immediate effect of making critical factors such as the pH, temperature, and solvent used for measuring the spectrum irrelevant. In the absolute chemical shift approach, maintaining control of or allowing for the variations introduced by these factors is a very important consideration.

If reliance is to be placed on the total numbers of sp^3 methyls, methylenes, and methines, then acquisition of a 2D-HSQC-DEPT spectrum may be required. In a collaboration with ACD/Labs, the MarinLit database includes both ^{1}H-and ^{13}C-NMR data for examination and has the capability of presenting an HSQC-DEPT plot of these combined data. For complex molecules, or where there are large numbers of methyls, methylenes, or methines, this is a very useful feature. Where possible, these spectroscopic data are the actual data and are not calculated.

8.5.2 Development of the AntiMarin Database

The database AntiMarin,[46] which numbers ~62 800 compounds, was a merger between MarinLit[9] (original Editors J. Blunt and M. Munro of the University of Canterbury, but since September 2013 owned and operated as a Web-based version by the Royal Society of Chemistry) with ~27 000 compounds (May 2014) from marine sources and AntiBase[45] (Editor H. Laatsch of the University of Göttingen) with ~38 600 compounds from microbial/algal sources. Although different in construction and appearance, MarinLit and AntiBase are congruent and more or less cover the same desirable features that make these databases essential and friendly working tools for the natural product chemist.

8.5.3 Extension of ^{1}H NMR Searching to the *Dictionary of Natural Products*

The DNP[10] (Editor J. Buckingham) is the most comprehensive compilation of natural product data and structures available. This database, originally supplied in print form but now available as a DVD or with online access, is updated every 6 months and contains the structures and chemical, physical, and biological data on more than 241 000 compounds. This is greater than the estimated ~176 000 actual natural products as many derivatives are also included in the compilation. In an arrangement between MarinLit and DNP, a form of the ^{1}H-NMR features searching database has been available to all subscribers to DNP.

8.6 Why ^{1}H-NMR Dereplication is Discriminatory

Interpretation of UV spectra can often suggest a chromophore within a molecule, but without available comprehensive databases to match spectra and TR values against there is little further information directly available until mass data are added. High-resolution MS data yield molecular formulae that can be matched against formula data in almost all general and specialist databases and can sometimes answer the questions, is it new, is it known? MS/MS data can also give the specialist mass spectrometrist the opportunity to ascertain substructures for searching against general and specialist databases.

The pattern recognition ^{1}H-NMR approach to dereplication is discriminatory and can quickly lead to a conclusion on novelty. Perhaps more importantly, this approach does not require specialist NMR knowledge, and comprehensive general and specialist databases are available for searching against parts, or against all known natural products. The databases are MarinLit for compounds of marine origin, AntiMarin for marine, algal, and microbial literature, and DNP covers all natural products.

8.6.1 Searchable Fields and ^{1}H-NMR Dereplication

A good example of searchable fields is the types of methyl groups present in a structure as methyls are usually one of the most easily recognized functional groups. The databases with searchable fields for ^{1}H-NMR features recognize nine different types of methyl group – singlet, doublet, triplet, *O*-Me, *N*-Me, *S*-Me, acetyl, aromatic methyl, and vinyl methyl. The selection of searchable fields is not limited and any combination can be searched for in addition to all the other search features in the databases along with a range within each type, but even just considering the number of methyl groups present is discriminatory. For example, as of this writing, there are 28 609 structures out of 228 970 in DNP that contain zero methyl groups. A ^{1}H-NMR spectrum of a compound that contains no methyl groups is very obvious, and by just that one observation the number of possible candidates has been reduced by ~88%. The distribution of methyl groups in DNP by number is shown in Figure 8.3, so a simple count of the methyl groups of any type observable in a ¹H-NMR spectrum rapidly reduces the number of possible candidates.

A keener demonstration of the discriminatory power of this pattern recognition approach arises when considering possible combinations of the nine possible types of methyl recognized in DNP. For example, for any two combinations of the nine types of methyl groups there are 45 possible combinations to spread the database across, or 165 for any three combinations from the nine, and so on.

8.6.2 Data Entry

The ¹H-NMR features databases that are incorporated into each of the MarinLit, AntiMarin, and DNP searches were originally constructed by

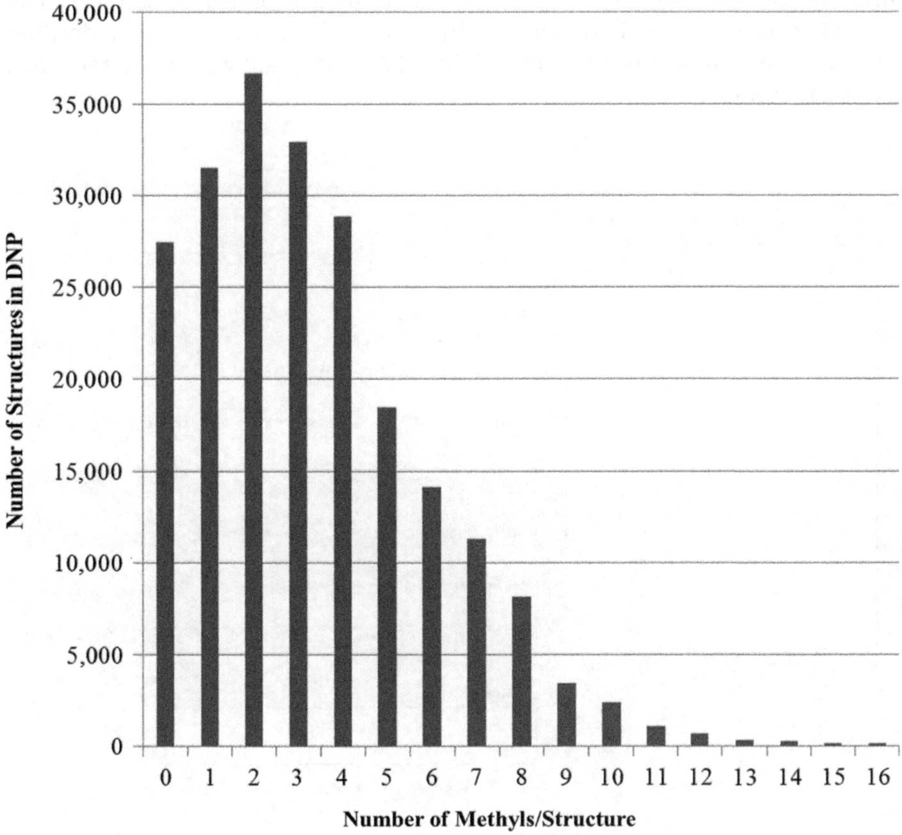

Figure 8.3 Distribution of the number of methyl groups/structure in the DNP ¹H-NMR database.

John Blunt at the University of Canterbury with just layout details of the versions differing from one another. A great attribute of the query entry page is its simplicity (see Figure 8.4 for AntiMarin). All essential numerical details for a search are entered as precise numbers (1, 7, 359.4567, *etc.*) or ranges (<5, >15, 0–7, *etc.*) in the appropriate boxes, formulae are entered in the normal fashion ($C_xH_yO_z$, then other elements in alphabetical order) while entries into the Name and Source boxes are in regular text.

Once the search is loaded, the query can be searched against the database with the results being shown in a comparable page for each successful match. In Figure 8.4, the search shown is for all compounds originating from a *Streptomyces* sp. that have a molecular mass in the range *m*/*z* 300–400, a total of four or five methyls of which three are methyl singlets, one a methyl doublet and zero or one methoxy groups, and has two >CH–O– groups. The results of this search gave five answers that matched (out of ~63 000) and each of these results can be examined one at a time. The record shown, Figure 8.5, is for albocycline M-2 from *Streptomyces bruneogriseus*, which has a molecular mass of 324.412, five methyl groups in total, of which three are singlets and one a doublet with one methoxy group. Albocycline M-2 has, as was required, two >CH–O– groups.

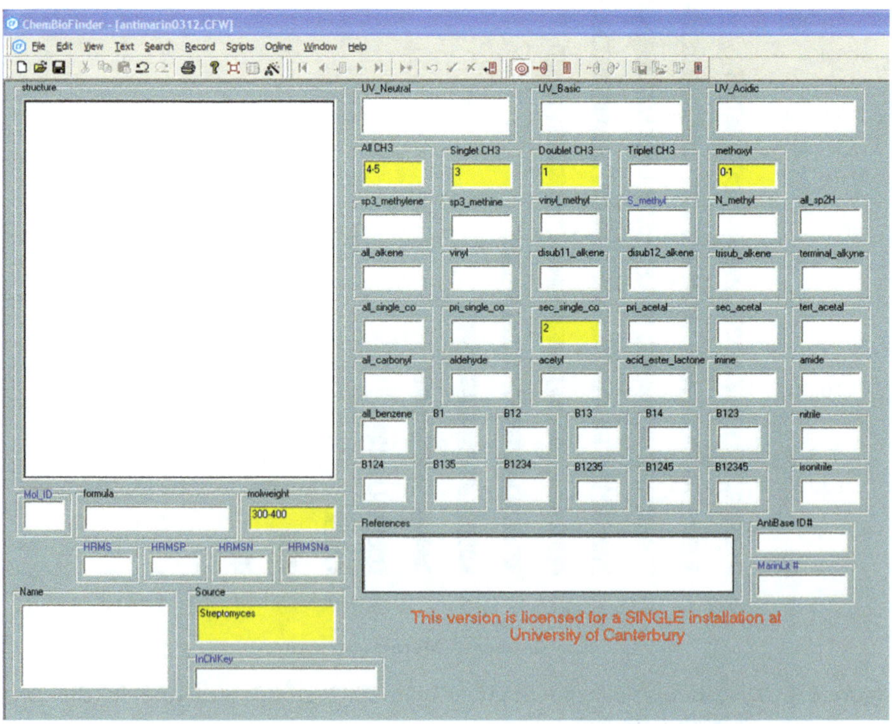

Figure 8.4 The AntiMarin "Query" page for entering ^{1}H-NMR search profiles.

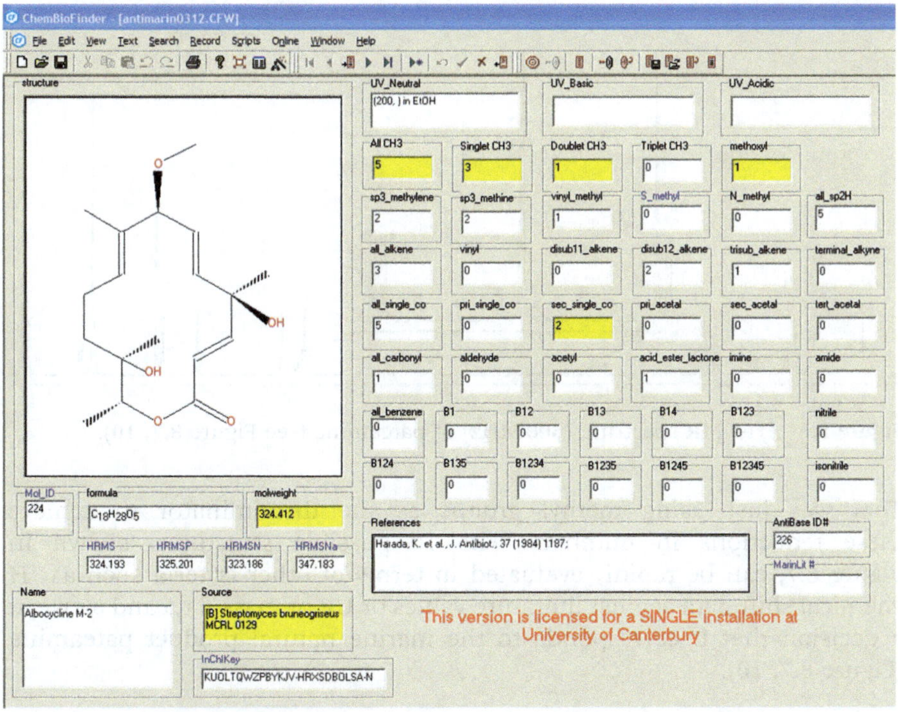

Figure 8.5 The first of five answers from the AntiMarin search depicted in Figure 8.4. One page per result.

8.6.3 Examples of the ^{1}H NMR Approach to Dereplication

8.6.3.1 *Methyl Chemical Shifts Only*

This first example illustrates the simplest possible approach to dereplication. The ^{1}H-NMR spectrum shown in Figure 8.6 (m/z 555.3131) has a total of seven methyl groups and on examination, and with no ambiguity at all, these can be grouped as two doublets (that is a methyl attached to a CH), three vinyl methyls, and two *N*-methyls.

Searching the available ^{1}H-NMR databases gave the following progression:

	MarinLit	AntiMarin	DNP
7 Me (any type)	1460	2995	11131
2 Me (d)	3073	6965	23339
3 Me (vinyl)	3357	2021	4986
2 *N*-Me	535	1667	3191

But,

7 Me/2 Me (d)/3 Me vinyl/2 *N*-Me	10	10	10

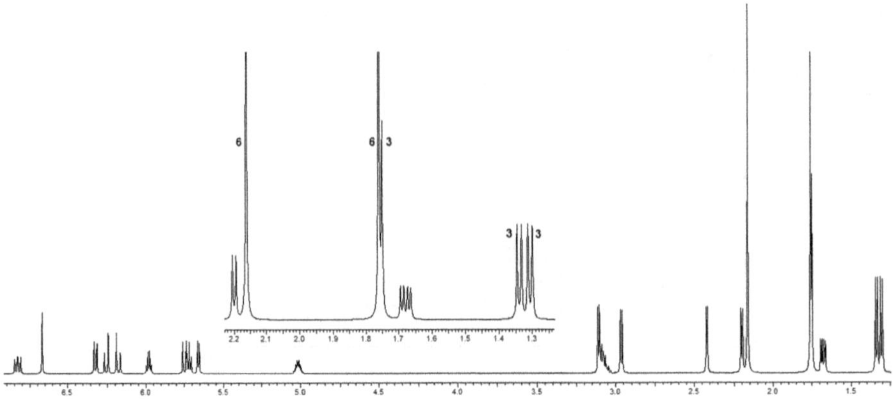

Figure 8.6 ^{1}H-NMR spectrum (500 MHz) of pateamine (see Figure 8.7, **10**).

That was just using methyl groups as the discriminator to achieve these reductions in numbers. The 10 possible structures, shown in Figure 8.7, can be rapidly evaluated in terms of other criteria such as ^{1}H chemical shift data against literature values or UV or mass data and arrive at a decision that it corresponds to the marine natural product pateamine (Figure 8.7, **10**).[63]

8.6.3.2 Combinations of Functional Groups

Searching combinations of functional groups is a powerful dereplication strategy and other easily recognized groups such as formyl, carbinol, alkene, or substituted benzenes could have been added into the search profile to increase the level of discrimination. Consider the ^{1}H-NMR spectrum shown in Figure 8.8.

Examination of the obvious features of the spectrum shows that it has eight methyl groups (five singlets, two doublets, and one methoxy group), two >CH–O–, and a 1,4-disubstituted benzene (1,4-B). Searching on these features across the ^{1}H-NMR databases gave the following results:

	MarinLit	AntiMarin	DNP
8 Me/5 Me (s)/2 Me (d)/1 *O*-Me	13	34	142
8 Me/5 Me (s)/2 Me (d)/1 *O*-Me/2 >CH–O–	0	6	23
8 Me/5 Me (s)/2 Me (d)/1 *O*-Me/2 >CH–O–/1 1,4-B	0	0	1

A simple search of just the methyl group patterns gave a marked reduction in the number of possible structures in each database. By adding in the two secondary carbinol type functionalities, the numbers in AntiMarin and DNP were reduced even further and there was no match of any such pattern in the marine natural products database, MarinLit. Searches with outcomes as low as 6 and 23 can readily enough be searched directly

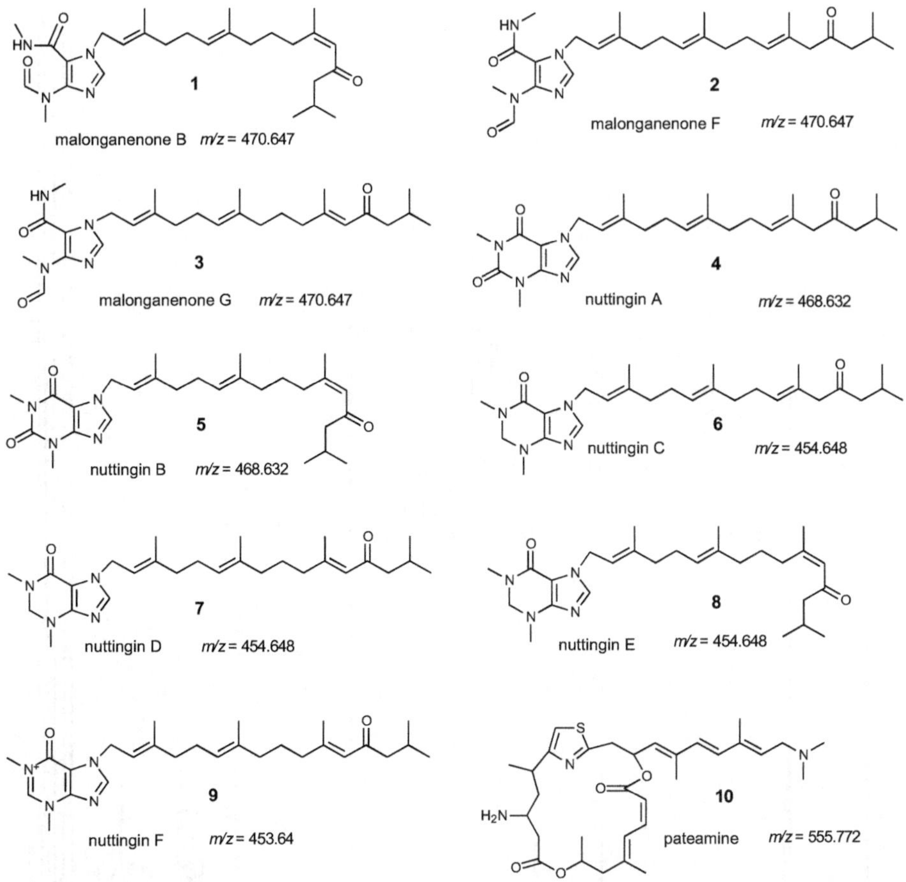

Figure 8.7 The 10 structures that matched the search profile: 7 Me/2 Me (d)/3 Me vinyl/2 *N*-Me.

against the literature to find a match. However, in this case, by adding in a 1,4-disubstituted benzene the possibility that the compound matched anything of algal or microbial origin (AntiMarin) was eliminated. Without having even considered any input of mass or molecular formula data, this combination of searchable ^{1}H-NMR features led to one unique choice from ~241 000 possibilities, a very powerful illustration of the discriminatory power of the ^{1}H-NMR approach to the dereplication of natural product extracts. The match in the database was to the triterpenoid 2-*O*-Me ether, 3-[4-hydroxy-(*E*)-cinnamoyl]-12-ursen-28-oic acid (Figure 8.9, **11**), also known as guajanoic acid, which is of terrestrial origin, being obtained from a Pakistani collection of *Psidium guajava*.[64] The final and necessary step in the process would be to compare the mass and NMR data with those published to establish that the compound under examination was actually guajanoic acid or an isomer.

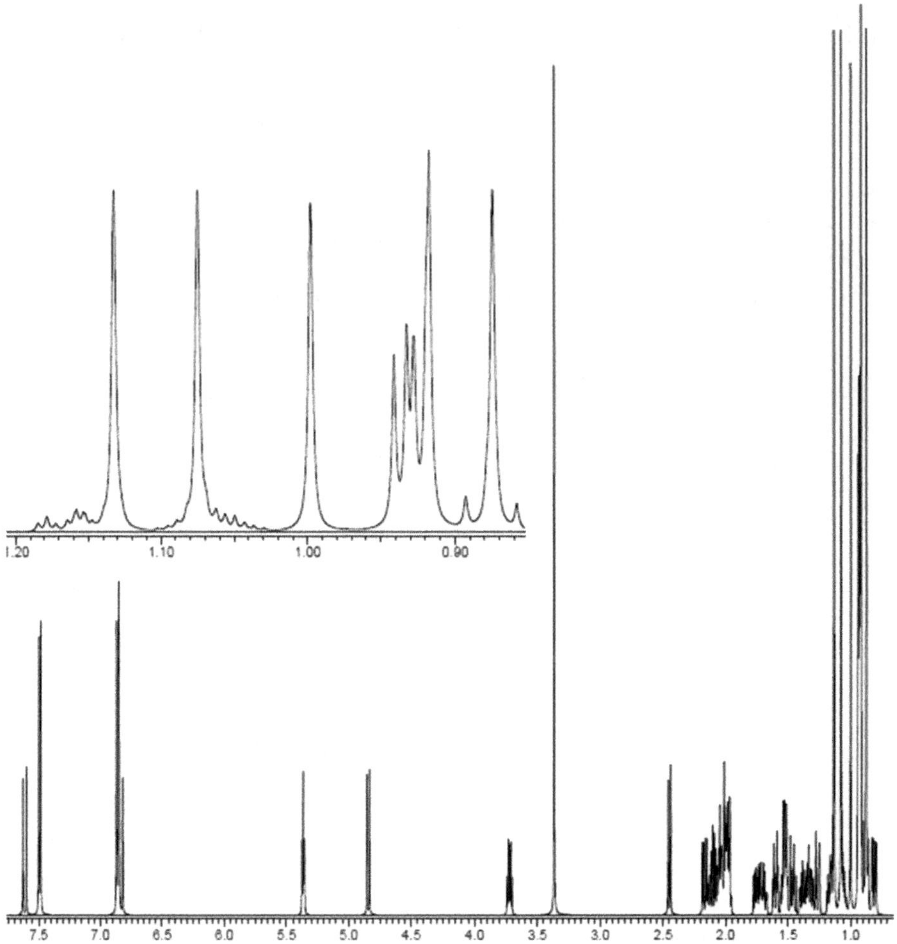

Figure 8.8 The ^{1}H-NMR spectrum (500 MHz) of the triterpene guajanoic acid (see Figure 8.9, **11**). The inset is an expansion of the high-field region.

8.6.3.3 *Adding Mass Data to a Search*

If a molecular formula or mass data are available, ideally they should form part of the initial probing of the ^{1}H-NMR databases. The four natural product databases, MarinLit,[9] AntiBase,[45] AntiMarin,[46] and DNP,[10] can each be searched using molecular mass and molecular formula criteria. Molecular formula data provide the opportunity for unique matches but, because the variable number of significant figures noted for the high-resolution data varies across the databases, it is more reliable to use a molecular mass range. For example, in the case of guajanoic acid (Figure 8.9, **11**), $C_{40}H_{56}O_6$, molecular mass *m*/*z* 632.869, it is a better strategy to search over the range *m*/*z* 632–633 to avoid any possible mismatching of data. For the

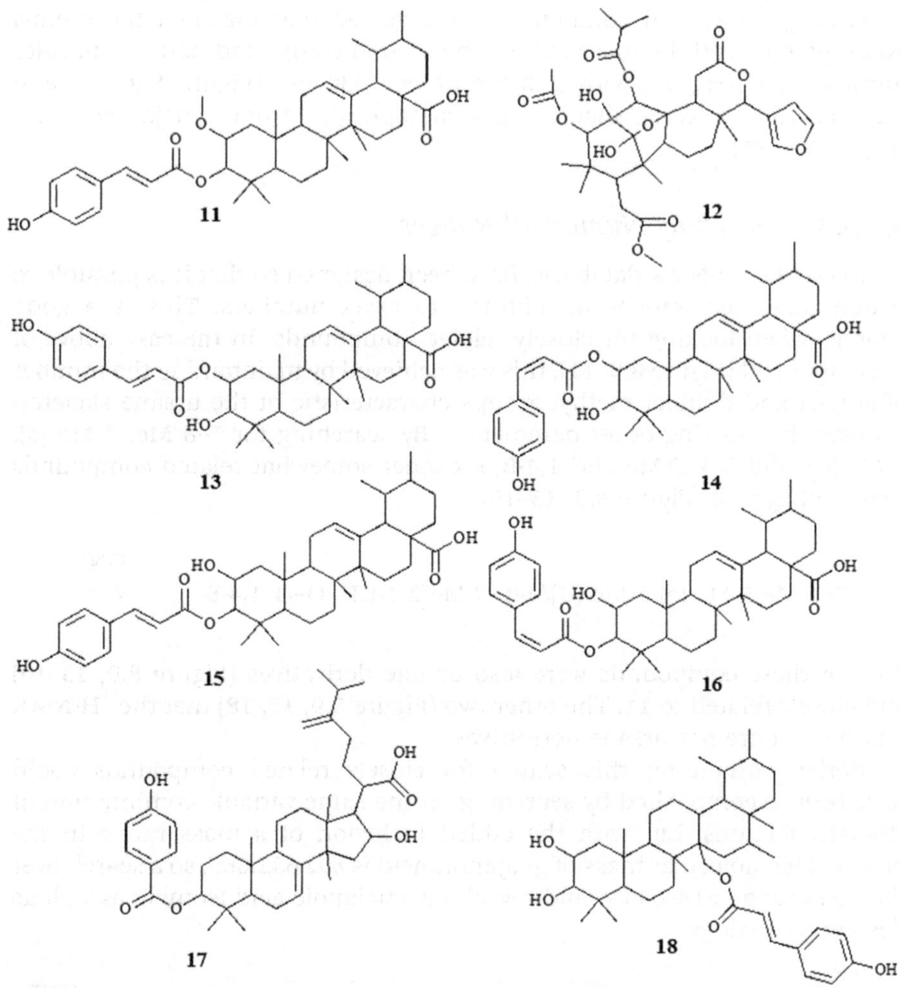

Figure 8.9 A selection of triterpenoids related to guajanoic acid (**11**) that resulted from a variety of ^{1}H-NMR search profiles.

guajanoic acid case, when the molecular formula or molecular mass data were added to the initial search profile, the following numbers were obtained:

	MarinLit	AntiMarin	DNP
8 Me/5 Me (s)/2 Me (d)/1 *O*-Me	13	34	142
8 Me/5 Me (s)/2 Me (d)/1 *O*-Me/$C_{40}H_{56}O_6$	0	0	1
8 Me/5 Me (s)/2 Me (d)/1 *O*-Me/*m*/*z* 632–633	0	0	2

When the molecular formula was used in combination with just the methyl group data, only one hit resulted across the three databases. Using the mass range, a second compound was detected that matched the methyl group pattern and the mass range. The second compound had a molecular formula of $C_{33}H_{44}O_{12}$ with a mass of *m/z* 632.695 (Figure 8.9, **12**) and was readily distinguishable spectroscopically from guajanoic acid (Figure 8.9, **11**).

8.6.3.4 Searching Numerical Ranges

As noted, the ^{1}H-NMR databases have been designed so that it is possible to search numerical ranges in addition to exact numbers. This is a good strategy when looking for closely related compounds. In the case above of guajanoic acid (Figure 8.9, **11**), this was achieved by maintaining the number of singlet and doublet methyl groups characteristic of the ursane skeleton constant but varying other parameters. By searching for 7–8 Me, 5 Me (s), 2 Me (d), and 0–1 *O*-Me and 1,4-B, six other somewhat related compounds were highlighted (Figure 8.9, **13–18**).

	DNP
7–8 Me/5 Me (s)/2 Me (d)/0–1 *O*-Me/2 >CH–O–/1 1,4-B	7

Four of these compounds were also ursane derivatives (Figure 8.9, **13–16**) and closely related to **11**. The other two (Figure 8.9, **17**, **18**) met the ^{1}H-NMR criteria, but are not ursane derivatives.

Another variant on this search for closely related compounds could have been accomplished by searching on the same variable combination of ^{1}H-NMR features, but with the added inclusion of a mass range in the profile. The molecular mass of guajanoic acid is *m/z* 632.869, so a search over the mass range 618–633 would reveal any guajanoic acid isomers as well as desmethyl analogs.

	DNP
7–8 Me/5 Me (s)/2 Me (d)/0–1 *O*-Me/2 >CH–O–/1 1,4-B/*m/z* 618–633	5

Four of the five compounds identified (Figure 8.9, **13–16**) had *m/z* 618.842 and are desmethyl analogues/isomers of guajanoic acid (Figure 8.9, **11**). The two compounds that were eliminated in this more refined search (Figure 8.9, **17**, **18**) lay outside the stipulated mass range.

8.6.3.5 Null Searches

Earlier, it was noted that 28 609 structures out of the ~241 000 in DNP contained zero methyl groups (Section 8.6.1), so recognition of the absence of any easily searchable group in a ^{1}H-NMR spectrum of a sample, a null search, is also an effective tool in the dereplication process. In the case of zero methyl groups, it immediately eliminates 88% of the compounds from

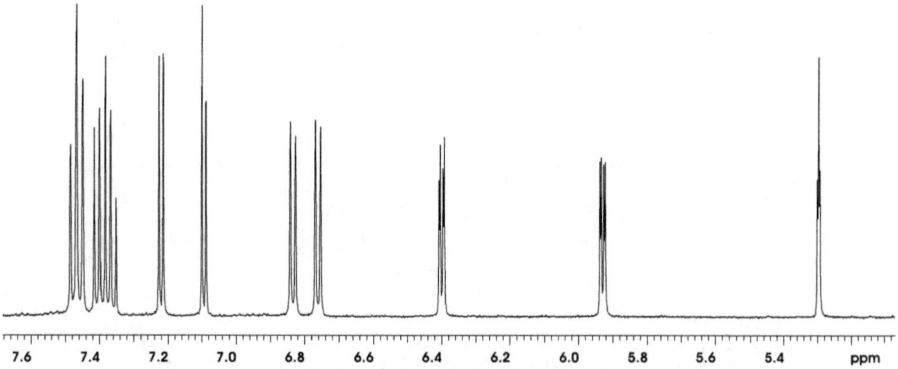

Figure 8.10 The ^{1}H-NMR spectrum (500 MHz) of spiro-mamakone A (see Figure 8.11, **19**). Each multiplet (δ 5.35–7.22) integrated for one proton. The multiplets (δ 7.35–7.50) integrated for four protons.

consideration. When the ^{1}H-NMR spectrum of the compound shown in Figure 8.10 was examined, the search profile was able to include both zero methyls and zero sp^3 methylenes. Further analysis of the ^{1}H-NMR spectrum and a look at the ^{1}H integral values suggested 10–11 sp^2 hydrogens as the one proton doublet at δ 5.35 could be assigned as either an sp^2 hydrogen or a >CH–O– group on the basis of chemical shift values.

The search profile across AntiMarin and the DNP, using variable ranges, could be

	AntiMarin	DNP
0 Me/0 –CH_2–	1969	5894
0 Me/0 –CH_2–/10–11sp^{2}H	191	522

This is a dramatic reduction in complexity using a dereplication search profile based solely on the lack of methyls and sp^3 methylenes and a count of possible sp^2 hydrogens. Observation and counting, not interpretation, were the requirements. The null search was effective in narrowing down the search.

If 2D-COSY data had also been available, a 1,2-disubstituted alkene [δ 7.10 (d)/7.22 (d); 1,2-alkene], a 1,2-disubstituted alkenol [δ 5.35 (d)/ 5.84 (dd) /6.40 (d)] and two 1,2,3-trisubstituted benzenes [δ 6.76 (d)/6.84 (d) and 7.34–7.49 (2d and 2t)] would have been recognized. Using these more definitive data, the search profile could have been

	AntiMarin	DNP
0 Me/0 –CH_2–	1969	5894
0 Me/0 –CH_2–/2 1,2-alkene	104	219
0 Me/0 –CH_2–/2 1,2-alkene/2 1,2,3-B	7	7
0 Me/0 –CH_2–/2 1,2-alkene/2 1,2,3-B/*m*/*z* 320–321	2	2

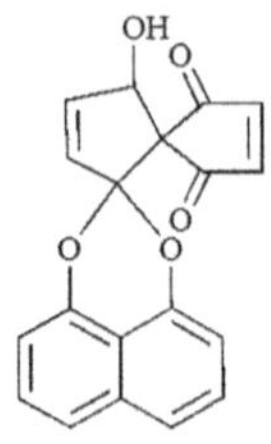

19

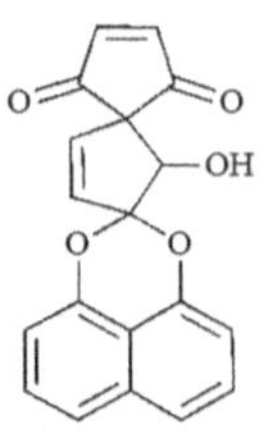

20

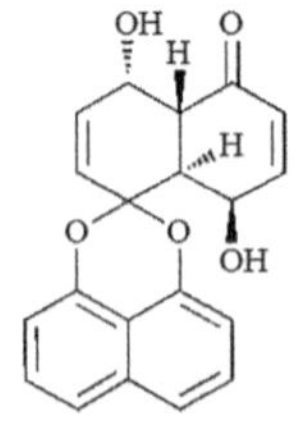

21

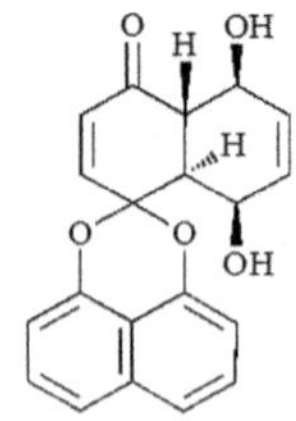

22

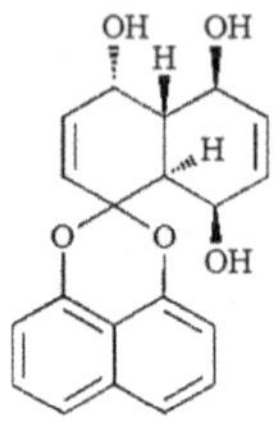

23

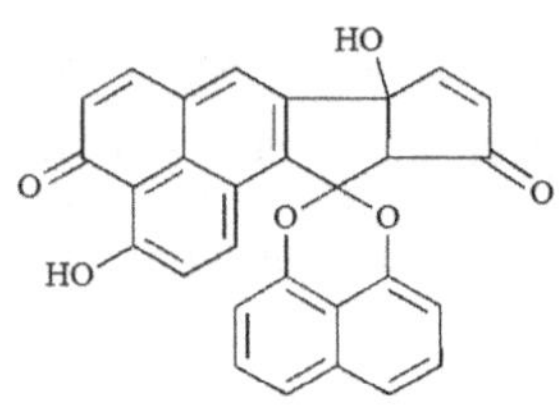

24

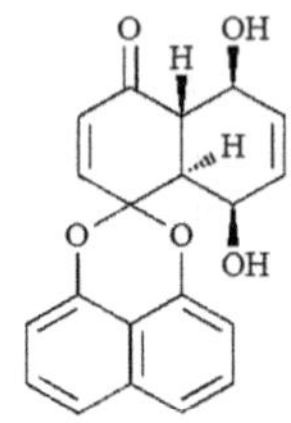

25

Figure 8.11 The seven spiro-bisnaphthalenes that matched the AntiMarin or DNP searches for 0 Me/0 $-CH_2-$/2 1,2-alkene/2 1,2,3-B.

The seven compounds selected after the third iteration all belonged to the spiro-bisnaphthalene family (Figure 8.11, **19–25**) and included spiro-mamakone A (**19**). By searching on molecular mass in a fourth iteration, only two compounds (**19, 20**) remained. The actual compound in question was spiro-mamakone A (**19**).[65] As the two compounds had identical mass and molecular formulae, the final choice between them would have to rely on comparison of the actual NMR spectral data and other physical properties.

8.6.3.6 The Role of Multiplicity-edited GHSQC Data

The acquisition of HSQC-DEPT data in addition to ^{1}H-NMR data is another strategy that conveys many advantages, as it allows ready distinction between, say, lower field >CH–O– hydrogens and higher field sp^2 hydrogens. which can be confusing if just the ^{1}H-NMR spectrum is available (see the example above using spiro-mamakone A). The 2D-HSQC-DEPT array is very useful for deconvoluting peaks overlapped in the ^{1}H-NMR spectrum and for obtaining a count of sp^3 methylene groups. Earlier, when the pateamine example was considered (Section 8.6.3.1; Figure 8.6), the results obtained based on the methyl group count required input of mass data to establish the identity of the molecule. However, if the HSQC-DEPT spectrum had been obtained (Figure 8.12; this spectrum is the calculated spectrum extracted from MarinLit), it would have been possible to distinguish unambiguously the two >CH–O– groupings (δ 5.11, 6.24) from the eight sp^2 protons.

The ^{13}C chemical shifts from the HSQC-DEPT spectrum also add confidence to the assignment of the two *N*-Me groups (δ_C ~45) and confirm five

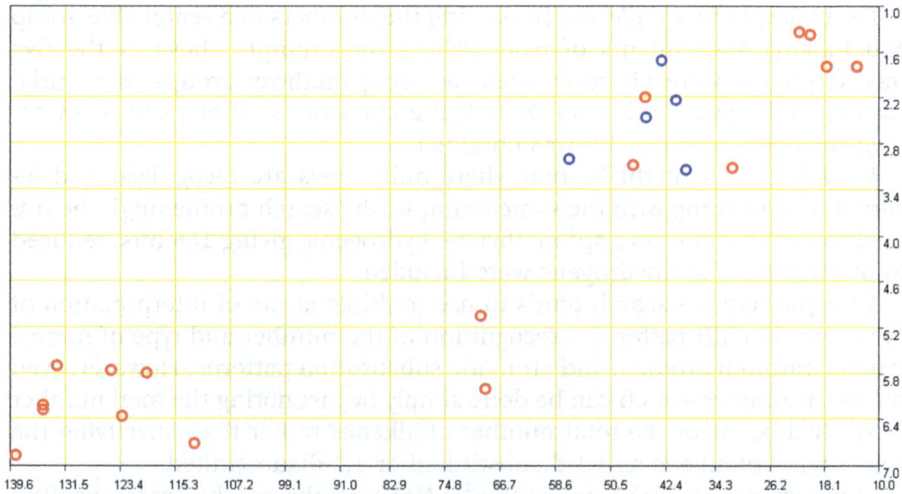

Figure 8.12 The calculated HSQC-DEPT spectrum for pateamine (Figure 8.7, **10**). Methyl and methine correlations shown in red, methylene in blue.

sp^3 methylenes. If these three other functionalities are taken in combination with the methyl groups, the number of possible hits in each case is reduced from 10 to one [corresponding to pateamine (Figure 8.7, **10**)].

	MarinLit	AntiMarin	DNP
7 Me/2 Me (d)/3 Me vinyl/2 *N*-Me	10	10	10

But,

7 Me/2 Me (d)/3 Me vinyl/2 *N*-Me/2 >CH–O	1	1	1
7 Me/2 Me (d)/3 Me vinyl/2 *N*-Me/8 sp^2 H	1	1	1
7 Me/2 Me (d)/3 Me vinyl/2 *N*-Me/5 –CH_2–	1	1	1

8.7 ^{1}H-NMR Pattern Matching Search Strategies

The strategies for formulating a pattern-matching ^{1}H-NMR search profile can range from the obvious to the subtle, from the simple to the complex. An *obvious* search could be just using the number of methyl groups of all recognizable types. Such a search would certainly reduce the number of potential candidates, but if used alone could still result in thousands of hits if, for example, there were five methyl groups. In DNP that would give 19 350 hits, but if the search criteria were combined with a mass range of *m*/*z* 328–329, then the hits decrease to only 76.

The concept of a *simple* search is using the numbers of a searchable group and making assignments of type. Taking the example above, if the five methyl groups were all recognized as being methoxy groups, the initial search would give 365 hits in DNP. If the mass range of *m*/*z* 328–329 were then included, only one hit was obtained.

A *subtle* search might be one where null values are recognized and incorporated. Keeping with the same example, the search profile might be five methoxy groups and zero sp^3 methylene hydrogens, giving 166 hits, reduced to five hits if six sp^2 hydrogens were included.

A *complex* type of search tends to use multiple layers of interpretation of the chemical shift patterns – recognition of the number and type of methyl groups, carbinol protons, and aromatic substitution patterns. However, even the most complex search can be done simply by just noting the total number of sp^2 hydrogens or the total number of alkenes rather than specifying the actual types of alkene as 1,1-disubstituted or 1,2-disubstituted.

With experience in the use of the ^{1}H-NMR databases, the search profiles generated tend to start as *obvious* or *simple* and evolve into the *complex* or *subtle*, which optimizes the search and minimizes the potential for dropping possible hits by being too subtle.

8.8 Chemical Shift-matching Approach to Dereplication

Possibly the earliest work in the area of chemical shift matching was research published in 1976 that used the wider dispersion of the ^{13}C chemical shift range to gain the resolution necessary to analyze and quantify complex mixtures of monosaccharides obtained as aqueous extracts directly from a natural product source.[66] Unlike the then standard GLC-based method, no derivatization was necessary and the method was direct and accurate with each data collection taking <5 min with the results then being analyzed automatically. To achieve these outcomes, careful attention was placed on aspects such as sample concentration, temperature, pH, and acquisition conditions. Care was taken in selecting the appropriate pulse width and acquisition time to account for variations in the longitudinal relaxation times (T_1), which if large enough could impede the accuracy of the method. About 25 years later, comparable approaches were taken in ensuring the accuracy of the ^{1}H-NMR approaches to metabolomics for the detection and quantitation of primary metabolites in body fluids, as exemplified by the work of Chenomx,[61] but also with an increasing number of online databases of NMR spectra obtained for metabolites.[67,68]

Using appropriate databases, such shift-matching approaches can also be successfully applied to the analysis of samples arising from the dereplication of crude natural product extracts. Three of the specialist databases (see Table 8.2) are appropriate. These are the ACD/Labs NMR,[40,41] AntiBase,[45] and MarinLit[9] databases. The ACD/Labs assigned ^{1}H and ^{13}C

NMR databases are currently the richest sources of assigned structures with associated chemical shifts and, although not limited solely to natural product structures, the content is without compare even when compared with dedicated natural product resources such as the *Dictionary of Natural Products*. While such dedicated resources contain rich collections of natural product structures, the collection is not chemical shift searchable in a form as useful to NMR spectroscopists as that provided in the ACD/Labs NMR database.

8.8.1 The ACD/Labs NMR Database

The Version 12 ACD/Labs NMR combined ^{13}C and ^{1}H NMR database contains over 322 000 compounds, >214 000 of them with ^{1}H data and >200 000 with ^{13}C data. This equates to over 2.5 million experimental ^{13}C NMR chemical shifts and over 105 000 coupling constants. The ^{1}H NMR data include over 1 758 000 experimental chemical shifts and over 624 000 coupling constants. Only a subset of the database contains natural products, as the database is compiled from around 200 separate data sources, the majority of them journal articles but also books, dissertations and theses, and online collections. Since the data are collected from such a diverse set of resources, they are not obtained under any particular conditions. Even in a single record there can be a significant distribution in the listed shifts, as the data are obtained at various temperatures, in different solvents, at different field strengths, and with different levels of impurities. Water levels especially can contribute dramatically to ^{1}H NMR shifts, as can concentration due to aggregation. Although the collection is valuable, it should be used with these considerations on data quality in mind. In terms of natural products, an indication of the sourcing includes >13 000 sets of chemical shift data from the *Journal of Natural Products* and >300 entries from the NMR Database of Lignin and Cell Wall Model Compounds.[69] Since natural products are reported in many journals, an estimate of the number of natural product compounds in the database would be in the range 40 000–50 000.

8.8.1.1 The ACD/Labs NMR Database Search Interface

Various options can be set in the search interface. These include the Looseness Factor, the Minimum Number of Query Shifts, and the Hit Quality Index. A schematic of the search interface is shown in Figure 8.13.

The "Looseness Factor" is the deviation allowed around each chemical shift during the search. The "Minimum Number of Query Shifts" to match is the number that should be found during a search. For example, if, in the "Enter Query Shifts" field, the shifts 1.2, 2.3, 2.7, 3.1, and 7.2 were entered and the minimum to match is 2, then the program will find all the records containing at least two chemical shifts from this list. It should be noted that it is possible to use the signal multiplicities for searching by adding the corresponding letter directly after the chemical shifts value. The option to

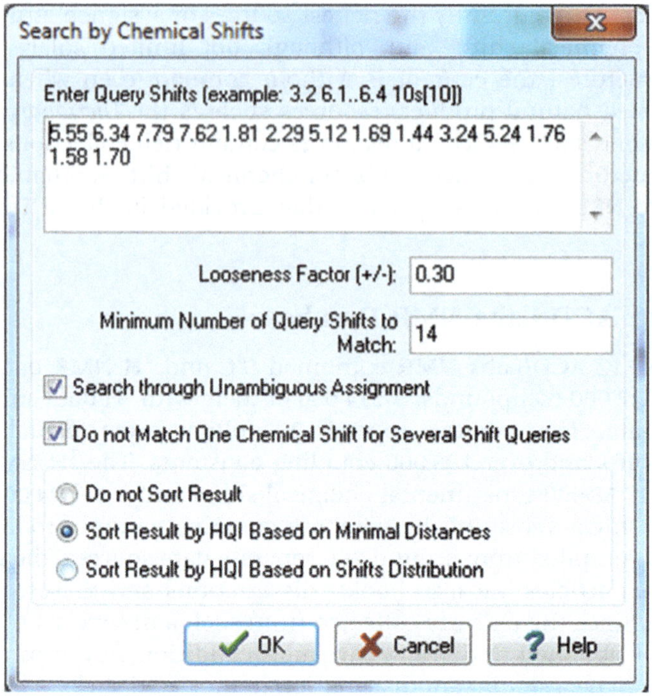

Figure 8.13 Schematic of the ACD/Labs NMR DB search interface.

sort the results by the HQI (Hit Quality Index) based on minimal distance orders the results such that the best matches are listed first, *i.e.* with the highest HQI.

8.8.1.2 *The ^{1}H Chemical Shift-matching Approach*

Using the ACD/NMR database of ^{1}H and ^{13}C chemical shifts, an example of shift searching can be demonstrated by considering the following dataset:

Molecular mass: 340.2.
^{1}H NMR: 5.55, 6.34, 7.79, 7.62, 1.81, 2.29, 5.12, 1.69, 1.44, 3.24, 5.24, 1.76, 1.58, 1.70.

Using the list of shifts as input, selecting a looseness factor of 0.3 ppm, and selecting the option to match all 14 chemical shifts (see Figure 8.13) 28 hits were retrieved. The hits were ordered by HQI based on the minimal deviations between the input chemical shifts and those contained within the database. Only one hit had a mass matching the experimental value and the result is shown in Figure 8.14.

The compound is identified as gaudichaudianic acid and the reference is included in the database. For additional reference, the ^{13}C data for the

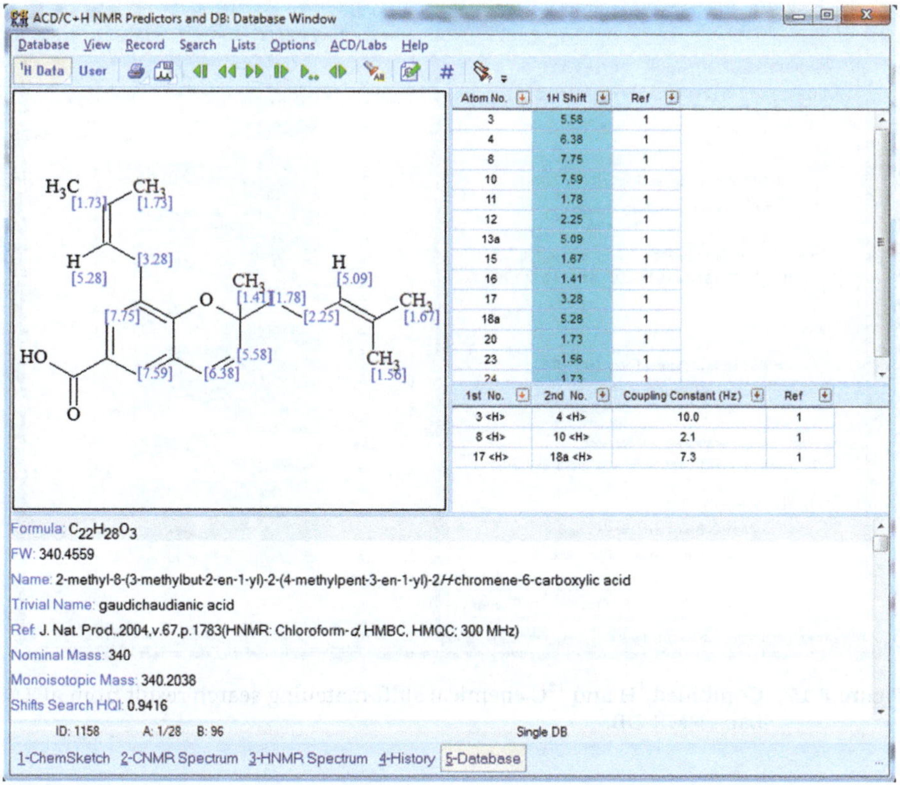

Figure 8.14 Chemical shift matching search result from ACD/Labs NMR DB.

compound are also available in case the data have been measured using either direct or indirect detection methods.

8.8.1.3 The ^{1}H and ^{13}C Chemical Shift-Matching Approach

When the following set of ^{13}C shifts:

> ^{13}C NMR: 80, 129, 122, 121, 156, 129, 132, 121, 127, 42, 23, 124, 132, 26, 27, 28, 122, 133, 26, 172, 18, 17.

is added into the combined search of chemical shifts, the hit list reduces from 28 hits using ^{1}H chemical shifts only to a single hit in the database as shown in Figure 8.15.

The ACD/Labs NMR database can also be searched in a variety of other ways using measured NMR properties. These include by ^{13}C NMR shifts only, combined ^{1}H and ^{13}C shifts, by coupling constants, and by correlations between ^{1}H–^{1}H and ^{1}H–^{13}C shifts.

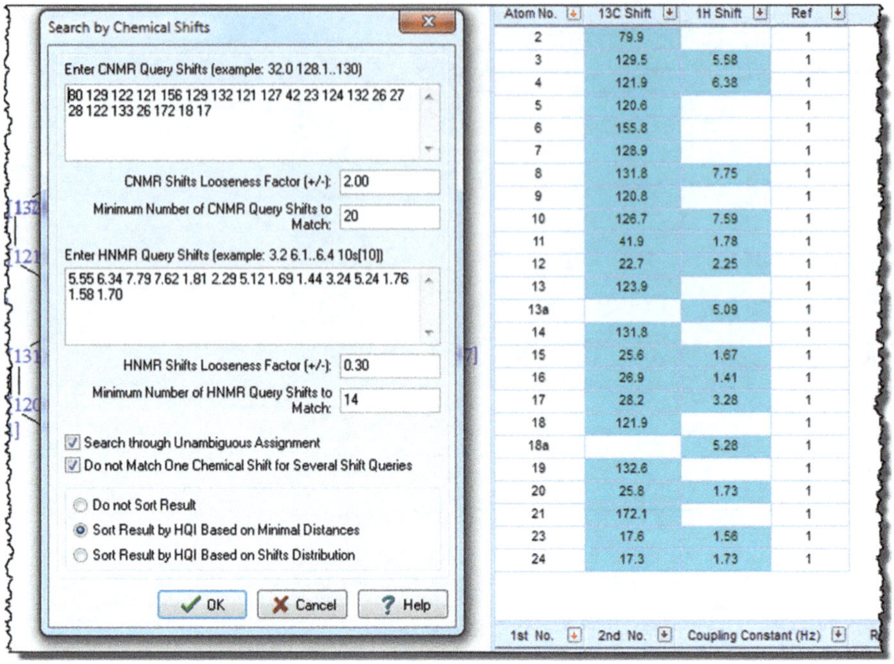

Figure 8.15 Combined ^{1}H and ^{13}C chemical shift-matching search result from ACD/Labs NMR DB.

8.8.2 MarinLit and AntiBase Databases and ^{13}C Chemical Shift Matching

The MarinLit and AntiBase databases can also be used for ^{13}C chemical shift matching. Under the Compound Search section in MarinLit, it is possible to enter carbon chemical shift data and search those data for a match or partial match against all marine natural products. The data can be entered with or without the number of attached protons to each carbon nucleus. The complete data set for a marine monoterpene is as follows and given in Figure 8.16:

> ^{13}C NMR($^{\#}H$): 69(0), 64(1), 35(2), 130(0), 124(0), 49(2), 30(3), 18(3), 131(1), 118(1).

Each carbon shift is searched against a combined actual and ACD/Labs calculated CNMR database with a user-set chemical shift tolerance. From this simple search, 12 compounds in the database matched these 10 values of carbon shift, with assigned proton counts and using a ± 5 ppm tolerance. Within MarinLit, it is possible to combine a pattern-match search with a chemical shift match or mass data. If the chemical shift data are now combined with a search containing two methyl singlets, or a mass range of

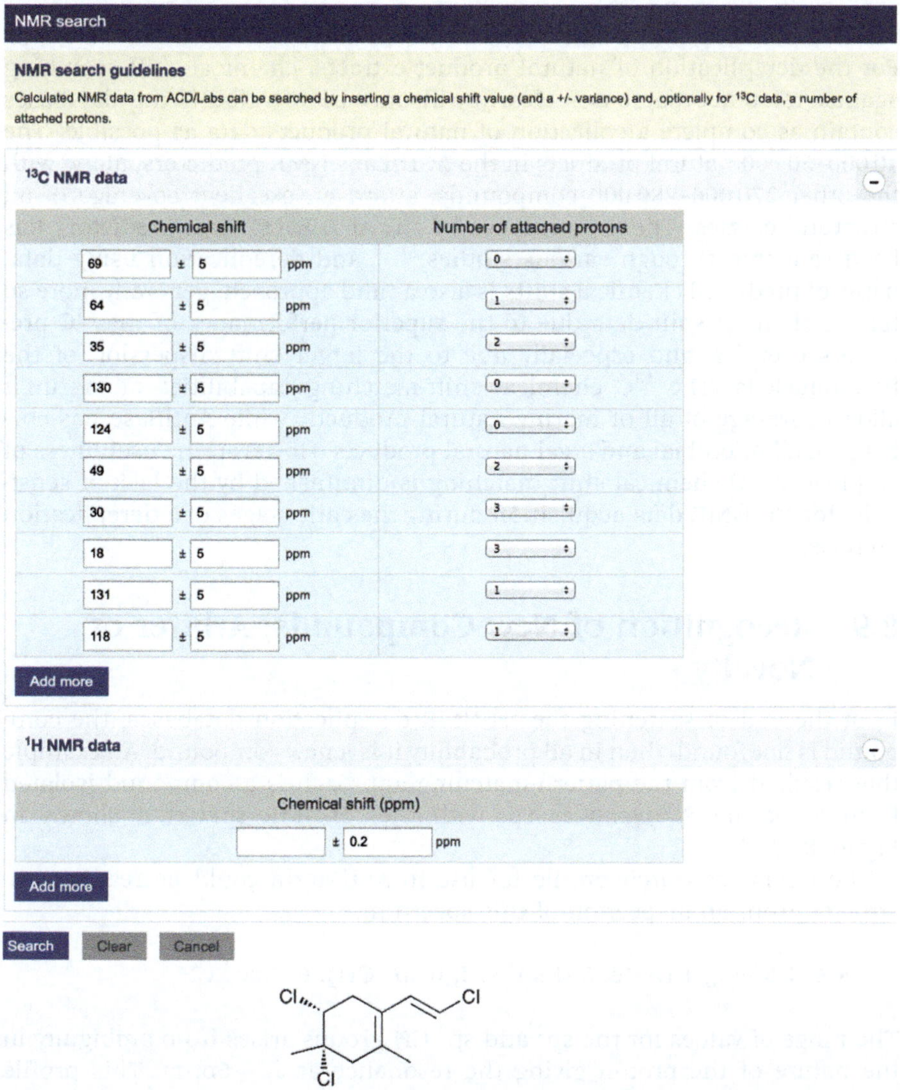

Figure 8.16 Part of the Compound Search entry page in MarinLit with ^{13}C shift data and the search for two methyl singlets. The structure shown is the compound (plocamene B) that matches the search requirements.

m/z 238 $\pm$ 0.5, then just one hit, plocamene B,[70] is obtained (see Figure 8.16). MarinLit is also able to carry out a comparable ^{1}H chemical shift-matching search.

A similar approach to that described here for MarinLit can be implemented in the SciDex version of AntiBase, which has calculated ^{13}C chemical shift data for most of the ~38 600 compounds of microbial or algal origin.

8.8.3 The Chemical Shift-matching Databases

For the dereplication of natural product extracts, chemical shift searching against reference databases clearly will only be of value if the databases contain as complete a collection of natural products data as possible. The 40 000–50 000 natural products in the ACD/Labs NMR predictors, along with the other 270 000–280 000 compounds, afford an excellent coverage of all structural classes. The performance of ^{13}C ACD/Labs NMR predictors has been validated through various studies,[71,72] and dereplication using databases of predicted chemical shifts is also a valid approach, generally more so for ^{13}C than ^{1}H shift data due to the superior performance of the ^{13}C predictors over ^{1}H, and especially due to the larger shift dispersion of the heteronucleus. The ^{13}C chemical shift-matching capabilities of MarinLit allows coverage of all of marine natural products, while AntiBase has coverage of all microbial and algal natural products. However, the usefulness of employing ^{13}C chemical shift matching is diminished by the lack of sensitivity for ^{13}C NMR data acquisition during the early stages of a dereplication exercise.

8.9 Recognition of New Compounds: Arbiter of Novelty

If, at the end of searching the ^{1}H-NMR dereplication databases, the compound is not found, then in all probability it is a new compound. An example that resulted from the pattern-matching approach is a compound isolated from a marine *Streptomyces* sp. with the ^{1}H-NMR spectrum shown in Figure 8.17.[73]

The following search profile for use in AntiMarin could be readily constructed from an inspection of this spectrum:

2 Me, 1 Me(s), 1 *O*-Me, 3–4 sp^3 CH, 0 sp^3 CH$_2$, 6–7 sp^2 CH.

The range of values for the sp^2 and sp^3 CH groups arises from ambiguity in the nature of the proton giving the resonance at $\delta_H \sim 6$ppm. This profile would have given no hits in AntiMarin, suggesting that this was a new microbial compound. Structural elucidation revealed the structure for kiamycin as shown in Figure 8.17. That this was a new compound could only be verified after searching one of the larger databases such as CAS Registry[35–37] or Reaxys.[39] Although these large databases do not include searchable ^{1}H-NMR data in the sense of pattern recognition and chemical shift matching, they are very comprehensive in their coverage of the chemical literature and should be considered the final arbiter of novelty. Once novelty has been established, the time spent on analysis of the full NMR data sets and mass data is fully justified as it is only with the establishment of a new structure that dereplication is complete.

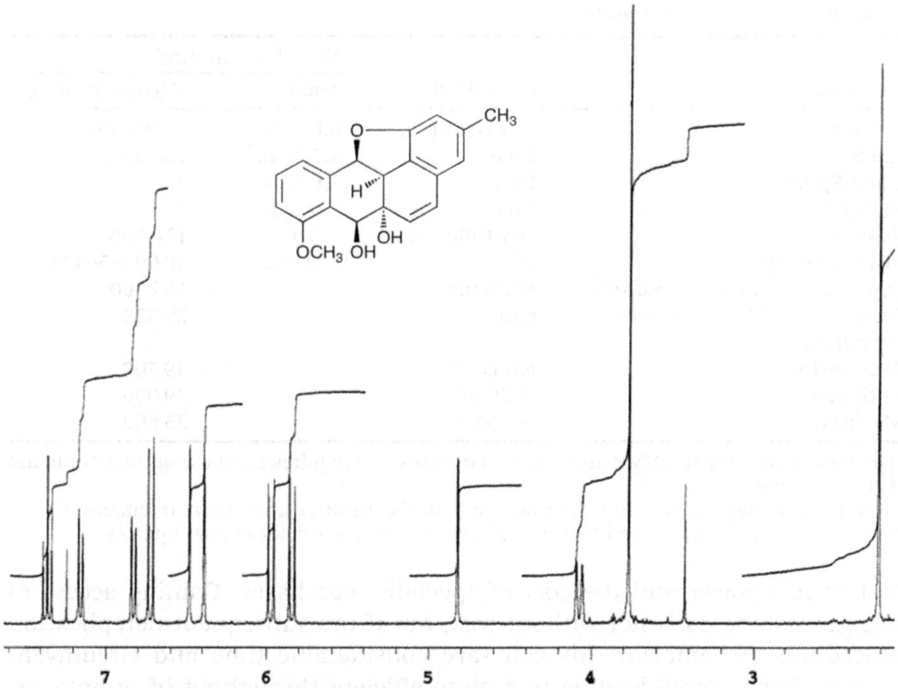

Figure 8.17 ^{1}H-NMR (300 MHz) spectrum of kiamycin.
Spectrum courtesy of Prof. Hartmut Laatsch.

8.10 The Costs Associated With Dereplication

The dereplication of natural product extracts is not without cost, but is probably the single most vital activity required for advancement, discovery of new compounds, and the exploration of known compounds and compounds from new ecological niches. The requirements for effective dereplication, in addition to effective manpower, include access to high-field NMR and mass spectrometers, HPLC equipment, and databases for data mining. The provision and maintenance of spectrometers and other equipment would normally arise from company, faculty, or institutional decisions rather than from individuals or research groups, who would normally be required to pay running costs only. Databases have an important role in advancing research in the natural products area. While the larger databases such as SciFinder, or other versions of CAS, are expensive, access is usually provided on an institutional basis. The specialized natural product-oriented databases are considerably less expensive and initially are more relevant for the natural product chemist. The relative costs of the relevant databases are given in Table 8.3.

For the individual researcher, the financial considerations for effective dereplication center on the cost of supplies and meeting running costs, the

Table 8.3 Costs of the databases.

Database	Cost (US$)	No. of compounds Total	Natural products
SciFinder	>50 000 p.a.	6.6×10^7	~260 000
CSLS	Free	4.7×10^7	Extracts
ChemSpider	Free	3.0×10^7	?
PubChem	Free	3.0×10^7	?
Reaxys	>40 000 p.a.	$>10^7$	170 000
ACD Labs DB	?[a]	322 000	40 000–50 000
Dictionary of Natural Products	6600 p.a.		165 500
Dictionary of Marine Natural Products	625		29 525
AntiMarin	No cost[b]		59 500
AntiBase	~3000[c]		39 000
MarinLit	~3000[c]		25 600

[a]Purchase of the TOTAL CNMR and HNMR databases also requires purchase of the HNMR and CNMR predictors.
[b]AntiMarin is "free" to current subscribers to both the AntiBase and MarinLit databases.
[c]These are the initial costs, and they are followed by lesser costs for annual updates.

cost of manpower and the cost of specialist databases. Gaining access to relevant databases is not without cost, but if the subsequent dereplication procedures are efficient this can save considerable time and circumvent wasted effort, overall leading to a more efficient throughput of samples by the researchers. In 1994, David Corley, a pharmaceutical chemist, estimated that "... in our laboratory that for each natural product dereplicated, at an average cost of $300 of online time (using STN databases), a savings of $50 000 is incurred in isolation and identification time".[74] With manpower, collection expenses, and operating costs comprising a major part of any natural product chemist's budget, the costs associated with accessing even the most specialized database are a small consideration in comparison with the savings possible with efficient dereplication of samples. ^{1}H-NMR-based specialist databases available since 1994 allow for an even more rapid and comprehensive dereplication process.

8.11 Conclusion

In considering taxonomic, biological, UV, MW/MF, and ^{1}H-NMR databases that are available for dereplication purposes, it is unlikely that just one technique alone will suffice, but of the possible approaches, the interpretation of ^{1}H-NMR data is the one most likely to provide a definitive outcome. There are two compelling reasons for this conclusion. First, there is access to ^{1}H-NMR databases that cover *all* natural products in the case or pattern matching (DNP, AntiMarin, and MarinLit) or a large section of natural products, and an extensive database of other compounds for the chemical shift-matching approach (ACD/Labs NMR databases). This is most certainly not the case for the matching of UV spectra. Although there are UV databases that might be able to cover many aspects of natural products, these are discreet databases and not available outside the institutions that developed them. A similar situation holds for the application of MS to dereplication. Details of specialist MS/MS^n-oriented databases have been published and these, in hand with the likes of SciFinder, Reaxys, and the NIST database, give access to the natural product MS data. The use of these data is, unfortunately, based almost entirely on molecular mass and molecular formula matching with little recognition of fragments, structural isomers, or stereoisomers. Regardless of the approach taken for ^{1}H-NMR dereplication, be it chemical shift recognition or pattern matching, there is strong database support.

The second reason focuses on the quality of the information. ^{1}H-NMR data are rich in structural information that on interpretation lead directly to structural elucidation. That does not hold for either the UV or the MS approach to dereplication. The recognition of a chromophore is helpful but not conclusive in arriving at a structure, and although fragmentation patterns in EIMS can be diagnostic, most dereplication MS techniques use soft ionization approaches yielding the MH^+, $[M - H]^-$, or adduct ions such as MNa^+ ions and not fragment ions. An MS/MS^n approach provides information on the mass of fragments produced but is not as helpful as the direct structural information that can be extracted from NMR data. There are, however, opportunities to use algorithmic fragmentation across such databases and then perform matching. Commercial MS fragmentation packages such as ACD/Labs MS Fragmenter[75] and Thermo Scientific's Mass Frontier[76] could be used to populate such databases or published algorithms could be utilized.[77]

In selecting the best approach to dereplication, it will be cost, not expediency or efficiency, that is the final arbiter of choice. The generation of UV data is the cheapest approach whereas the MS and ^{1}H-NMR approaches both require levels of investment, usually by the institution or company, which it might not be practicable to make.

References

1. A. J. der Marderosian, *Pharm. Sci.*, 1969, **58**, 1.
2. F. Sertürner, *Journal der Pharmacie fuer Aerzte und Apotheker*, 1805, **13**, 229.
3. F. Sertürner, *Ann. Phys.*, 1817, **55**, 56.
4. D. L. Hawksworth and M. T. Kalin-Arroyo in *Global Biodiversity Assessment*, ed V. Heywood, Cambridge University Press, Cambridge, UK, 1995, p. 107.
5. A. D. Chapman in *Numbers of Living Species in Australia and the World*, 2nd edn, Australian Biological Resources Study, Canberra, 2009.
6. R. M. May, *Science*, 1998, **241**, 1441.
7. L. Tangley, in *US News and World Report*, Aug 18, 1997. See http://www.usnews.com/usnews/culture/articles/970818/archive_007681.htm. Accessed April, 2012.

8. J. Blunt, J. Buckingham and M. Munro in *Handbook of Marine Natural Products*, ed. E. Fattorusso, W. H. Gerwick, O. Taglialatela-Scafati, Springer, Dordrecht, Heidelberg, New York, London, 2012, p. 3.
9. MarinLit. See http://pubs.rsc.org/marinlit/.
10. *Dictionary of Natural Products*, ed. J. Buckingham, Chapman & Hall/CRC, Boca Raton, USA, 2013.
11. F. VanMiddlesworth and R. J. P. Cannell in *Methods in Biotechnology*, Vol 4, ed. R. J. P. Cannell, Humana Press, Totowa NJ, 1998, p. 279.
12. A. Langlykke, Foreword, in *CRC Handbook of Antibiotic Compounds*, ed. J. Berdy, CRC, Boca Raton, FL, 1980.
13. Webster's Online Dictionary, http://www.webster-dictionary.org/. Accessed May, 2014.
14. L. J. Hanka, S. L. Kuentzel, D. G. Martin, P. F. Wiley and G. L Neil, *Cancer Res.*, 1978, **63**, 69.
15. J. A. Beutler, A. B. Alvarado, D. E. Schaufelberger, P. Andrews and T. G. McCloud, *J. Nat. Prod.*, 1990, **53**, 867.
16. J. Antonio and T. F. Molinski, *J. Nat. Prod.*, 1993, **56**, 54.
17. J. A. Beutler, T. C. McKee, R. W. Fuller, M. Tischler, J. H. Cardellina II, T. G. McCloud, K. M. Snader and M. R. Boyd, *Antiviral Chem. Chemother.*, 1993, **4**, 167.
18. J. H. Cardellina II, M. H. Munro, R. W. Fuller, K. P. Manfredi, T. C. McKee, M. Tischler, H. R. Bokesch, K. R. Gustafson, J. A. Beutler and M. R. Boyd, *J. Nat. Prod.*, 1993, **56**, 1123.
19. M. Månsson, R. K. Phipps, L. Gram, M. H. G. Munro, T. O. Larsen and K. F. Nielsen, *J. Nat. Prod.*, 2010, **73**, 1126.
20. R. A. Gosselin, *Lloydia*, 1962, **25**, 24.
21. D. J. Newman and G. M. Cragg, *J. Nat. Prod.*, 2012, **75**, 311.
22. See http://www.protasis.com/MicroFlowNMR/index.htm. Accessed May, 2014.
23. See http://www.bruker.com/products/mr/nmr/probes/cryoprobes.html. Accessed May, 2014.
24. G. Lang, N. A. Mayhudin, M. I. Mitova, L. Sun, S. van der Sar, J. W. Blunt, A. L. J. Cole, G. Ellis, H. Laatsch and M. H. G. Munro, *J. Nat. Prod.*, 2008, **71**, 1595.
25. J. Bitzer, B. Kopcke, M. Stadler, V. Hellwig, Y-M. Ju, S. Seip and T. Henkel, *Chimia*, 2007, **61**, 332.
26. ChemSpider. http://www.chemspider.com. Accessed May, 2014.
27. CSLS. http://cactus.nci.nih.gov/. Accessed May, 2014.
28. PubChem. http://pubchem.ncbi.nlm.nih.gov/. Accessed May, 2014.
29. NMR Shift DB. See http://nmrshiftdb.nmr.uni-koeln.de. Accessed May, 2014.
30. Naproc-13. See http://c13.usal.es/. Accessed May, 2014.
31. SuperNatural. See http://bioinformatics.charite.de/supernatural/. Accessed May, 2012.

32. SDBS (Spectral Database for Organic Compounds). See http://riodb01.ibase.aist.go.jp/sdbs/cgi-bin/cre_index.cgi?lang = eng. Accessed May, 2014.
33. BindingDB. http://www.bindingdb.org. Accessed May, 2014.
34. Chemical Abstracts Service. http://www.cas.org. Accessed May, 2014.
35. CAS Registry. See https://www.cas.org/content/chemical-substances. Accessed, May, 2014.
36. STN. See http://www.stn-international.de. Accessed, May, 2014.
37. SciFinder. See http://www.cas.org/products/scifinder. Accessed, May, 2014.
38. SpecInfo. See http://cds.dl.ac.uk/cds/datasets/spec/specinfo/specinfo.html. Accessed, May, 2014.
39. Reaxys. See https://www.reaxys.com/reaxys/session.do. Accessed, May, 2014.
40. ACD/Labs. See http://www.acdlabs.com. Accessed, May, 2014.
41. ACD Spectral Libraries. http://www.acdlabs.com/products/adh. Accessed May, 2014.
42. NaprAlert. See http://www.napralert.org/. Accessed, May, 2014.
43. NIST 11. See http://www.sisweb.com/software/ms/nist.htm. Accessed, May, 2014.
44. *Dictionary of Marine Natural Products*, ed. J. W. Blunt, M. H. G. Munro, Chapman & Hall/CRC, Boca Raton, USA, 2008.
45. AntiBase. See http://wwwuser.gwdg.de/~hlaatsc/antibase.htm.
46. AntiMarin: a combination database formed from AntiBase and MarinLit. See http://wwwuser.gwdg.de/~hlaatsc/antibase.htm and/or http://www.chem.canterbury.ac.nz/marinlit/marinlit.shtml.
47. GVK Biosciences Natural Product DB. See https://gostardb.com/gostar/.
48. The Marine Group, University of Canterbury's UV data was acquired on a Dionex HPLC using Chromeleon software.
49. T. O. Larsen, M. A. E. Hansen in *Bioactive Natural Products: Detection, Isolation and Structural Determination*, 2nd edn, ed. S. M. Coalgate, R. J. Molyneaux, CRC Press, 2007, p. 221.
50. K. F. Nielsen and J. Smedsgaard, *J. Chromatogr. A*, 2003, **1002**, 111.
51. K. F. Nielsen, M. Månsson, C. Rank, J. C. Frisvad and T. O. Larsen, *J. Nat. Prod.*, 2011, **74**, 2338.
52. J. Lei and J. Zhou, *J. Chem. Inf. Comput. Sci.*, 2002, **42**, 742.
53. S. Moss, G. Bovermann, R. Denay, J. France, C. Guenat, L Oberer, M. Ponelle and H. Schroder, *Chimia*, 2007, **61**, 346.
54. The National Centre for Plant and Microbial Metabolomics. See http://www.metabolomics.bbsrc.ac.uk/currentactivities.htm. Accessed May, 2014.
55. CH-NMR-NP. See https://www.las.jp/CH-NMR-NP/English/English_help.html. Accessed May, 2014.
56. See http://www.cosmoscience.org/pdfs/Session%20IV_Presentation%20I_Zink.pdf. Accessed May, 2012.

57. Catalogue of Life. See http://www.catalogueoflife.org. Accessed May, 2014.
58. WoRMS. See http://www.marinespecies.org. Accessed May, 2014.
59. J. L. Little, A. J. Williams, A. Pshenichnov and V. J. Tkachenko, *J. Am. Soc. Mass Spectrom*, 2012, **23**, 179.
60. A. Fredenhagen, C. Derrien and E. Gassmann, *J. Nat. Prod.*, 2005, **68**, 385.
61. CHENOMX. See http://www.chenomx.com/software/. Accessed May, 2014.
62. J. Bradshaw, D. Butina, A. J. Dunn, R. H. Green, M. Hajek, M. M. Jones, J. C. Lindon and P. J. Sidebottom, *J. Nat. Prod.*, 2001, **64**, 1541.
63. P. T. Northcote, J. W. Blunt and M. H. G. Munro, *Tet. Lett.*, 1991, **32**, 6411.
64. S. Begum, B. S. Siddiqui and S. I. Hassan, *Nat. Prod. Lett.*, 2002, **16**, 173.
65. S. A. van der Sar, J. W. Blunt and M. H. G. Munro, *Org. Lett.*, 2006, **8**, 2059.
66. J. W. Blunt and M. H. G. Munro, *Aust. J. Chem.*, 1976, **29**, 975.
67. NMR metabolomics database of Linkoping. See http://www.liu.se/hu/mdl/main/. Accessed May, 2014.
68. Biological Magnetic Resonance Data Bank. See http://www.bmrb.wisc.edu/metabolomics/query_metab.php. Accessed May, 2014.
69. NMR Database of Lignin and Cell Wall Model Compounds. S. A. Ralph, J. Ralph, L. L. Landucci. November 2004. See http://ars.usda.gov/Services/docs.htm?docid = 10491, Accessed May, 2014.
70. P. Crews and E. Kho, *J. Org. Chem.*, 1975, **40**, 568.
71. K. A. Blinov, C. Steinbeck, M. E. Elyashberg and A. J. Williams, *J. Chem. Inf. Model.*, 2008, **48**, 550.
72. Y. D. Smurnyy, K. A. Blinov, T. S. Churanova, M. E. Elyashberg and A. J. Williams, *Chemom. Intell. Lab. Syst.*, 2009, **97**, 91.
73. Z. Xie, B. Liu, H. Wang, S. Yang, H. Zhang, N. Ji, S. Qin and H. Laatsch, *Mar. Drugs*, 2012, **10**, 551.
74. D. G. Corley and R. C. Durley, *J. Nat. Prod.*, 1994, **57**, 1484.
75. See http://www.acdlabs.com/products/adh/ms/ms_frag/. Accessed May 2014.
76. See http://www.thermoscientific.com/ecomm/servlet/productsdetail?productId = 11961841&storeId = 11152&ca = massfrontier. Accessed May 2014.
77. See http://onlinelibrary.wiley.com/doi/10.1002/rcm.2177/abstract. Accessed May, 2014.

CHAPTER 9

Application of Computer-assisted Structure Elucidation (CASE) Methods and NMR Prediction to Natural Products

M. E. ELYASHBERG,*[a] ANTONY J. WILLIAMS*[b] AND K. A. BLINOV[a]

[a] Advanced Chemistry Development, Moscow Department, 117513 Moscow, Russian Federation; [b] ChemConnector Inc., Wake Forest, NC 27587, USA
*Email: elyas@acdlabs.ru; tony27587@gmail.com

9.1 Introduction

The characterization of unknown chemical structures forms the basis of natural product chemistry. In previous chapters, different NMR spectroscopy techniques for organic molecule structure elucidation have been described. To elucidate the structures of large and complex natural products, a set of 2D-NMR spectra in combination with mass spectrometric (MS) data are usually required. The application of X-ray crystallography is also very attractive since it allows the determination of not only the structure but also a 3D model of the molecule. Unfortunately, there are numerous challenges that hamper the elucidation of a structure using X-ray analysis, including insufficient sample size and difficulty in obtaining a crystal of the appropriate quality. Therefore, it is a rather common situation that a combination of the most informative 2D-NMR experiments [usually HSQC (with or

Modern NMR Approaches To The Structure Elucidation of Natural Products,
Volume 1: Instrumentation and Software
Edited by Antony J. Williams, Gary E. Martin and David Rovnyak

Published by the Royal Society of Chemistry, www.rsc.org

without multiplicity editing), COSY, HMBC and ROESY(NOESY)] provides the necessary data to allow determination of the structure and the relative stereochemistry of newly isolated natural products. The careful logical and deductive analysis of all of the available 1D- and 2D-NMR data to infer the structure is time consuming and commonly requires the efforts of a skilled spectroscopist. The number of potential structural hypotheses that should be considered is frequently very large, which encourages the potential application of software capable of mimicking human expert reasoning.

The advantages of automating the procedure of spectrum structural information processing were originally realized in the 1960s.[1–4] As a result of the efforts of many research groups, a general ideology regarding computer-assisted structure elucidation (CASE) was developed over the following 20 years and a series of artificial intelligence systems were elaborated. These systems are now referred to as expert systems (ESs). These first-generation systems were based on 1D-NMR, MS and IR spectra used independently or in different combinations. They were capable of assisting in the structure determination of relatively small organic molecules with up to 20 skeletal atoms (see relevant books[5–7] and reviews[8–11]). The analysis of large and complex natural product molecules was impossible with the aid of these programs. In the 1990s, when 2D-NMR techniques became routinely available, scientists directed their efforts at the adaptation of ESs to utilize 2D-NMR data. As a result, a new generation of CASE systems was developed for which the molecular size limits were extended to 100 or more skeletal atoms. A comprehensive review of 2D-NMR-based expert systems was published by our group[12] that clearly demonstrates that a contemporary 2D-NMR-based CASE system is a versatile analytical tool capable of assisting the spectroscopist to solve complex structural problems in natural product chemistry. As discussed in this chapter, an ES can dramatically reduce the time necessary for a spectroscopist to elucidate the structures of *new* natural products and can significantly increase the reliability of structure determinations.

We describe the main principles on which CASE systems are based and demonstrate the applications of ESs to natural product structure elucidation. In order to investigate the uses of such systems, we use as an example the most advanced ES, Structure Elucidator[13–16] (StrucEluc), developed by our group. Familiarization with this ES will help researchers to understand the strategy of software utilization and obtain the information necessary to master this approach quickly so as to be able to use a CASE system effectively in their research.

9.2 Axiomatic Theory of Structure Elucidation

In the initial stages of the development of CASE systems, it was shown[5,17,18] that the methodology of *structural-group spectral analysis* of a given unknown can be interpreted in terms of a *partial axiomatic theory*, reflecting interrelations between molecular fragments and their characteristic spectral features. Later, this approach was extended to the whole methodology of molecular structure elucidation.[19,20] The methodology is reduced to the logical inference of the most probable structure from a set of statements ("axioms" and hypotheses) reflecting the interrelations between a set of observed spectral features (peaks in IR, mass and 1D and 2D-NMR spectra) and the analyzed structure. This methodology was *implicitly* used well before computer methods appeared. Independent of computer-based methods, the path to a target structure is the same, while CASE expert systems can mimic many, but not necessarily all, approaches of a human expert.

The implementation of the axiomatic approach into the algorithms contained within CASE systems supplies the systems with the following notable and unique abilities: (1) all statements regarding the interrelation between spectra and a structure ("axioms") are expressed *explicitly* and can be surveyed by the chemist; (2) all logical consequences (structures) following from the set of initial "axioms" are completely deduced without any exclusions; (3) the process of CASE is transparent and is generally very fast, providing tremendous savings in both time and labor for the chemist; (4) if the chemist has several alternative sets of axioms related to a given structural problem, then using an ES allows for the rapid generation of all structures from each of the sets and identification of the most probable structure by comparison of the solutions obtained.

In the following, we describe the main kinds of statements used when employing an ES for structure elucidation. These can be conventionally divided into the categories described below.

9.2.1 Axioms and Hypotheses Based on Characteristic Spectral Features

Axioms are those statements that can be considered true based on prior experience. To elucidate the structure of a *new* unknown compound, the chemist first uses known *characteristic spectral features* in NMR and, historically, IR spectra [spectrum–structure correlations (SSCs)], established as a result of the efforts of several generations of spectroscopists. Statements regarding SSCs play the role of *axioms* in the theory of structure elucidation. The general form of typical axioms belonging to this category can be presented as follows:

> *If a molecule contains a fragment A_i then the characteristic features of fragment A_i are observed in certain spectrum ranges $[X_1]$, $[X_2]$, …, $[X_m]$ that are characteristic of this fragment.*

We illustrate axioms of this type with simple examples. If a molecule contains a CH_2 group then a vibrational band is known to occur around 1450 cm^{-1} in the IR spectrum. If a molecule contains a CH_3 group then two bands around 1450 and 1380 cm^{-1} appear. These axioms can be presented formally in the following way using the symbols of implication (→) and conjunction (∧) conventional in symbolic logic:

$$CH_2 \rightarrow [1450\,cm^{-1}];\ CH_3 \rightarrow [1380] \wedge [1450\,cm^{-1}] \tag{9.1}$$

Analogously, for characteristic ^{13}C NMR chemical shifts, the following implications are also example axioms:

$$(C)_2C{=}O \rightarrow [200\ ppm];\ (C)_2C{=}S \rightarrow [200\ ppm] \tag{9.2}$$

When characteristic IR and NMR spectral features are used for the detection of fragments that can be present in a molecule under investigation, then the chemist usually forms statements for which a typical "template" is as follows:

If a spectral feature is observed in a spectrum range $[X_j]$, then the molecule contains at least one fragment of the set $A_i(X_j)$, $A_k(X_j)$, . . ., $A_l(X_j)$, where A_i, A_k, . . ., A_l are fragments for which the spectral feature observed in the range $[X_j]$ is characteristic and the fragments form a finite set.

This statement is a *hypothesis*, not an axiom, because: (1) the feature X_j can be produced by some fragment that is not known as yet and (2) the feature X_j can appear due to some intramolecular interaction of known fragments. Therefore, if an absorption band is observed at 1450 cm^{-1} in an IR spectrum, then the molecule can contain either CH_2 or CH_3 groups, both of them (band overlap at 1450 cm^{-1} is allowed) or the 1450 cm^{-1} band, which can be present as a result of the presence of another unrelated functional group. This statement can be expressed formally using the symbol for logical disjunction (∨): $1450\,cm^{-1} \rightarrow CH_2 \vee CH_3 \vee \alpha$, where α is a "sham fragment" denoting an unknown cause of the feature origin. For our ^{13}C NMR examples, we may obviously formulate the following hypothesis:

$$200\ ppm \rightarrow (C)_2C{=}O \vee (C)_2C{=}S \tag{9.3}$$

It is very important to keep in mind that if $A_i \rightarrow X_j$ is true, then the inverse implication $X_j \rightarrow A_i$ can be true or not true. In other words, the presence of a characteristic spectral feature X_j in a spectrum does not yet imply the presence of a corresponding fragment A_i. A true implication is $X_j \rightarrow A_i$. This implication means that if the characteristic spectral feature X_j does *not occur* in a spectrum, then the corresponding fragment A_i is absent from the molecule under investigation. The latter statement can be considered as another equivalent formulation of the basic axiom.

9.2.2 Axioms and Hypotheses of 2D-NMR Spectroscopy

The use of 2D-NMR spectroscopy is known to be a method that, in principle, permits inferring a molecular structure from the available spectral data *ab initio* without using any SSCs and additional suppositions. In practice, the structure elucidation of large molecules by the *ab initio* application of

2D-NMR data only (without 1D NMR SSCs) is generally impossible, although this might change in the future with the development of new 2D-NMR methods. The 1D and 2D-NMR data are usually combined synergistically to obtain solutions to real analytical problems in the study of natural products, synthetic products, impurities and many other classes of molecules. The number of hydrogen atoms responsible for the propagation of structural information across the molecular skeleton and the number of skeletal heteroatoms are the most influential factors for the time being.

When 2D-NMR data are used to elucidate a molecular structure, then the chemist (or ES) deduces conceivable structures from the molecular formula and a set of *hypotheses* matching the data from 2D-NMR spectroscopy. When we deal with a *new* chemical entity we must interpret a *new* 2D-NMR spectrum or spectra. In this case we have no possibility of relying on "axioms" valid for the *given* spectrum–structure matrix so *hypotheses* that are considered as the most plausible are formed. These hypotheses are based on the general regularities that are the significant *axioms of 2D-NMR spectroscopy*. We will attempt to express these axioms in an explicit form and classify them. The most important and common are axioms of homonuclear ^{1}H–^{1}H and heteronuclear ^{1}H–^{13}C and ^{1}H–^{15}N spectroscopy.

A *necessary condition* for the application of 2D-NMR data to CASE is the chemical shift assignment of all proton-bearing carbon nuclei (*i.e.* all CH_n groups where $n = 1$–3). This information is extracted from the HSQC (or alternatively HMQC) data using the following axiom:

- If a peak (δC-*i*, δH-*i*) is observed in the spectrum, then the hydrogen atom H-*i* with chemical shift δH-*i* is attached to the carbon atom C-*i* having chemical shift δC-*i*, anisochronous methylenes accepted.

The main sources of structural information are COSY (or TOCSY) and HMBC correlations that allow the elucidation of the backbone of a molecule. We refer to "standard" correlations[21] as those that satisfy the following axioms reflecting the experience of NMR spectroscopists:

- If a peak (δH-*i*, δH-*k*) is observed in a COSY spectrum, then a molecule contains the chemical bond (C-*i*)–(C-*k*), assuming that H-*k* is on a carbon, which is formally presented as

$$(\delta\text{H-}i,\ \delta\text{H-}k) \rightarrow [(\text{C-}i)\text{–}(\text{C-}k)] \tag{9.4}$$

- If a peak (δH-*i*, δC-*k*) is observed in a HMBC spectrum, then atoms C-*i* and C-*k* are separated in the structure by *one* or *more* chemical bonds:

$$(\text{C-}i)\text{–}(\text{C-}k) \text{ or } (\text{C-}i)\text{–}(\text{X})\text{–}(\text{C-}k),\ \text{X} = \text{C, O, N, } \ldots$$

  In the general case, the corresponding implication can be presented as

$$(\delta\text{H-}i,\ \delta\text{C-}k) \rightarrow [(\text{C-}i)\text{–}(\text{C-}k) \vee (\text{C-}i)\text{–}(\text{X})\text{–}(\text{C-}k)] \tag{9.5}$$

Note that both fragments shown in the right side of implication (9.5) can be present simultaneously in a molecule if and only if both of them are included in a three-membered ring. In other cases, an implication

$$(\delta\text{H-}i,\ \delta\text{C-}k) \rightarrow [(\text{C-}i)\text{–}(\text{C-}k)\ \nabla\ (\text{C-}i)\text{–}(\text{X})\text{–}(\text{C-}k)] \tag{9.5a}$$

is valid, where the symbol "∇" denotes *exclusive* disjunction, which has the following interpretation: only one of two fragments can exist in a molecule.

By analogy, the main axiom associated with employing the nuclear Overhauser effect (NOE) for the purpose of structure elucidation can be formulated in the following manner:

- If a peak (δH-*i*, δH-*k*) is observed in a NOESY (ROESY) spectrum, then the distance between the atoms H-*i* and H-*k* through space is no more than 5 Å.

It is important to note that there is a principal difference between logical interpretations of 1D- and 2D-NMR axioms. For instance, for COSY there exists another equivalent form of the main axiom:

- If a molecule does not contain the chemical bond (C-*i*)–(C-*k*), then no peak (δH-*i*, δH-*k*) is observed in a COSY spectrum.

In this case, the interpretation of logical implication (9.4) allows us to conclude that the absence of a peak (δH-*i*, δH-*k*) says *nothing* about the existence of the chemical bond (C-*i*)–(C-*k*) in the molecule: the bond may or may not exist. Consequently, the absence of the COSY peak (δH-*i*, δH-*k*) cannot be used to reject structures containing the bond (C-*i*)–(C-*k*), which is in agreement with chemical common sense. Analogous conclusions are also applicable to HMBC and NOESY/ROESY spectra.

Although it is known that the listed axioms hold in the overwhelming majority of cases, there are many exceptions and these correlations are referred to as *non-standard correlations* (NSCs).[21] Since standard and non-standard correlations are not easily distinguished, the existence of NSCs is the main hurdle to inferring logically the molecular structure from the 2D-NMR data. If the 2D-NMR data contain both undistinguishable *standard* and *non-standard* correlations, then the total set of "axioms" derived from the 2D-NMR data will obviously contain contradictions. This means that the correct structure cannot be inferred from these axioms and in this case the structural problem either has no solution or the solution will be incorrect: the set of suggested structures will not contain the "genuine" structure.

Unfortunately, as yet there are no routine NMR techniques that distinguish between 2D-NMR signals belonging to standard and non-standard correlations. In some fortunate cases, the application of time-consuming INADEQUATE[22] and 1,1-ADEQUATE[22] experiments, and also H2BC[23,24] experiments, is expected to help to resolve contradictions, but these techniques are also based on their own axioms, which can be violated. Nevertheless, it has been shown[25,26] that application of HMBC in combination with 1,1-ADEQUATE data acquired by means of CryoProbe technology could dramatically alleviate CASE in the presence of NSCs.

9.2.3 Structural Hypotheses Necessary for the Assembly of Structures

When chemical shifts in 1D- and 2D-NMR spectra are assigned and all COSY and HMBC correlations are transformed into *connectivities* between skeletal atoms in the molecular framework, then feasible molecular structures should be assembled from "strict fragments" (suggested on the basis of the 1D-NMR, 2D-COSY, MS and MS/MS fragment ion data and IR spectra, in addition to those postulated by the researcher) and "fuzzy fragments" determined from the HMBC data. To assemble the structures, it is necessary to make a series of logically consistent decisions, equivalent to constructing a set of hypotheses ("axioms"). At least the following choices should be made:

- Allowable chemical composition(s): CH, CHO, CHNO, CHNOS, CHNOCl, *etc.*
- Possible molecular formula (formulae) as selected from a set of possible accurate molecular masses.[27–29] The postulation of a molecular formula is crucial for assembling structures.
- Possible valences of each atom having variable valence: N (3 or 5), S (2, 4 or 6), P (3 or 5). If ^{15}N and ^{31}P spectra are not available then, in principle, all admissible valences of these atoms should be tried.
- Hybridization of each carbon atom: sp; sp^2; sp^3; *not defined.*
- Possible neighborhoods with heteroatoms for each carbon atom: *fb* (forbidden), *ob* (obligatory), *nd* (not defined).
- Total number of hydrogen atoms attached to carbons that are the nearest neighbors to a given carbon (determined, if possible, from the signal multiplicity in the ^{1}H NMR spectrum or from a multiplicity-edited 2D-NMR spectrum).
- Maximum allowed bond multiplicity: 1 or 2 or 3. The main challenge relates to the triple bond. Strictly, it can be solved reliably only based on vibrational spectra (IR/Raman).
- List of fragments that can be assumed to be present in a molecule according to chemical considerations or based on a spectral fragment search in some fragment database (DB). The presence of the most significant functional groups (C=O, OH, NH, C≡N, C≡C, C≡CH, *etc.*) can be suggested from both IR and Raman spectra.
- List of fragments that are *forbidden* within the given structural problem. These include fragments unlikely in organic chemistry: for example, a triple bond in small rings, *trans* double bonds in small rings or an O–O–O connectivity. IR and Raman spectra can also hint at the specification of forbidden fragments and the axiom $\overline{X}_j \rightarrow \overline{A}_i$ is usually a fairly reliable basis for making a particular decision. For example, if no characteristic absorption bands are observed in the region 3100–3700 cm^{-1}, then an alcohol group will be absent from the unknown.

It should be evident that at least one poor decision based on the points listed above would likely lead to a failure to elucidate the correct structure.

If we generalize all axioms and hypotheses forming the partial axiomatic theory of a given molecule structure elucidation, then we will arrive at the following properties of initial information that should be logically analyzed:

- Information is *fuzzy by nature*, *i.e.* there are either two or more carbon–carbon bonds between pairs of H-*i* and C-*k* atoms associated with a two-dimensional peak (δH-*i*, δC-*k*) in the HMBC spectrum.
- Not all possible correlations are observed in the 2D-NMR spectra owing to steric factors, *i.e.* information is *incomplete*.
- The presence of NSCs frequently results in *contradictory* information.
- The number of NSCs and their lengths are *unknown* and signal overlap leads to the appearance of *ambiguous* correlations. Information is otherwise *uncertain*.
- Information can be *false* if a mistaken hypothesis is suggested.
- Information contained within the "structural axioms" reflects the opinion and bias of the researcher and the information is, therefore, *subjective* and typically based on synthetic or biosynthetic arguments.

Consequently, a 2D-NMR-based CASE system should be capable of processing *fuzzy*, *contradictory*, *incomplete*, *uncertain*, *subjective* and even *false* spectrum–structural information. The StrucEluc system was developed to meet these requirements.

9.3 General Principles of the CASE Systems

In order to obtain maximum information about the structure of an unknown molecule, different kinds of spectra, normally MS, NMR, IR, Raman and UV/VIS, are used. In this case, the molecule under analysis acts like a specific cipher machine that codes structural information into each kind of spectrum using its own code. The goal of a researcher is to crack these codes and extract the maximum structural information achievable. Figuratively, the CASE problem can be formulated in the following way: create a decoding machine capable of interpreting *as completely as possible* the structural information contained in the available spectra for an unknown. The structural information is known to be coded into different types of spectra at different levels of complexity. For instance, MS can produce spectra containing a lot of structural information, but extraction of the details can be very complicated. Nevertheless, MS can deliver the accurate molecular mass, which often translates directly to a molecular formula – a key parameter necessary for molecular structure elucidation. Ultimately, the molecular mass is a carrier of structural information.

An IR spectrum can provide valuable information about the presence or absence of certain functional groups but communicates very little about their environment in the molecule. The richest structural information can be extracted from NMR spectra, since the environment of a given magnetically active nucleus (^{1}H, ^{13}C, ^{15}N, *etc.*) can be revealed through the chemical shift and the spin–spin couplings with neighboring nuclei. NMR (especially 2D-NMR) spectra are therefore considered as a primary source of structural information.

The main idea on which the CASE approach is based can be easily explained starting from the nature of isomerism. Figure 9.1 displays the structures of a series of known small organic molecules and the numbers of potential structural isomers *N* calculated by our group.[30] The figure shows that even the simplest structures can theoretically have hundreds of billions and even trillions of isomers. The *N* value associated with the structures of medium-sized organic molecules can be estimated as about 10^{20}–10^{30} isomers (on the order of Avogadro's number). Although the number of isomers is huge, those corresponding to a given molecular formula do make up a *countable* (at least in principle) and *finite* set. We can conclude that the general CASE strategy utilizes processes to eliminate "superfluous" isomers from the full isomer set by imposing different structural constraints produced from the molecular spectra and *a priori* information (sample origin, chemical rules, *etc.*). A successful result depends on the screening and rejection of *N* – 1 structural formulae that do not comply with the experimental data and systematic constraints applied. It is important to note that the described strategy of structure elucidation allows one to relate this problem to the class of so-called inverse problems.[31,32]

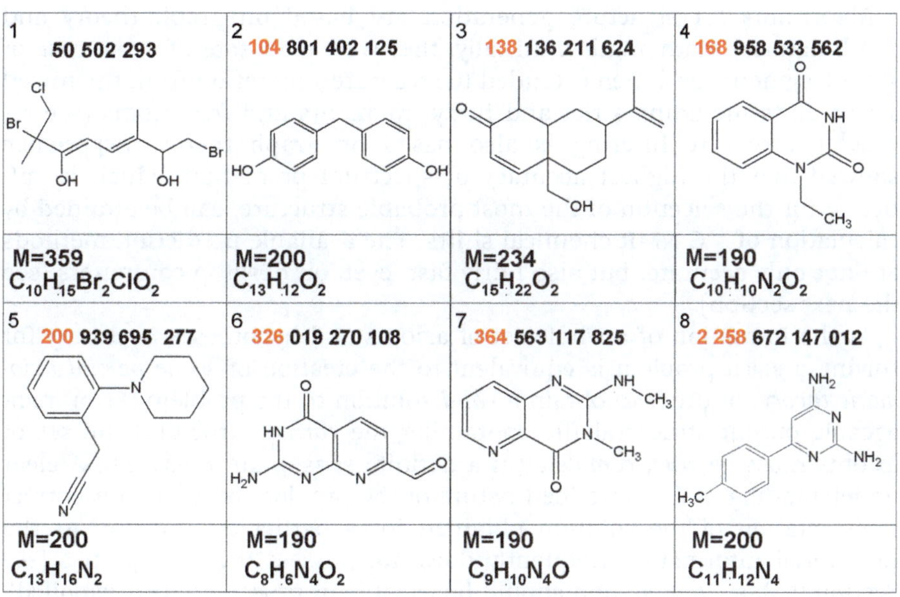

Figure 9.1 The structures of some small organic molecules and the theoretical numbers of isomers (*N*) corresponding to their molecular formulae.[30]

Direct mathematical isomer generation and isomer screening using different constraints are practically impossible as the number of isomers and the associated processing time would be so huge that the problem could not be solved, a situation referred to as a "combinatorial explosion." To reduce the dimension of the problem, it is necessary to introduce *molecular fragments* that "absorb" a significant number of the skeletal atoms. Therefore, to create a CASE system, it is necessary to elaborate algorithms capable of solving the following four main problems in series:

1. Detection of appropriate fragments from the molecular formula (MF) and available spectra.
2. Generation of all isomers from fragments and free atoms provided that all constraints are satisfied.
3. Filtering out isomers whose structures contradict the observed characteristic spectral features and chemical rules.
4. Selection of the most probable structure using prediction of spectra and properties for all structural hypotheses.

For 1D-NMR and IR spectra, fragment detection can be automated using the "characteristic features" concept. This concept has been shown[5,17,18] to be amiable to automation on the basis of mathematical logic *via* establishing the main axioms and hypotheses used for the spectral interpretation. In the case of 2D-NMR, axioms and hypotheses of 2D spectroscopy allow one to derive a set of *strict diatomic* fragments from COSY correlations and a set of *fuzzy* 2–3 fragments from HMBC data.

Algorithms for structure generation are based on graph theory and combinatorial mathematics. Initially, they were elaborated for the case of strict fragments and then extended to structure generation from the mixed set of units including strict and fuzzy fragments and free atoms (see review[12]). Structure filtering is also based on graph theory. Experience showed that the highest accuracy of spectrum prediction, which is sufficient for the selection of the most probable structure, can be attained by calculation of ^{13}C NMR chemical shifts. The available prediction methods are not only accurate, but also fairly fast, even on desktop computers (see the next section).

As the formation of a whole set of axioms and hypotheses necessary for solving a *given* problem is equivalent to the creation of some *partial axiomatic theory*, in order to obtain a *valid* solution to the problem (*i.e.* a manageable output structural file containing the *correct* structure) the set of axioms must be *true*, *complete* (in a definite sense) and *consistent*. A clear understanding of the described nature of the problem is crucial for correct interpretation of the solution obtained for a particular problem. As the real initial information frequently does not possess these properties (see Section 9.2), a CASE system should be capable of processing this "spoiled" information to provide a valid solution to the problem. We will show that StrucEluc adheres to this condition.

As discussed previously, the procedure of structure elucidation is essentially reduced to imposing different structural constraints on a set of conceivable structures. In general, structural constraints can be arbitrarily set in two forms: affirmative (positive) and negative. Among the affirmative constraints are the specification of the hybridization of carbon atoms, the obligatory neighborhoods of some carbon atoms with heteroatoms, the enumeration of fragments that can (or must) be present in the molecule, the specification of the permissible sizes of cycles, *etc.* Negative constraints form a system of prohibitions: the prohibition of neighborhoods with certain heteroatoms, the prohibition of the presence of particular fragments, sizes of rings, specific bond orders, *etc.* The requirement of the best match between the calculated spectrum of the expected structure and the experimental spectrum can be considered as the most rigid constraint. Calculated spectra impose constraints not only on characteristic spectral features, but also on all spectral features without exception. ^{13}C NMR spectra are known to be more informative than ^{1}H NMR spectra. However, their combined use yields a synergistic effect and is especially pronounced in 2D-NMR spectra. Both ^{13}C and ^{1}H calculated spectra are used for selecting the "best" structure.

It is worth noting that "negative" structural constraints implied by characteristic spectral features are commonly the most informative ones. Indeed, as was mentioned above (Section 9.2.1), both implications $A \rightarrow X_j$ and $\overline{X}_j \rightarrow \overline{A}_i$ are true, whereas implication $X_j \rightarrow A_i$ may be either true or false. For example, the absence of signals in the region 150–200 ppm in the ^{13}C NMR spectrum suggests with a high probability that the carbonyl group is absent in the molecule, whereas the presence of a signal in this region can also be accounted for by the presence of other groups (C=N, C=S, C=C–O, *etc.*). This circumstance is effectively used at the output file filtering stage. Molecular fragments along with their characteristic spectral ranges in NMR spectra form a set of filters. These fragments are searched for in each generated structure and the structures containing fragments that are not confirmed by the spectra are excluded from the output structural file.

The four stages of CASE enumerated above and suggested in the 1970s have essentially remained valid until today, despite the fact that the algorithms have been continuously varied and improved during the last 40 years and 2D-NMR spectra have become the main source of structural constraints (instead of SSCs).

9.4 Methods of NMR Spectral Prediction

Depending on the rigorous nature of the structural constraints imposed by the experimental data, the output file of generated structures from the ES may contain tens, hundreds or even tens of thousands of structural formulae. A correct structure cannot easily be distinguished by taking into account changes in the characteristic spectral features of the functional groups and fragments existing in the probable structures. Therefore, the selection of the most probable structure is carried out by comparing experimental to predicted spectra and this step is generally the conclusion of the ES workflow.

1D-NMR spectral prediction has been available to the chemical community for a number of years. ^{1}H and ^{13}C spectra are the primary analytical techniques utilized by chemists for structure verification. ^{1}H NMR is used with at least a 20 : 1 ratio over direct detection ^{13}C spectroscopy[12]. The development of NMR prediction tools has therefore focused on ^{13}C and ^{1}H nuclei, although chemical shift calculation for ^{15}N, ^{31}P and ^{19}F nuclei can also be performed.

In general, the methods of chemical shift prediction can be divided into two categories: quantum mechanical (QM) and empirical. QM methods are slow (at least several hours per structure) and are not amenable to full automation. Obviously they cannot be applied to the NMR spectrum prediction of large structural files, which is common for ES or for large molecules. Empirical methods combine high speed of calculation with fairly high accuracy; therefore, empirical approaches are used in ES for the selection of the most probable structure. The relative performance of empirical and QM methods was considered in comparison in our work.[33]

The prediction of NMR chemical shifts to facilitate the batch analysis of spectra has been reported by a number of workers.[34–37] Applications have been developed to perform analysis on combinatorial plates of data.[38] High-throughput analysis of both 1D- and 2D-NMR has also been validated.[36,39]

There are three widely used procedures for predicting NMR spectra. The first consists of the construction of linear empirical models based on additivity rules (*incremental approach*).[40–43] The second assumes the application of prediction algorithms that employ data collected within spectral databases (*fragmental approach*).[44–49] These methods are present in a series of commercially available programs.[44,48,50,51] The third method for ^{13}C NMR chemical shift prediction uses *artificial neural networks* (ANNs)[52] and has been reported, for example, for a series of studies.[53–58]

Grant and Paul[40] suggested the first additive linear model for calculating the chemical shifts of carbon atoms in aliphatic hydrocarbons using increments accounting for environmental effects up to four atoms away. Following the initial groundbreaking work, linear models were extended to some classes of organic compounds.

Fürst and Pretsch[41] used large databases containing structures and their associated assigned ^{13}C NMR spectra to construct linear models that could be applied to many classes of chemical compounds. The models contain both configuration- and conformation-dependent parameters and take into consideration the configuration of C=C bonds and also the presence of axial and equatorial substituents in cyclohexyl rings. The software developed also allows the modification of parameters, both reference values and increments, and the input of new additivity rules.

In the fragmental approach, databases containing chemical structures and their assigned carbon chemical shifts form the foundation data set for the derivation of prediction algorithms. For every carbon atom in each chemical structure contained in the database, atom-centered fragments (ACFs) with a prescribed number of concentric layers are generated according to the HOSE code (Hierarchical Ordering of Spherical Environments[59]). These fragments and their corresponding chemical shifts are stored as an ordered list for use in the prediction algorithms. To predict the spectrum of a candidate structure, the program selects all possible ACFs present in a structure, performs a search for their analogs in the database and ascribes the chemical shifts taken from the reference fragments to the carbon atoms being predicted. If an ACF is not found in the database, then the program interpolates the chemical shifts using the most similar structural environments available. The results obtained by using this approach are generally in good agreement with the experimental data when using a large database containing a diversified set of structures. Frequently, the difference between individual predicted and experimental chemical shifts lies within 1 ppm limits, delivering high prediction accuracy.

Despite the impressive accuracy reported for this approach, it does suffer some drawbacks. The absence of stereochemical information in databases containing only connection tables without explicit stereochemistry defined can strongly affect the prediction quality. As has been shown,[60] neglecting stereochemical effects can increase the deviations between predicted and experimental chemical shifts to more than 10 ppm. This problem was circumvented to some extent by Schutz *et al.*,[61] who introduced three-dimensional descriptors to modify the HOSE code. It should be noted that the spectral properties of the reference fragments used to derive the prediction may appear to be unrelated in certain cases, but this is simply the nature of the approach.

A number of commercially available ^{13}C chemical shift prediction software packages based on the fragment database approach have become available in recent years. The most popular products to date are those of ACD/Labs (Advanced Chemistry Development),[48] Chemical Concepts,[44] Upstream[51] and Sadtler.[62,63] The authors are familiar with the ACD/Labs product suite and these products are used as examples in further discussions.

When a new structure is drawn in the structure drawing interface of ACD/CNMR, the program automatically splits the structure into a set of unique fragments that are then compared with the structural fragments from the internal database.

- If a fragment from the drawn structure *coincides* with a fragment contained within the database, the program will use its experimental δ_C value as part of the final set of chemical shifts for the structure. For such δ_C values, the program will not show confidence intervals in the table of chemical shifts. The program utilizes a reference structure up to 16 spheres in depth for a particular carbon atom. As a result, the size of the fragment is defined by the size of the largest fragment common to both the predicted and the reference structure, the fragments being centered on the given carbon atom.

- If some fragments from the structure *cannot be found* in the internal database, then the program will search for the most similar fragments in the database. First, the program composes sets of fragments from the database that are structurally similar to each of the fragments generated from the analyzed structure. Second, the program estimates δ_C values for the fragments using secondary algorithms and compares them with the estimated δ_C values of fragments selected from the database. This second step allows the program to narrow down to a set of similar fragments from the database. Third, the program calculates both the average values (δ_{Av}) of the experimental data and produces estimated δ_C values after application of the second criterion described above. The resulting δ_C value is calculated using both the estimated δ_C value of the given fragment and the average δ_{Av} values. The obtained δ_C values are used to compose the final set of chemical shifts for the structure. After composing the final lists of chemical shifts, ACD/CNMR composes and diagonalizes the spin Hamiltonian matrices to generate the exact number, location, intensities and assignment of the spectral lines associated with the structure.

The array of chemical shift, coupling constants and line width parameters describing an NMR spectrum are influenced by many external factors, including solvent, concentration, temperature, relaxation times, concentration of paramagnetics, shimming and observation frequency, to cite just a few. Many of these parameters are simply too complex to take account of during a prediction, but certainly solvent dependence can be accounted for to a certain extent. ACD/CNMR Predictor[48] provides the ability to perform solvent-specific prediction. The user can select from a list of common NMR solvents and predict a solvent-specific NMR spectrum. The stereochemistry of a particular structure is crucial in determining the molecular properties, and when the stereochemistry of an atom is included in the submitted chemical structure, the information is utilized during the prediction process.

Beginning in the early 1990s, the attention of chemists was drawn to the possibilities of promising new mathematical tools developing in computer-based chemistry, *e.g.*, ANNs. There was a rapid increase in the number of studies on the application of ANNs to the interpretation, classification and prediction of spectral data, including NMR chemical shift prediction.

A neural network can be considered as a simplified computer model of the human brain, consisting of several layers of neurons that send signals to other neurons as a function of the input signals received. Such networks have a "black box" nature and possess the common ability to construct empirical models of the systems for which theoretical dependences between the input and output are too complicated or are unknown. Models are obtained as a result of network training. In the course of training, the network is represented in the form of input–output pairs related by a simulated transformation. A network trained in this manner is able to predict the output signals from input signals not originally contained in the training set.

The training procedure may be time consuming (tens of hours), but a network, once trained, generates a prediction result almost instantaneously. For instance, a network can be trained to generate structural information (output) retrieved from a spectrum (input) or to predict a spectrum (output) from structural information (input). The theory of ANNs and examples of their application in chemistry and spectroscopy have been described.[52]

ANNs are trained to predict the NMR spectra of compounds belonging to classes defined *via* a training set, including encoded structures and their associated spectra. In the course of training, the network uses reference structures as input information and the output signals are compared with the NMR spectra of these structures. The training process is complete if the deviations of the predicted spectra from the reference set are less than a chosen threshold. Algorithms of ^{13}C chemical shift prediction were first elaborated by Meiler *et al.*[53]

Two fast calculation algorithms[64–66] were developed by the authors, one of which is based on additivity rules and the other employing ANNs. These algorithms provide a calculation speed of 3000–10 000 ^{13}C chemical shifts per second with an average deviation between calculated and experimental chemical shifts of $d = 1.6$–1.8 ppm. The maximum calculation speed is achieved using the incremental approach. For a file containing tens of thousands of structural isomers, the calculation time by either of the two methods is no longer than a few minutes. Both algorithms are implemented in the StrucEluc system and their high speed and accuracy have strongly influenced the CASE strategy.[16] The third algorithm included in the set of system predictors is based on a fragmental method,[12] for which a database containing 355 000 structures with assigned ^{13}C and ^{1}H chemical shifts is used. Although the fragmental method is not as fast as the other two, it allows the user to obtain a detailed explanation of how each predicted chemical shift was calculated. For each atom within the candidate structure, the related structures used for the prediction can be shown with their assigned chemical shifts, allowing the user to understand the origin of the predicted chemical shifts. All three methods can be used for ^{1}H, ^{13}C, ^{15}N, ^{19}F and ^{31}P NMR chemical shift prediction and all of them are implemented within the StrucEluc software program.

9.5 Expert System Structure Elucidator

The expert system Structure Elucidator (StrucEluc) was developed towards the end of the 1990s. For the last decade, it has been in a state of ongoing development and improvement of its capabilities. The areas of focused development were determined by solving many hundreds of problems based on the elucidation of structures of new natural products. The different strategies for solving problems using StrucEluc, and also the large number of examples to which we have applied the system, have been reported in numerous publications and have been reviewed.[12,16] A very detailed description of the system can be found in a review,[12] and we will not repeat that analysis here. Rather, in this section we will give a short explanation of the algorithms underpinning the system and also specifying the various operational modes that provide a high level of flexibility to the program.

Generally, the purpose of the system is to establish topological and spatial structures, in addition to the relative stereochemistry of new complex organic molecules from high-resolution mass spectrometric (HRMS) and 2D-NMR data. Mass spectra are used to determine the most appropriate molecular formula for an unknown. The availability of an extensive knowledgebase within StrucEluc allows the application of spectrum–structural information and experience accumulated by chemists and spectroscopists when solving the task of CASE.

9.5.1 Knowledgebase of the StrucEluc System

The knowledge of the system can be divided into two segments containing factual and axiomatic knowledge. The factual knowledge consists of a database of structures (420 000 entries) and a fragment library (1 700 000 entries) with the assigned ^{1}H and ^{13}C NMR spectra (subspectra). There is also a library containing >355 000 structures and their assigned ^{13}C and ^{1}H NMR spectra used for the prediction of ^{13}C and ^{1}H chemical shifts from input chemical structures and using the fragment (HOSE code-based) approach.

The axiomatic knowledge includes correlation tables for spectral structural filtering by ^{13}C and ^{1}H NMR spectra and an Atom Property Correlation Table (APCT). The APCT is used to suggest automatically atom properties (hybridization, possibility of neighboring with heteroatoms, *etc.*). A list of fragments that are unlikely for organic chemistry (a permanent BADLIST) can also be related to axiomatic knowledge of the system.

The reliability of this axiomatic knowledge was thoroughly checked. Filtering of both correlation tables through the database subset containing 280 000 structures showed that 98% of structures passed through the verification procedures.[15] A general flow diagram of StrucEluc is shown in Schemes 9.1 and 9.2.

9.5.2 Molecular Connectivity Diagram (MCD)

The molecular formula or accurate molecular mass of the analyzed compound, the HSQC, HMBC and COSY spectra and the 1D ^{13}C and ^{1}H NMR spectra are generally used as initial data, if they are all available. If the ^{13}C NMR spectrum cannot be recorded because of concentration or time limitations, then the program will attempt to create a spectrum from the 2D-NMR data. When the molecular mass is input, the program determines the molecular formula or the most probable formulae and all of them can then be checked by the system. To establish the relative stereochemistry of the molecule, either NOESY or ROESY spectral data are used. The application of ^{13}C NMR prediction is also helpful for this purpose (see Section 9.5.5).

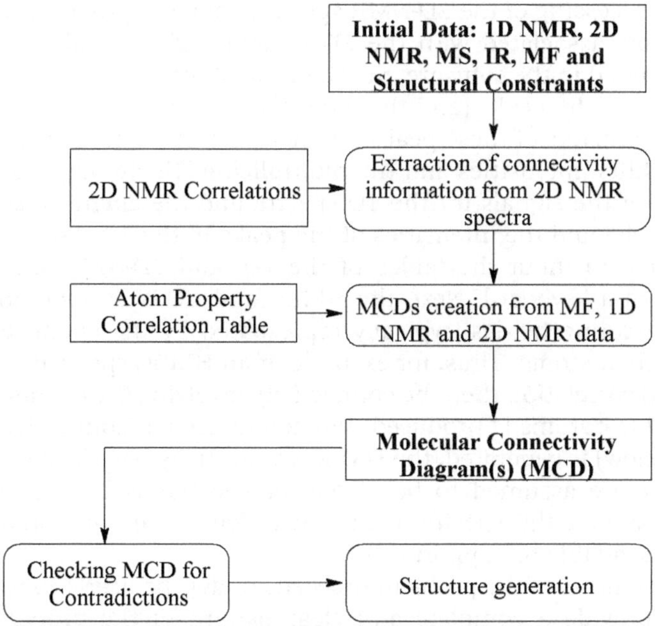

Scheme 9.1 The Common Mode of structure generation contained within the StrucEluc system. Depending on the results, the system continues the process as shown in Scheme 9.2.

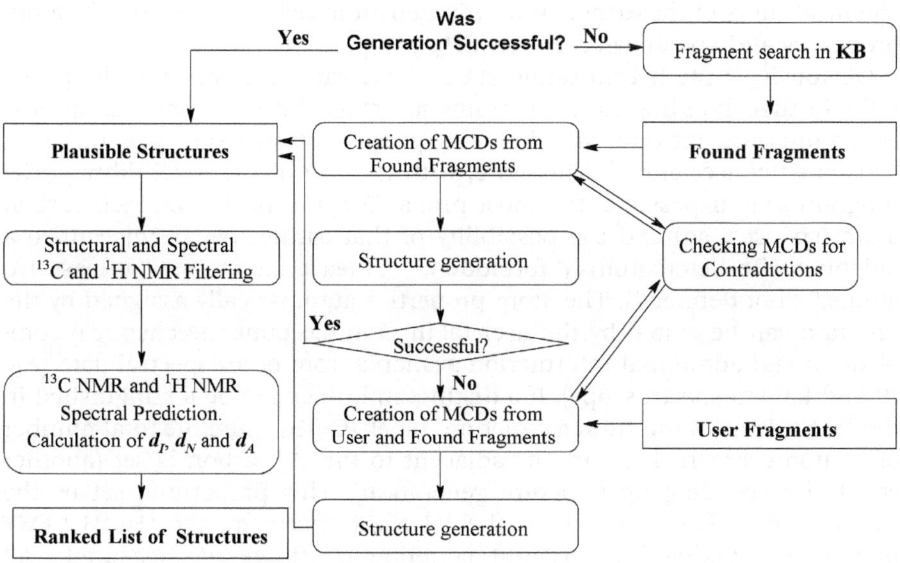

Scheme 9.2 The possible stages of the process of structure elucidation depend on the results of structure generation in the Common Mode. If the Common Mode fails, StrucEluc initiates the Fragment Mode of generation. The symbols d_I, d_N and d_A denote the average deviations between the experimental and predicted NMR spectra calculated by the different methods (see Section 9.5.3.2).

During processing of the 2D-NMR spectral data, the program analyzes the contour plots associated with the 2D spectra and determines, to specific criteria encoded in the software, the chemical shifts of the interacting nuclei represented by the peaks (and therefore the coordinates of the peaks). The spectral parameters of these peaks are then imported into tables containing chemical shifts, intensities and the multiplicities (including those for ^{1}H if measured) of the signals for the 1D spectra and the chemical shifts of the coupled nuclei and the intensities of the peaks in the 2D-NMR spectra. It is also possible to input the tables of the 1D- and 2D-NMR spectral peaks directly from a keyboard. Next, the HMBC and COSY correlations are converted into connectivities, typically represented by the chemical shifts of pairs of carbon atoms. Thus, for example, if an HMBC spectrum exhibits an (H-*i*)→(C-*k*) correlation then the connectivity involving the chemical shifts of the C-*i* and C-*k* atoms is produced. When a molecular connectivity diagram (defined below) is generated, the HMBC connectivity lengths between atoms C-*i* and C-*k* are assumed to be of one or two bonds by default, but the chemist may edit the specific connectivity lengths if some additional information available to support this.

Further solution of the problem proceeds under the user's control in most cases. To provide a complete and clear pattern of the properties of the skeletal atoms and the connectivities between them, the program places skeletal atoms together with hydrogen atoms attached to the skeletal atoms in a display window. We refer to this visual depiction as a *molecular connectivity diagram* (MCD) (see the example in Figure 9.2). The values of the chemical shifts of the carbon and hydrogen atoms are accompanied by atom properties and are shown for each CH_n group.

Obviously, if the hybridization state of the carbon atoms and the possibility of their bonding to heteroatoms are taken into account (*i.e.* specific constraints are introduced), then the process of structure generation is substantially accelerated. Therefore, with the use of the APCT library, the program sets, if possible, the most probable hybridization of each carbon atom (sp^3, sp^2, sp) and the possibility of that carbon being adjacent to a neighbor with heteroatoms ("forbidden," "at least one atom," "at least two atoms," "not defined"). The atom properties automatically assigned by the program can be edited by the user taking into account the chemical composition and additional information available from other spectral data (*e.g.* IR and Raman spectroscopy). If a distinct multiplet can be distinguished in the ^{1}H NMR spectrum from a structural block (C-*i*)H_n, then the total number of H atoms attached to carbons adjacent to the C-*i* carbon is set (another constraint speeding up structure generation). This property is set by the chemist after visual analysis of the ^{1}H NMR spectrum, the ^{1}H–^{1}H COSY pattern and taking into account coupling constants (if measured). All structural constraints presented in the molecular connectivity diagram are used during structure generation. Note that a group of carbon atoms showing a chain of COSY connectivities between them makes up a fragment (a connected subgraph), while each carbon atom taken together with others

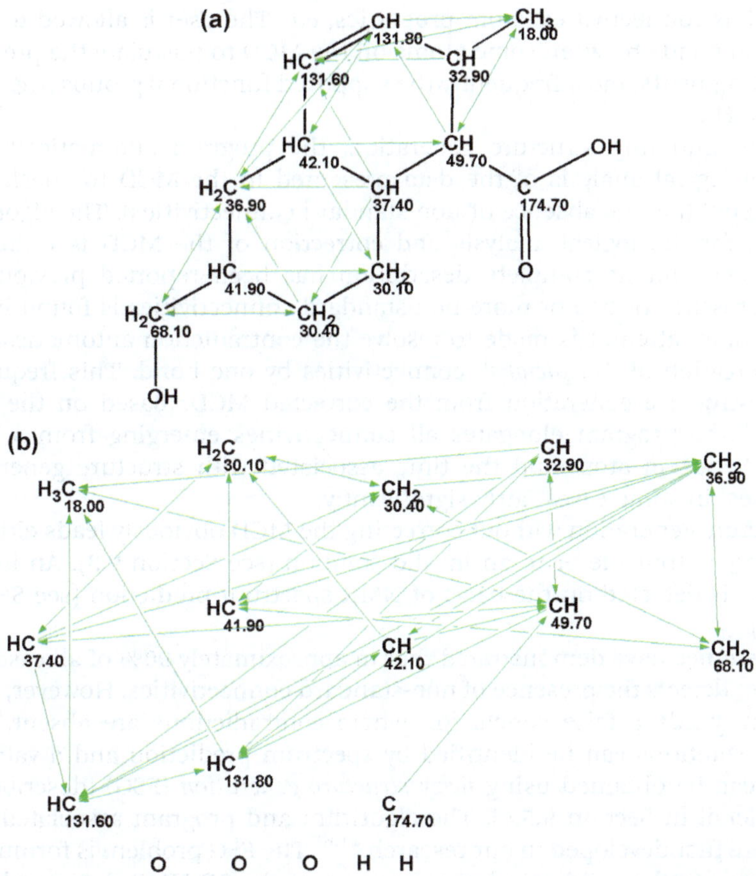

Figure 9.2 An example of a structure (a) and the associated Molecular Connectivity Diagram of HMBC connectivities (b). In the structure, the HMBC connectivities are shown by arrows. On the structure it is shown that the 131.6–18.0 and 36.9–131.8 connectivities are non-standard (extending out more than three bond correlations).

connected to it *via* HMBC connectivities forms a "fuzzy fragment." Such "fuzziness" emphasizes that the distances between a central C atom and its neighboring carbons, even though they are limited by not more than by two bonds by default, are not necessarily strictly defined.

9.5.3 Structure Generation and Verification

9.5.3.1 Common Mode of Structure Generation

In StrucEluc, two main modes are provided for structure generation: *Common Mode* and *Fragment Mode*. The Common Mode is used most frequently. In this mode, structure generation is performed from the structural blocks C, CH, CH_2, CH_3 and heteroatoms when constraints are imposed that are entered as connectivities, atom properties, *etc.* The user is allowed to draw chemical bonds between some atoms on the MCD to postulate the presence of any fragments, most frequently the supposed functional groups (*e.g.* C=O, O–H, N–H).

Before initiating structure generation, the program automatically performs a logical analysis of the data presented in the MCD to check their consistency (*i.e.* the absence of non-standard connectivities). The algorithm utilized for the logical analysis and correction of the MCD is rather sophisticated and its complete description has been reported previously.[21] If the presence of one or more non-standard connectivities is found by the program, an attempt is made to resolve the contradiction automatically by the elongation of "*suspicious*" connectivities by one bond. This frequently allows structure generation from the corrected MCD. Based on the algorithm,[21] the program elongates all connectivities emerging from a "suspicious" carbon atom and the time associated with structure generation increases, in some cases fairly significantly.

Structure generation without correcting the MCD obviously leads either to an empty output file or to an invalid solution (see Section 9.3). An invalid solution is detected on the basis of NMR spectrum prediction (see Section 9.5.3.2).

Our studies have demonstrated that in approximately 90% of all cases the program detects the presence of non-standard connectivities. However, if the program yields a false conclusion where contradictions are absent, then invalid solutions can be identified by spectrum prediction and a valid solution can be obtained using *fuzzy structure generation* (FSG) (described in more detail in Section 9.5.4). The algorithm and program associated with FSG were first developed in our research.[21,67] The FSG problem is formulated as follows: find a valid solution provided that the 2D-NMR data involve an unknown number m ($m = 1$–15) of non-standard connectivities and the length of each of them is also unknown. The efficiency of this proposed approach is discussed in Section 9.5.4.

Structure generation is controlled by options that impose constraints on the sizes of the rings within the molecule, the bond orders, lists of *obligatory* and *forbidden* fragments and include a check for the fulfillment of Bredt's rule, *etc.* The use of the APCT library for setting the atom properties (see above) significantly accelerates the structure generation process. In particular, for 80% of >300 problems that we have solved, the generation time was less than 1 min.

It is important to note that during the structure generation process, chemical bonds are set between atoms possessing definite chemical shifts and properties specified in the MCD (not between abstract atoms!). Therefore, in the generated structures, N (if ^{15}N NMR data are available), C and H atoms already have assigned chemical shifts. The spectral filtration of structures (see Sections 9.3 and 9.5.1) proceeds simultaneously with generation and three modes of filtration severity can be specified, taking into account the ambiguity of the boundaries of characteristic spectral ranges.

It has been found that filtration in even the most relaxed mode decreases the number of structures in the output file by a factor of 10 or even up to 100.[15]

9.5.3.2 Selection of the Most Probable Structure

For the correct elimination of duplicates and to choose the most probable structure, the prediction of ^{13}C NMR spectra and the calculation of the average deviations of the calculated spectra from the experimental data are used in the StrucEluc system. These procedures are performed in following three stages.

1. ^{13}C chemical shift calculation is performed for the full output file using the incremental algorithm[64,66] implemented in StrucEluc and the average deviations d_I between experimental and predicted chemical shifts are calculated. As noted above, even for a file containing tens of thousands of structural isomers, the calculation time is not longer than a few minutes. Next, redundant identical structures are removed. Since different deviations correspond to duplicate structures with different signal assignments, the structure with the minimum deviation is retained from each subset of identical structures (*i.e.* the "best representatives" are selected from each family of identical structures). NOESY correlations can also be used for selecting the best generated structures at this stage. The structure candidates are then ranked by ascending average deviation d_I.
2. A ^{13}C chemical shift calculation based on the ANN approach is applied to the reduced and ordered output structural file. Structures are reordered again in ascending order of d_N deviations, which refines the position of a correct structure in the output file. Our experience has shown that the correct structure frequently is in first place with the smallest chemical shift deviation or at least is among the first several structures at the beginning of the list.
3. A ^{13}C chemical shift calculation is carried out using the HOSE-based approach for n ($n = 10$–50) top structures of the file ranked in ascending order of d_N deviations. Then the calculated n structures are ranked again in ascending order of d_A deviations ($d_A = d_{HOSE}$) and further refinement of the position of the correct structure is carried out. As noted above, although the fragmental method is not as fast as the incremental and ANN methods, it does allow the user to obtain a detailed explanation of how each predicted chemical shift was calculated.

If the difference between the deviations calculated for the first- and second-ranked structures is small [$d(2) - d(1) < 0.2$ ppm] then the final determination of the structure is performed by the expert. In so doing, additional experiments may be required. Generally, the choice is reduced to between two or, less frequently, three structures.

In difficult cases, the ^{1}H NMR spectra can be calculated by the fragmental method for a detailed comparison of the signal positions and multiplicities in the calculated and experimental spectra. Solutions that may be invalid are revealed by a large deviation of the calculated ^{13}C spectrum from the experimental for the first structure of the ranked file. For instance, if $d_A(1) > 3$–4 ppm, then it is desirable to check the solution using fuzzy structure generation (see Section 9.5.4). The reduced $d_A(1)$ value found as a result of FSG should be considered as a hint regarding the presence of one or more non-standard connectivities. The correct solution is usually obtained using different modes of fuzzy structure generation.[67] The NOESY spectrum,[68] which imposes constraints on geometric distances between intervening protons, can also give valuable structural information (spatial constraints) in this step. We would expect, however, that either HSQC–TOCSY or ADEQUATE experiments might be more effective than NOESY in such cases. Note that StrucEluc is capable of generating ionic structures in addition to symmetric molecules from 2D-NMR data, for which the algorithm of structure generation was enhanced.

Let us consider an example demonstrating the application of the StrucEluc system to structure elucidation in its Common Mode. Ge *et al.*[69] isolated and determined the structure of an unusual new natural product named hopeanolin ($C_{42}H_{28}O_{10}$). To challenge StrucEluc, we used published 1D- and 2D-NMR data[69] to elucidate the "unknown" structure. The HSQC peak list and 80 HMBC and three COSY correlations were supplied to the program and the MCD was created. Atom hybridization was automatically set for all carbons except eight CH atoms and two quaternary C atoms with chemical shifts in the range 90–120 ppm: the program took into account that chemical shifts observed in this region can be assigned either to C=C or to O–C (O–C–O) carbons. Only one obvious constraint (hypothesis) was added by the user: the sp^2 carbon atom with a chemical shift of 171.4 ppm was marked as having at least one neighboring oxygen. No NSCs were detected in the 2D-NMR data by checking the MCD. The results of the structure generation and filtering were 259 structures generated in 2 min 10 s, 85 structures remained after filtering, and 36 structures were stored after removing duplicates. We denote this as $k = 259 \rightarrow 85 \rightarrow 36$, $t_g = 2$ min 10 s. The ^{13}C NMR chemical shifts were predicted for all structures using all of the fragment, incremental and neural net approaches. The four structures at the top of the structural file ranked with d_A deviation are shown in Figure 9.3, where the "best" structure, No. 1, which has rank $r = 1$ in the ordered file, is hopeanolin. The stereochemistry of this molecule is discussed in Section 9.5.5.1.

9.5.3.3 *The Fragment Mode of Operation*

If the structural restrictions imposed by the MCD are not sufficient for the generation of a reasonable number of plausible structures within an appropriate time, the utilization of molecular fragments has been shown to greatly facilitate the solution of the problem.[14,63] Fragments have been

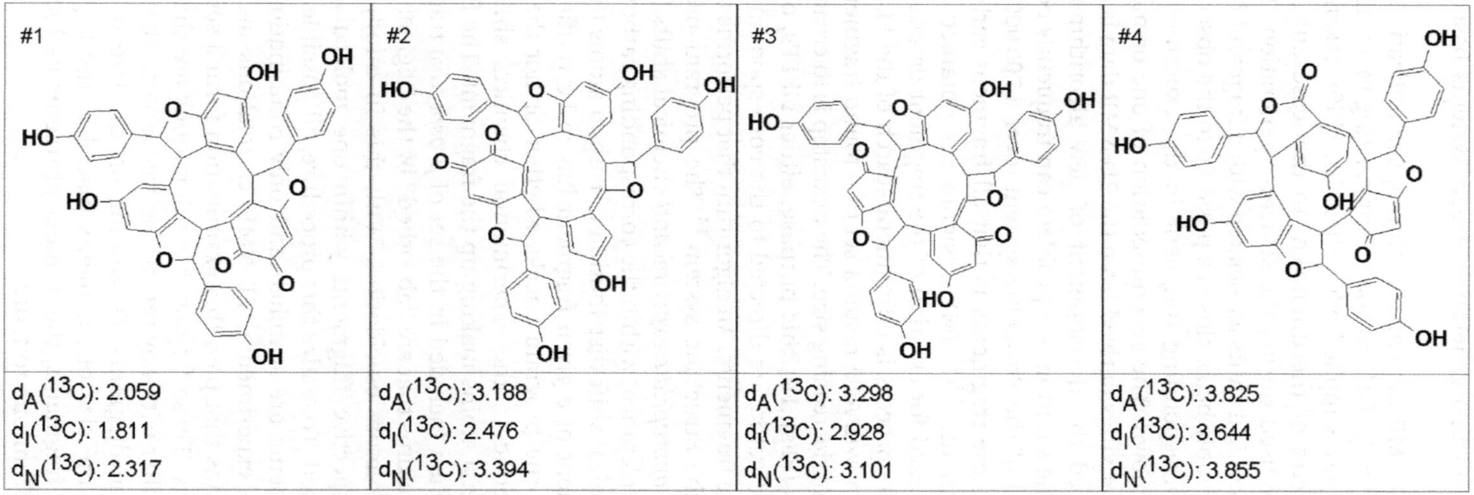

Figure 9.3 The four structures at the top of the ranked structural file. The first-ranked structure No. 1 is identical with the structure of hopeanolin determined by the authors.[69]

successfully used in all first-generation expert systems based on 1D-NMR spectra only.

However, when 2D-NMR data are employed, the utilization of molecular fragments is effective only if *all* carbon atoms existing in a fragment used in structure generation are supplied with chemical shifts taken from the *experimental* NMR spectra of the unknown. In this case only, the 2D-NMR connectivities can be used during the structure generation. Therefore, the *supposed* values of chemical shifts associated with a fragment involved in the elucidation will preferably be as close as possible to the observed values for the atoms of the corresponding fragment in the experimental ^{13}C NMR spectrum of the unknown. The accommodation of one or more fragments within a set of connectivities derived from the 2D-NMR data is a complicated problem that required the development of new algorithms. Appropriate fragments to aid in the solution of a problem can frequently be found in the fragment library (FL) of the StrucEluc system (over 1 700 000 entries). The main advantage of these fragments is that all fragment carbon atoms are already supplied with the ^{13}C NMR assignments obtained from the full structures that were used for creation of the fragment database.

The first step in the process is a fragment search of the FL using the ^{13}C spectrum of the unknown. As a result a set of L found fragments is selected and ranked in order of decreasing size. The next step is to create MCDs using the found fragments (FFs). For this purpose, either all FFs, or any number selected by the investigator, are directed to the corresponding block of the program to utilize the fragments. An algorithm that performs this procedure was developed for the StrucEluc system.[14] The program produces all rearrangements of the *appropriate experimental* chemical shifts (*i.e.* those that meet the postulated tolerance) within the corresponding carbon atoms of the fragment. Each chemical shift distribution of carbon atoms that produces a conceivable assignment of a given fragment has to be verified. During the verification process, the program checks whether or not the carbon atom assignments correspond to the experimental chemical shift correlations comprising the skeletal atoms making up the fragment. The fragments that survive the test are then included in the set of *prospective* fragments.

The more skeletal atoms that are "absorbed" by the fragments, the shorter is the process of structure elucidation. With this in mind, an algorithm combining the prospective fragments within one molecular connectivity diagram was developed. To realize this procedure, all possible combinations of prospective fragments are searched and only combinations that are in agreement with the experimental 2D-NMR correlations are chosen. The fragment combinations that pass this examination form a set of *prospective fragment combinations*. These fragment combinations are then "projected" onto the MCDs together with any remaining free atoms. The user can then visually analyze these diagrams. Depending on the size of the molecule being analyzed and the size of fragments placed at the beginning of the ranked list of found fragments, the number of fragments included into an MCD usually varies from one to four.

The conclusion of all further verification procedures is a check of all of the MCDs produced for the presence of contradictions. The program offers an option that deletes all MCDs that are identified as containing contradictions. Contradictory MCDs contain fragments with carbon atoms associated with assignments that contradict the standard length of the corresponding connectivities. The diagrams remaining after checking can then be used in the structure generation process.

In the process of analyzing a novel compound, it is entirely possible that there will be no *easily detectable* fragments in the database that will reduce the magnitude of the challenge. In such cases, it may help to introduce user-defined fragments (UDFs). The main qualitative difference between an FF and a UDF is that the FF already contains carbon atoms with chemical shifts that have been assigned somehow while the carbon atoms of the UDF have no carbon chemical shift assignments. Two ways to introduce user fragments into the program have been developed:

- calculate the carbon chemical shifts of the fragment;
- search the FL for fragments that *comprise* the user fragment.

It is likely that fragments from at least one of these two sources will be available for use by the program. Experience has shown[13,14,70] that an appropriate combination of FFs and UDFs frequently allows the solution of rather difficult problems.

9.5.3.3.1 Example. A new dimeric natural product, ashwagandhanolide (Figure 9.4a), was isolated by Subbaraju *et al.*[71]

Its molecular formula was determined as $C_{56}H_{78}O_{12}S$ on the basis of the molecular ion observed at *m*/*z* 975.5285. The structure of this compound was determined using 2D-NMR data together with additional information obtained from comparison of experimental spectra with the structures and the spectra of related molecules. In the original article,[71] only 35 HMBC correlations are reported and no COSY data were given. The number of correlations is small owing to severe overlap in the ^{1}H NMR spectrum. An attempt to solve the problem using StrucEluc in Common Mode showed that the processing time and the number of generated structures would be unmanageable. Therefore, a fragment search using the ^{13}C NMR spectrum was performed and 5524 fragments were found in the Fragment Library. As mentioned previously, the displayed Found Fragments are ranked in decreasing order of the number of carbon atoms. The first-ranked fragment, along with the structure of ashwagandhanolide, is shown in Figure 9.4b to illustrate our approach. Visual comparison of the molecular structure with the structure of the fragment confirms that fragment (b) is a substructure of structure (a) and its carbon chemical shifts are very close to the values measured for the full structure. The procedure of creating MCDs from the FFs was initiated and the program produced 960 MCDs with different shift

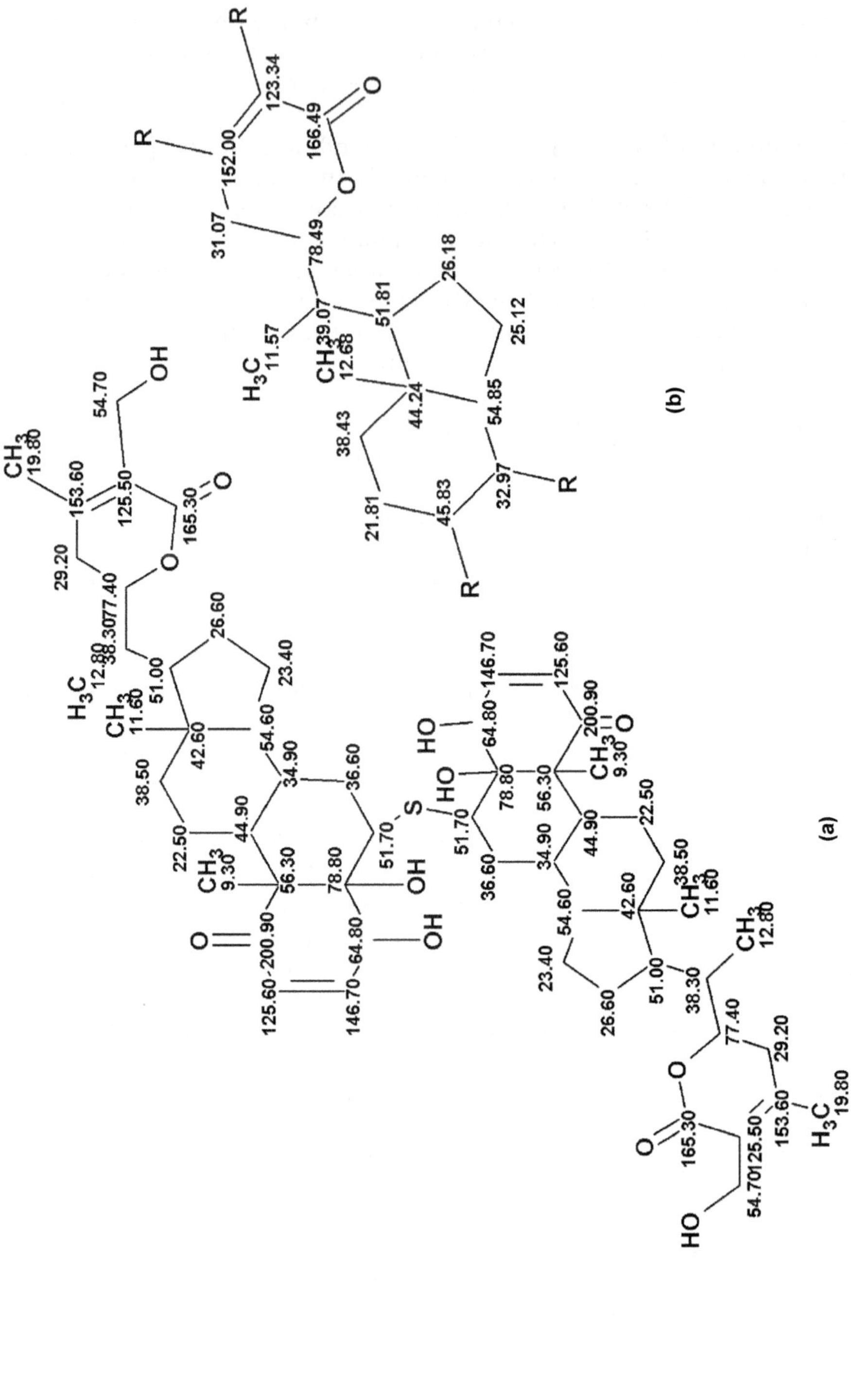

Figure 9.4 The structure of ashwagandhanolide (a) and a Found Fragment (b).

assignments. Checking MCDs for contradictions took 23 min and structure generation from 360 consistent MCDs resulted in $k = 360 \rightarrow 24 \rightarrow 6$, $t_g = 22$ s.

The three top structures in the ranked file are shown in Figure 9.5. The most probable structure, No. 1 characterized by $r = 1$, coincides with the structure of ashwagandhanolide determined by Subbaraju *et al.*[71]

9.5.3.4 *User Fragment Database Application*

Currently, about 90 million compounds have been identified by chemists and well over a quarter of a million new compounds are synthesized, isolated or identified each year. It is possible that many new compounds will have no analogs in the knowledgebase of an expert system. As a result, it is not always possible to find fragments in the system database that will help to elucidate the structure of a compound from an entirely new class of molecules. Investigations have shown[14] that the inability to utilize library fragments was most frequently due to the following issues: (a) the fragments appropriate for a given problem are missing from the knowledgebase; (b) appropriate fragments are found but the number of possible permutations of the carbon atom assignments in these fragments is so large (combinatorial explosion) that the structure generation process is too long; and (c) the molecule under investigation is proton deficient and, as a result, the 2D-NMR correlations do not provide sufficient constraints to produce a manageable output file within a reasonable time.

Our previous studies[13,14,72] have shown that if the methods described above are ineffective, then the creation of a *user database* could permit a solution. The StrucEluc system provides both the algorithms and the capabilities to create user databases and thereby to allow searches for fragments of related compounds. In particular, even if only one compound with a similar structure is known, it can be used successfully for the creation of a user database. With the help of user databases, the system can easily be adjusted for the elucidation of compound classes that are commonly investigated by a given laboratory. Examples of successful utilization of the user database for the structure elucidation of natural products belonging to the *Cryptolepis* family of indoloquinoline alkaloids have been presented in our previous publications[13,14,72] and are discussed in Section 9.6.

9.5.4 Structure Generation in the Presence of NSCs

Numerous computational experiments have allowed us to conclude that if the program detects the presence of NSCs but fails to resolve contradictions in the 2D-NMR data using the appropriate algorithms,[21] then FSG[67] should be used to solve the problem. Moreover, it is quite probable that structure elucidation from 2D-NMR data on the basis of FSG can be considered as a general CASE strategy because it is almost independent of the presence or absence of NSCs in the 2D-NMR data.

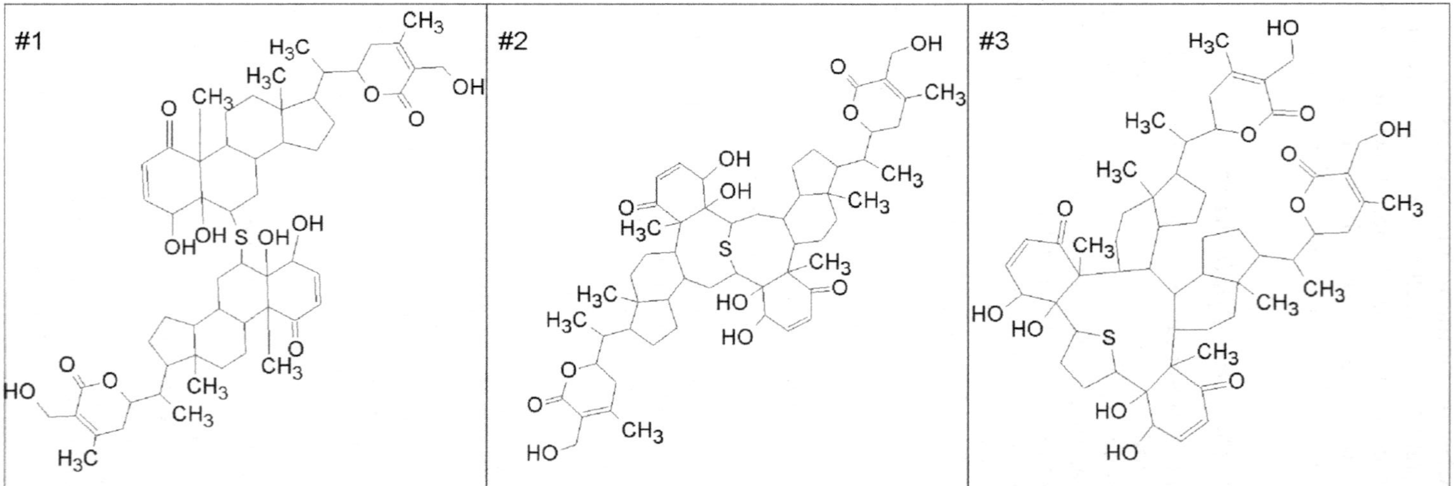

#1	#2	#3
$d_A(^{13}C)$: 2.187	$d_A(^{13}C)$: 2.397	$d_A(^{13}C)$: 2.910
$d_I(^{13}C)$: 2.827	$d_I(^{13}C)$: 3.014	$d_I(^{13}C)$: 3.227
$d_N(^{13}C)$: 2.767	$d_N(^{13}C)$: 2.774	$d_N(^{13}C)$: 3.186

Figure 9.5 The three top structures of the ranked output file. Structure No. 1 coincides with the structure of ashwagandhanolide as reported.[71]

Fuzzy structure generation can easily be controlled by parameters that make up a set of options. The two main parameters are *m*, the number of non-standard connectivities, and *a*, the number of bonds by which some connectivity lengths can be augmented. Unfortunately, in general, 2D-NMR spectral data cannot deliver definitive information regarding the values of these variables in any case, although the data obtained from both 1,1-ADEQUATE and 1,*n*-ADEQUATE can significantly reduce the uncertainties in the connectivity lengths. When these data are not available, both *m* and *a* can be determined only during the process of structure elucidation. We have concluded that, in many cases, choosing an erroneous value for *a* can be avoided and the solution of a problem can be considerably simplified if the lengthening of the *m* connectivities is replaced by their *deletion*. When set in the options, the program can delete the necessary connectivity responses that have to be augmented (by convention, the parameter *a* is set to *X* in these cases). Such an approach can be successful in those cases when the number of 2D-NMR connectivities is, in some sense, optimal. In this sense, we mean that the total number of connectivities (structural constraints), *N*, must be large enough to avoid the combinatorial explosion during the fuzzy structure generation process.

If the number of connectivities, *N*, is small, then further decreasing *N* by *m* in a connectivity combination can lead to an excessive decrease in the number of structural constraints required for solving the problem. In such a case, the problem may be difficult to solve because the 2D-NMR data structural constraints will only reduce the total number of possible isomers very slightly.

Independent of the use of augmentation or removal of connectivities, the crucial point in the application of FSG is the number of connectivity combinations that should be checked during structure generation. For instance, if $N = 60$ and $m = 5$, then the number of connectivity combinations, $n_{\text{math}} = C_N^m$, is equal to $\sim$5.5 million. Any attempt at structure generation has to be performed using each of these combinations. It is necessary to perform the generation of structures from each of the C_N^m data sets and obtain the output file as a unification of all of the intermediate results. Even though the StrucEluc structure generator is fast, the productivity is certainly insufficient in terms of coping with a combinatorial problem as outlined here.

To overcome this difficulty, the system includes an algorithm capable of reducing the number of combinations without the risk of losing the correct solution. This is attained as a result of logical analysis of the initial 2D-NMR data. If connectivity sets potentially containing NSCs are identified,[21] then groups of these connectivities are utilized to produce connectivity combinations. As a consequence, connectivities that are suspected to be non-standard are included in all resulting combinations and the initial number of combinations reduces (it was found that this number could be reduced by many factors[67]). In addition, the algorithm is capable of immediately detecting combinations of connectivities from which structure generation is impossible – a connectivity combination of this kind still contains at least one non-standard connectivity. These combinations are skipped during the structure generation process. As a result, FSG can be performed in a reasonable time even in those cases when n_{math} is very large (for instance, when $n_{\mathrm{math}} \approx 10^6$).

The algorithm developed by the authors provides six different FSG modes that are employed depending on the 2D-NMR correlation properties and the result of their logical analysis. The algorithm was developed and tested in the process of solving real problems. A set of more than 100 problems was selected where either the GHMBC or COSY spectra, or both, contained a total of 1–18 non-standard connectivities corresponding to a range of coupling constants $^nJ_{\mathrm{HH}}$ or $^nJ_{\mathrm{CH}}$ where $n = 4$–6. The structures under investigation were all natural products and the number of skeletal atoms in the molecules varied between 15 and 75. The experimental data were obtained from articles published mainly in the *Journal of Natural Products* or from collaborations with various laboratories.

As a result of these studies, all problems were classified into three sets as follows:

1. 53 problems were identified where NSCs were detected and the initial MCDs were successfully updated.
2. 34 problems were identified where the program revealed the presence of NSCs but failed to update the MCDs.
3. 13 problems were identified where the program failed to detect NSCs.

This classification describes all conceivable results that can be obtained from checking the MCDs. Depending on the results of checking the MCD, various modes or combinations of modes can solve the problem. Attempts to solve each problem were made using the different FSG modes to investigate possible approaches. The problems for which valid solutions could not be found during the first attempt were eventually solved after utilizing different fuzzy generation options. Logical data preprocessing frequently allowed for a significant reduction in the number of connectivity combinations that had to be tested during the FSG. For instance, in 20 problems the theoretical number decreased by 10^4–10^6 times but the real number of combinations still remained rather large. Nevertheless, the speed of the structure generator algorithm was fast enough to solve almost all problems.

It is difficult to describe the myriad of nuances associated with FSG, since these depend on each 2D-NMR data set associated with a given problem. A series of examples illustrating the strategies leading to valid solutions with the minimum number of user assumptions have been presented in our work.[67] Here we briefly describe one example.

In the analysis of cleospinol A[73] with molecular formula $C_{20}H_{32}O_2$ (**1**), the 2D-NMR data are comprised of 21 COSY and 55 HMBC correlations.

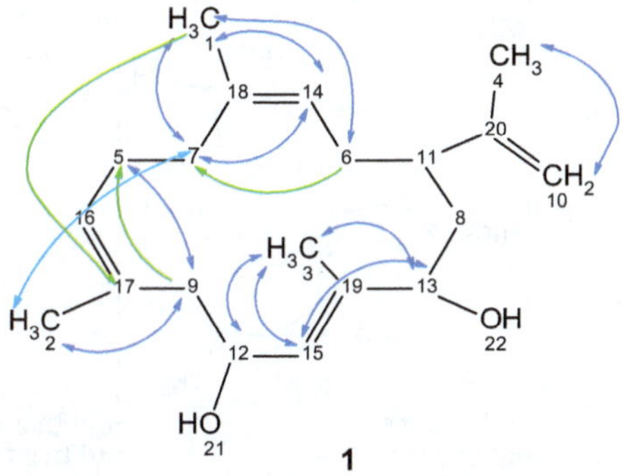

1

These data were used to evaluate the possibility of solving a problem in those cases where a large number of non-standard correlations were present. In this case, the 2D-NMR data contained the following combination of NSCs:

$$3\text{HMBC}[2a(1),\ 1a(3)] + 12\ \text{COSY}[8a(1),\ 3a(2),\ 1a(3)] = 15$$

This nomenclature describes the fact that there are three HMBC non-standard correlations, two of which must be lengthened by one bond and one by three bonds; the information about the 12 COSY correlations is interpreted analogously. The total number of NSCs is hence 15. The COSY connectivities are represented below on the structure by blue double-headed arrows whereas the HMBC correlations are defined by green single-headed arrows from the proton to the carbon to which it is long-range coupled.

The COSY, HMQC and HMBC spectral data associated with compound **1** were input to the program and the MCD was generated. A check of the MCD was accompanied by the automated removal of contradictions. When the computation started, the program displayed a message declaring that the contradictions had been detected and resolved; the minimum number of NSCs was estimated by the program to be seven at that point. Unfortunately, strict structure generation from the automatically edited MCD resulted in an empty output file. This result was interpreted as evidence of the presence of either additional undetected non-standard correlations or NSCs whose length must be increased by more than one bond.

FSG was initiated assuming only that the number of non-standard connectivities is not more than 15, *i.e.* options $\{m \leq 15,\ a = X\}$ were set. In this case, 18 281 379 connectivity combinations from 40 225 345 056 theoretically possible combinations were used for structure generation. The following result was obtained: $k = 769 \rightarrow 430 \rightarrow 245$, $t_g = 29$ min 9 s. The correct structure was ranked first (Figure 9.6) by all methods of spectrum prediction, $r_{all} = 1$.

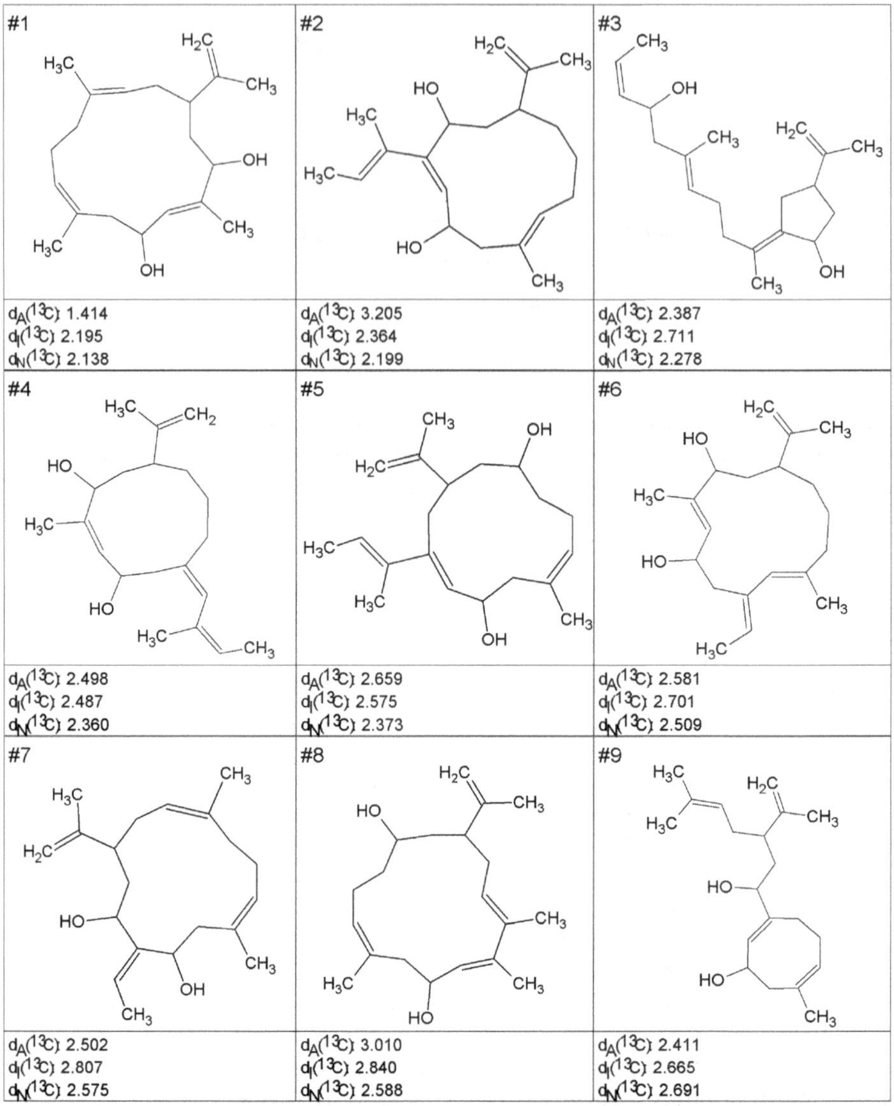

Figure 9.6 The first nine structures of the ranked output file found as a solution to the cleospinol structure elucidation.

The program therefore identified the correct solution even when 15 non-standard connectivities existed in the 2D-NMR data. This result was particularly noteworthy since the HMBC and COSY spectra both contained $^6J_{CH}$ and $^6J_{HH}$ correlations. Note that only $\sim 10^{-4}$ of the theoretically possible connectivity combinations were processed. In spite of the fact that $n_{real} >$ 18 million, the high-speed structure generator present in the StrucEluc program completed the process in a reasonable time.

9.5.5 Determination of Relative Stereochemistry of Identified Structures

The biological activity of natural products and drug molecules is known to be highly dependent on stereochemistry. Hence the final step in contemporary structure characterization efforts is to define the relative and if possible, absolute stereochemistry. NMR methods are well suited to the former, whereas the latter is generally obtained using chemical structure modification combined with NMR studies[74,75] or by X-ray crystallographic methods.[74] The NMR-based determination of relative stereochemistry utilizes the NOE effect. Typically NOESY (at 300–400 MHz) or ROESY (at 500–600 MHz) two-dimensional NMR experiments or their selective 1D analogs are used to provide the data for this analysis in rigid molecules.

Determining the relative stereochemistry for new organic compounds, especially natural products, has become a routine procedure that needs to be speeded up and automated as much as possible. Conditionally, it is conceivable to distinguish the following two stages in the traditional strategy of relative stereochemistry determination: (1) selecting a set of the most probable stereoisomers using similar reference structures for which the relative stereochemistry has been determined; (2) examining these stereoisomers by NOESY/ROESY spectra, molecular modeling and QM ^{13}C NMR prediction for finding the most preferable stereoisomer. Methods allowing computational modeling of stereochemistry have been developed and incorporated into StrucEluc system.

9.5.5.1 Selection of the Set of Most Probable Stereoisomers

As mentioned previously (Section 9.4), when ACD/NMR predictors were developed, information regarding the relative stereochemistry of reference molecules was taken into account. All three computational approaches (I, ANN and HOSE) are sensitive to different degrees to the orientation of the stereobonds in the structure under investigation. In our work,[76] we reported results of a study aimed at evaluating the possibility of using empirical NMR chemical shift prediction as a preliminary filter to select a set of the most probable stereoisomers that could be used for the subsequent determination of the actual configuration of a molecule. We found that, of the three methods, the fragmental approach was the most sensitive to stereochemistry and we showed that it can be used for the purpose of stereochemical investigations prior to using time-consuming QM methods.

The fragmental approach was examined using a series of new natural products reported in the literature and belonging to a number of different classes: steroids, alkaloids, terpenes, cembranoid diterpenes, *etc.* The stereochemistry of the compounds was reported in the corresponding publications. The structural formula of each structure examined was input into the Proposed Structure window of StrucEluc and all $N = 2^n$ (n = number of stereocenters) mathematically conceivable stereoisomers were generated and depicted by the program. In our computational experiments, the value of N varied between 64 and 4096 ($n = 6$–12). ^{13}C NMR chemical shift prediction was then performed and the stereoisomers were ranked in descending order of the average deviation d_A between the experimental and calculated spectra. The study showed that the correct stereoisomer was usually placed at the top of the ranked file and took between the first and third positions in the list, therefore allowing the program to serve as a filter capable of rejecting improbable stereoisomers. Note that NOE data were not even used at this stage. Subsequent visualization of the NOESY/ROESY connectivities on the structures allows for rapid determination of the most preferred member of the "best" stereoisomers set. QM-based geometry optimization and chemical shift calculations can be performed at this stage in order to facilitate a final decision.[76,77]

Maloney *et al.*[78] reported the structural characterization of a new cucurbitacin, **2**. Twelve stereogenic centers were determined and marked by the StrucEluc program automatically and 4096 stereoisomers (2048 enantiomeric pairs) were generated. ^{13}C NMR spectra were calculated for each enantiomeric pair using the fragmental approach in ~1.5 h and the ranking procedure promoted the correct stereoisomer, **2**, to the first position.

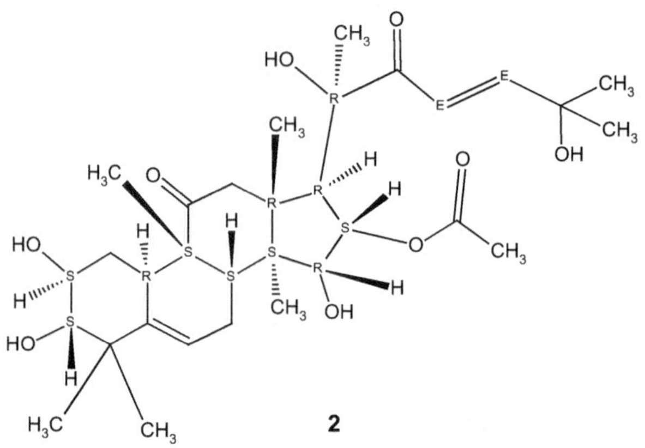

2

For the case of hopeanolin (Section 9.5.3.2), StrucEluc placed the correct stereoisomer in third position, but when observed NOESY correlations were displayed on stereoisomeric structures, the assessment of an expert promoted the correct stereoisomer to first position.

9.5.5.2 Simultaneous Determination of Relative Stereochemistry and 3D Modeling

The StrucEluc system was also enhanced by adding an algorithm[79] to allow for the *automated* determination of the relative stereochemistry and 3D modeling of a molecular structure using constraints imposed by the NOE effect. The program extracts NOE information from either the NOESY and/or ROESY spectra and determines the relative stereochemistry and 3D model accordingly. This process can be carried out for several of the most likely structures produced by StrucEluc during a structure elucidation or performed on a chemical structure proposed by the chemist.

The utility of NOESY/ROESY spectra for relative stereochemistry determination is based on a direct correlation between both the cross-peak *volume* integration and the internuclear *distance*. This dependence is used by the program for energy minimization.

Minimization algorithms deal with *numerical values* and, in this case, these numerical values are extracted from a set of NOEs overlaid on a 3D structure and examined for goodness of fit. The function describing this goodness of fit is called a *penalty function*. The better the solution, the lower is the value of the function. The function must exhibit the lowest value for the best-matching stereoisomer.

In our work,[79] an appropriate function was suggested that could be minimized by calculating *all* stereoisomeric structures or by using a stochastic *genetic algorithm*[80] to limit the number of stereoisomers that need to be investigated. To improve the convergence of the genetic algorithm, alternative, efficient methods of parameter optimization were compared. Since the algorithm used only information about stereocenter configuration and not information such as chair or boat ring conformations, conformational rigidity is essential to obtain accurate results. This requirement currently limits the algorithm to the fused ring portions of molecules. Meanwhile, relatively simple problems with 2–6 chiral centers may be solved in a straightforward manner by the enumeration of all stereoisomers and the calculation of the corresponding penalty function values. The process is fast and this approach is preferred over the use of genetic algorithms for molecules with smaller numbers of stereocenters.

The advantage of employing the genetic algorithm was demonstrated on two complex natural products, taxol ($C_{47}H_{51}NO_{14}$) and brevetoxin B ($C_{50}H_{70}O_{14}$).[81,82] The most challenging is the structure of brevetoxin B (**3**).

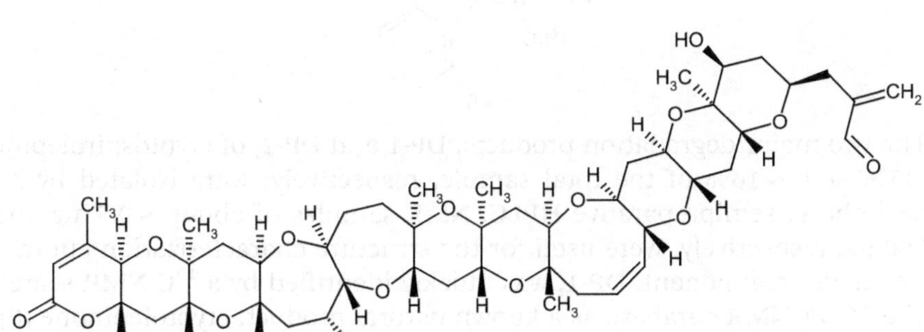

3

This remarkable structure includes 11 rings, 23 stereogenic centers and three carbon–carbon double bonds. The processing time necessary for running all ~8.4 million stereoisomers corresponding to this structure was estimated to be about 1 month. The application of a genetic algorithm allowed us to determine correctly the brevetoxin B stereochemistry and the 3D geometry model of the molecule in a processing time of 2 h 50 min on a PC. This demonstrates the power of the approach we have described to facilitate the identification of relative stereochemistry in a complex molecule containing multiple stereocenters.

9.6 Challenging StrucEluc

In this section, we present examples of challenges solved with the aid of StrucEluc. The first two examples discussed relate to the elucidation of complex alkaloids of the cryptolepine (**4**) series.[72,83]

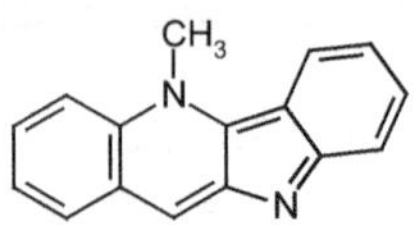

4

9.6.1 Structure Elucidation of a Cryptospirolepine Degradant

Martin *et al.*[72] employed a combination of cryogenic NMR probe technology[84] and the StrucEluc system in the characterization of unknown degradants of a complex spiro monocyclic alkaloid, cryptospirolepine (**5**). A 2.5 mg sample of this compound had been stored in a sealed 5 mm NMR tube in DMSO-d_6 for ~10 years, which allowed the compound to degrade.

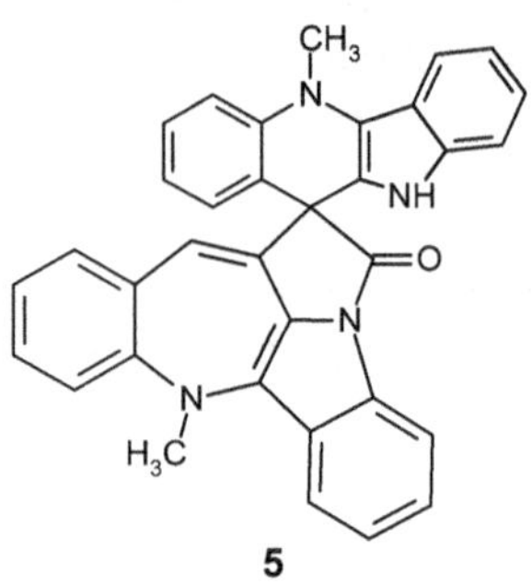

5

The two major degradation products, DP-1 and DP-2, of cryptospirolepine (~35% and ~16% of the total sample, respectively) were isolated by reversed-phase, semipreparative HPLC. NMR samples of about ~0.5 mg and ~200 μg, respectively, were used for the structure characterization effort.

The major component, DP-1, was quickly identified by a ^{13}C NMR search in the ACD/CNMR database as a known natural product, cryptolepinone (**6**).

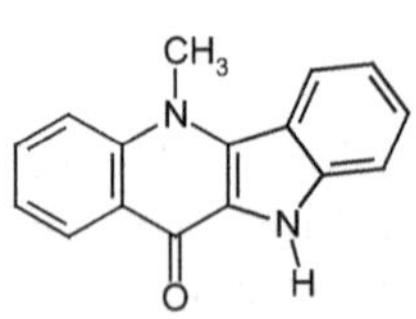

6

Mass spectrometry performed on the second isolate (DP-2) gave a molecular ion, $MH^+ = 479$, which suggested a molecular formula of $C_{32}H_{22}N_4O$. A 1D ^{13}C NMR spectrum was not available, as was very common in natural product structure elucidation for very small samples at that time. It should be noted that nowadays a decent ^{13}C NMR spectrum of 0.5 μmol of strychnine can be obtained overnight using a 1.7 mm Micro-CryoProbe.[85] The ^{13}C shift inputs were thus created from the HSQC and HMBC spectra. Eighteen peaks were identified in the HSQC (2 CH_3 and 16 CH) data and 13 peaks were extracted from the HMBC to give a total of 31 peaks. According to the molecular formula, the molecule contained 32 carbon atoms. It was concluded that one quaternary carbon atom did not show an HMBC peak and one was added to the spectrum with a chemical shift of 130 ppm, in the middle of the aromatic interval (an "axiom"). The number of peaks in the HMBC spectra acquired in standard and phase-sensitive mode were different, 32 and 45, respectively. These additional responses are likely due to improved resolution in the congested regions of the spectrum, although possibly longer range couplings are being detected. To avoid contradictions caused by the presence of NSCs, the extra peaks observed in the second HMBC experiment were attributed to a range of potential couplings and concluded to be $^{2-4}J_{CH}$ (another "axiom").

Attempts to solve this problem in both the Common and Fragment Modes quickly showed that structure generation would be extremely time consuming, which was interpreted as a hint to apply a User Fragment Database (UFDB) formed from the known structures of the cryptolepine series. A UFDB containing 342 fragments was created specially for the identification of alkaloids belonging to the cryptolepine series,[14] for which eight compounds of this class were used. Searching the ^{13}C NMR spectrum in the UFDB resulted in 44 fragments; 776 MCDs were created and each MCD contained four found fragments. No constraints on the generated structures were imposed. The result of structure generation was $k = 1572 \rightarrow 228 \rightarrow 8$, $t_g = 12$ s. As the structures were ranked by the deviation values, the best structure was found in first position as shown in Figure 9.7.

All three methods of ^{13}C NMR prediction pointed to structure No. 1 as the best one. This allowed Martin *et al.*[72] to conclude that the structure of compound DP-2 is 7.

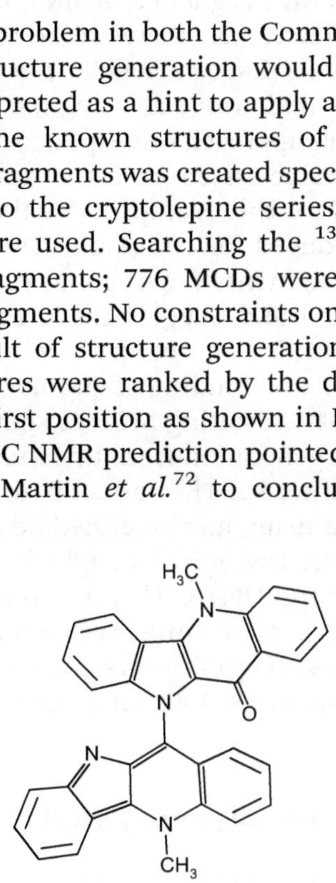

7

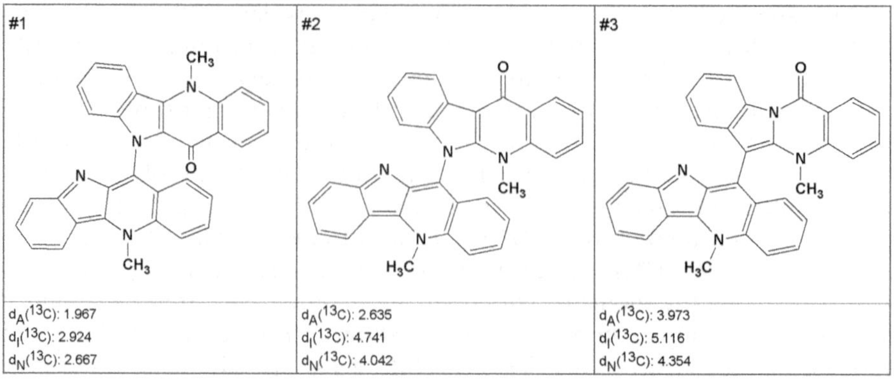

Figure 9.7 The first three structures of the ranked file deduced as the solution to the structure analysis of DP-2.

They also considered how StrucEluc could assist in solving this problem when both traditional and computer-based approaches are combined.[72] It is common for an experienced spectroscopist to detect molecular fragments simply by visual analysis of 1D- and 2D-NMR data. The approach is based on experience, knowledge and the insight of a highly qualified researcher, and the structural information extracted can therefore be invaluable. Providing spectroscopists with software tools that can facilitate the assembly of the molecular structure in an interactive mode while allowing them to modify their hypotheses is of obvious value. This approach was expected to have a synergistic effect.

The ability of the StrucEluc system to act as an "assistant" to the elucidation process was tested for this example. Visual analysis of the molecular connectivity diagram produced for compound DP-2 allowed the experts to clearly see three 1,2-Ar fragments ("hexagonal stars" produced by HMBC connectivities inside rings) and suggest the presence of the fourth one. HMBC connectivities identified the connection of the two aromatic fragments *via* the *N*-CH_3 group and binding of another *N*-CH_3 group to the third aromatic ring fragment. The resulting MCD is shown in Figure 9.8.

The results of structure generation from this manually created MCD were $k = 496\,528 \rightarrow 26$, $t_g = 13$ min 15 s. The correct structure was again identified as the most probable one using all three methods of ^{13}C NMR prediction. The application of the spectroscopist's insight had a beneficial effect and allowed progress without the UFDB. This example indicates that a highly qualified expert is capable of determining very complex structures relying on theirr knowledge and the capacity of the system for deducing all, without any exception, logical consequences following from postulated suggestions (axioms and hypotheses).

9.6.2 Solution of a Cryptolepine Family "Puzzle"

When the 2D-NMR data were being acquired for cryptospirolepine (**5**) by Martin's group,[86] data were also accumulated in late 1991 for another

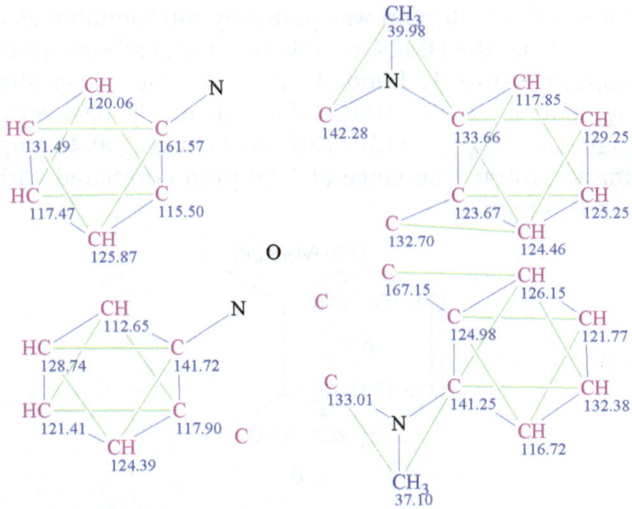

Figure 9.8 The molecular connectivity diagram of DP-2 displaying fragments deduced by the expert. Ambiguous connectivities are not shown.

alkaloid fraction from *Cryptolepis sanguinolenta* that was given the notebook designation TC-6. A data set consisting of proton and carbon reference spectra, COSY, ROESY, ^{1}H–^{13}C HMQC and HMBC spectra in MeOD was acquired. A structure consistent with all of the available data was not assembled in 1991–92 when these data were first examined. The data generated were associated with a very small sample amount, very scant knowledge about the cryptolepine indoloquinoline alkaloids at the time and the experimental capability of the instruments then available. Consequently, TC-6 was reinvestigated about 10 years later by Blinov *et al.*[83] using new instrumentation and StrucEluc was applied again in a mode of tight interaction with the spectroscopist.

The retained reference sample of this alkaloid was 95% pure with a molecular weight of 448 Da. Major fragmentation was simple, with the molecule essentially splitting into two "halves," producing fragment ions at 217 and 232 Da. The accurate mass was measured as 448.1683 Da, which is within 1.2 ppm of the theoretical mass of the empirical formula of $C_{31}H_{21}N_4$.

Despite a relatively congested proton NMR spectrum at 400 MHz, the COSY spectrum still readily allowed the protons of the four individual four-spin systems to be identified and ordered. These included ordered sets of resonances (ppm) as follows:

8.88 – 8.23 – 7.79 – 7.86
8.86 – 7.59 – 7.85 – 7.57
8.68 – 7.52 – 7.58 – 7.11
8.31 – 7.76 – 7.53 – 7.80

In addition, a CH_3 singlet was observed at 5.28 ppm in the ^{1}H NMR spectrum that can be attributed to an indoloquinoline *N*-methyl group.

A singlet resonating at 7.90 ppm was plausibly interpretable as an isolated aromatic proton. Using the HMQC correlation data, carbons were associated with their respective directly bonded protons, which suggested the substructural fragments **8–11**. The HMQC data, obviously, also correlated with the *N*-CH_3 singlet at 5.28 ppm with a carbon resonating at 43.1 ppm and the isolated aromatic proton resonance at 7.90 ppm correlated with a carbon resonating at 115.8 ppm.

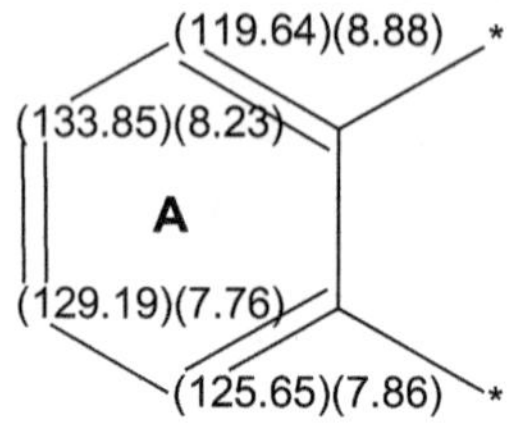

8

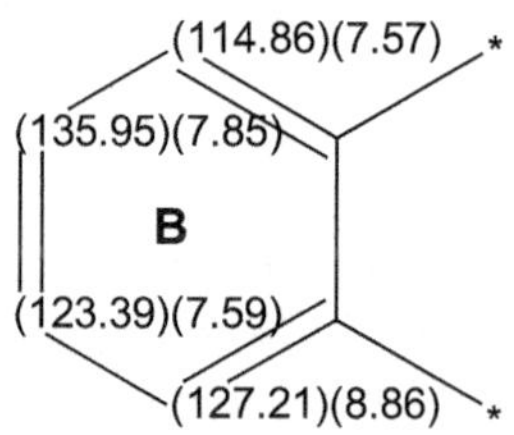

9

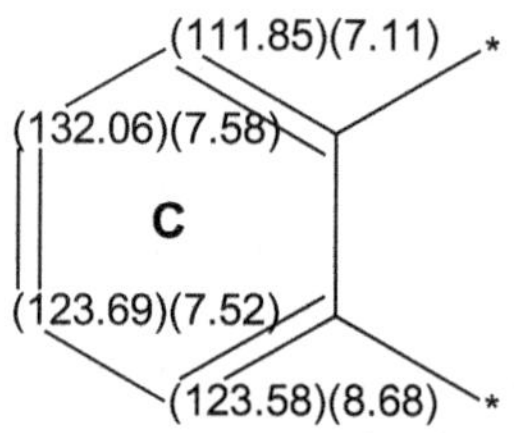

10

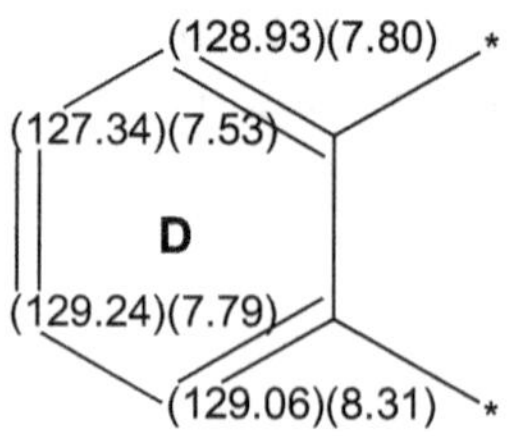

11

When the molecular formula and all of the NMR data were fed into StrucEluc, the MCD shown in Figure 9.9 was created. Because of the highly congested region in the vicinity of 129 ppm in the ^{13}C spectrum, there was some potential for ambiguity in the assignments. All ambiguous correlations are displayed with dotted lines, which allows a spectroscopist to analyze the whole picture visually and edit correlations step-by-step in accord with chemical "common sense."

To remove ambiguous correlations, a heuristic approach suggested by spectroscopic "common sense" was used. We describe in detail the first step of the process. Figure 9.9 shows that there are two very close chemical shifts in the ^{1}H NMR spectrum at 7.86 and 7.85 ppm. The first of them was assigned to ring A and the second to ring B. It is obvious that these chemical shifts may be interchanged since they are observed in a very congested region of the ^{1}H NMR spectrum.

Furthermore, in ring A, a distinct HMBC correlation of standard length from C(125.65) to H(8.23) was observed. At the same time, there are no correlations from C(125.65) to H(7.59), H(7.57) and H(8.86), which belong to ring B. This observation suggests that C(125.65) is related to ring A and consequently ambiguous correlations associated with this atom and ring B can be deleted. In ring B, a correlation from C(135.95) to H(8.86) is observed whereas there are no correlations from C(135.95) to protons H(7.76), H(7.79), H(7.86) and H(8.88), which are related to ring A. Hence C(135.95) is included

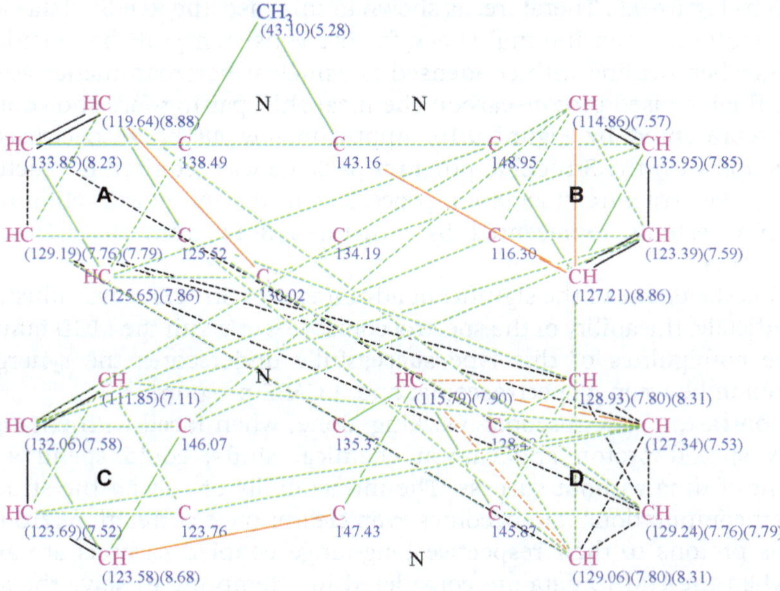

Figure 9.9 The MCD showing all potentially ambiguous correlation pathways as dashed lines. The solid lines denote correlations that were initially thought to be correct. Vicinal connectivities are denoted by solid black lines. Two- and three-bond heteronuclear correlations are shown using solid or dashed green lines (the latter are possibly ambiguous correlations). Suggested longer range correlations ($^nJ_{CH}$, $n \geq 4$) are shown in orange.

in ring B. With these associations established, the ambiguous correlations from H(7.86) associated with carbons situated in rings B and D, and also the ambiguous correlations from H(7.85) associated with the carbon atoms contained in ring A, were removed. The ambiguous correlations inside the rings were then transformed into unambiguous correlations.

Working in an analogous fashion, some minor revisions of proton–carbon pairings involving resonances with closely similar chemical shifts were performed and ambiguous connectivities were removed for all four four-spin systems.

A ROESY correlation was observed from the *N*-methyl resonance at 5.28 ppm to the aromatic proton resonating at 8.88 ppm. The *single* ROE correlation observed from the *N*-CH_3 is important in that it excludes cryptolepine (**9.4**) from consideration as a possible fragment of the TC-6 structure. The reason is that the *N*-methyl group in systems containing cryptolepine exhibits correlations to *both peri* aromatic protons, which would in turn be required again if one of these indoloquinoline systems were a constituent of the TC-6 structure. Consequently, the presence of a single ROESY correlation from *N*-CH_3 can be considered as a very distinctive feature of the target molecule. An important cross-ring ROESY correlation was observed between a clearly resolved aromatic proton resonating at 7.90 ppm (*single* for this proton) and the proton resonating at 7.80 ppm, consistent with the MCD shown by Figure 9.9. Therefore, as shown in this case, the ROESY data can be very important as an internal check for the consistency of the elucidation process when dealing with condensed polynuclear heteroaromatic systems.

The final, revised proton–carbon chemical shift pairings are shown in the MCD represented by Figure 9.10. Approximately 48 h of spectroscopist interaction with the StrucEluc program package was required to reach this point in the structure elucidation process from the initial extraction of the four-spin systems represented by structures **8–11** from the COSY and HMQC data.

At this stage, one of the significant advantages of StrucEluc was illustrated – specifically, the ability of the spectroscopist to work with the MCD family to resolve ambiguities of this type successfully underscores the synergistic interaction between a spectroscopist and a CASE program.

In contrast, a spectroscopist working alone, when faced with entangled, closely spaced proton and carbon chemical shifts, could spend a vast amount of time without success. The intractability of solving the structure without computational aid becomes even clearer once correlations from the various protons to their respective long-range coupled carbons are added and when the HMBC data are considered in attempting to solve the structure. In part, this sort of confusion was probably responsible for the frustrated initial attempts to elucidate the structure of this molecule manually.

From the MCD shown in Figure 9.10, the structure generation process was initiated and the following result was obtained: $k = 353 \rightarrow 266$, $t_g = 10$ s. ^{13}C chemical shift calculations with subsequent file sorting allowed the program to distinguish the set of top-ranked structures presented in Figure 9.11.

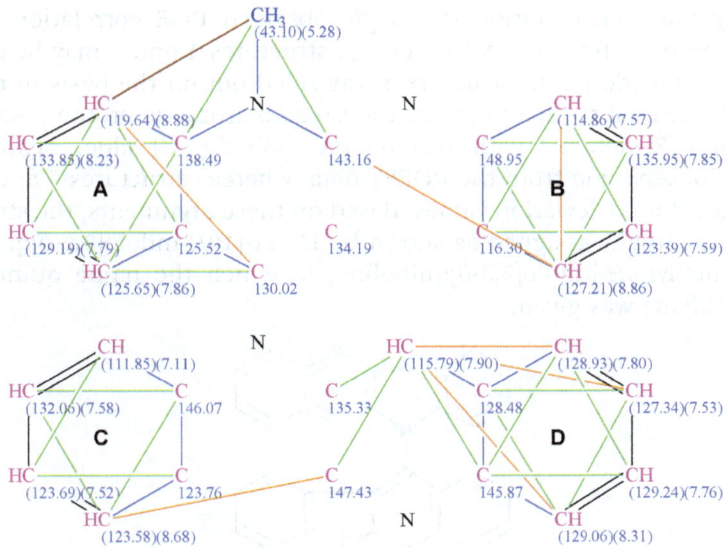

Figure 9.10 The final MCD obtained by continued pairwise successive removal of ambiguities associated with all four ring systems.

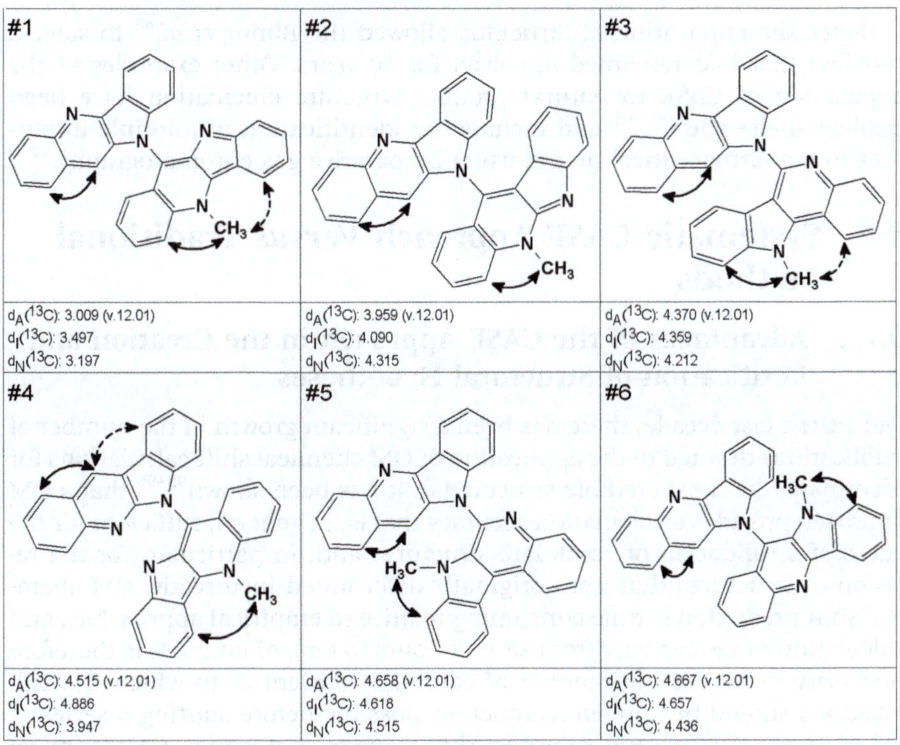

Figure 9.11 The first six of 266 non-identical structures generated by StrucEluc and sorted on the basis of $d_A(^{13}C)$. Arrows show experimental (solid) and expected (dotted) ROESY correlations from the CH_3 group and from the isolated aromatic proton at 7.90 ppm.

Taking into consideration the *single* observed ROE correlation in the ROESY spectrum from the N-CH_3 group, structures 1 and 3 may be eliminated from consideration. Structure 4 was ruled out on the basis of the observed single ROE correlation from the isolated aromatic proton resonating at 7.90 ppm. Structure 2, with the more favorable $d_A(^{13}C)$ value, is consistent with this observation from the ROESY data, whereas structures 5 and 6 can be rejected due to deviation values. Based on these arguments, the structure of TC-6 was finally assigned as shown by **12**, 11-(10*H*-indolo[3,2-*b*]quinolin-10-yl)-5-methyl-5*H*-indolo[2,3-*b*]quinoline, to which the name quindolinocryptotackieine was given.

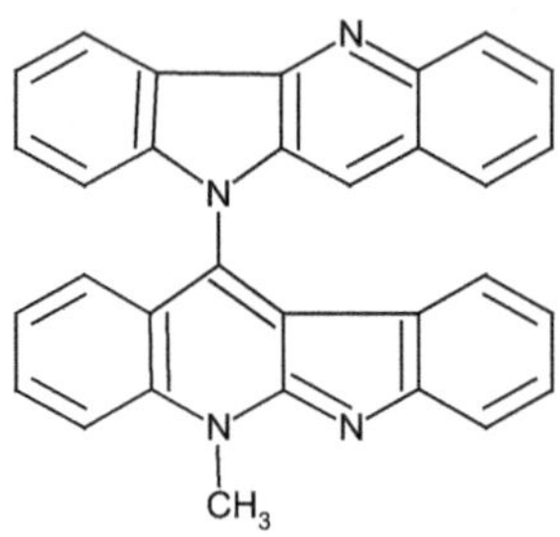

12

Hence the application of StrucEluc allowed the Blinov *et al.*[83] to solve a problem that had remained unsolved for 10 years. Other examples of the application of CASE to natural product structure elucidation have been published elsewhere[87–91] and include the identification of multiple impurities in a pharmaceutical matrix using preparative gas chromatography.[92]

9.7 Systematic CASE Approach *Versus* Traditional Methods

9.7.1 Advantages of the CASE Approach in the Creation and Verification of Structural Hypotheses

During the last decade, there has been a significant growth in the number of publications devoted to the application of QM chemical shift calculations for identifying the most credible structure(s). It has been shown[93–98] that a QM approach provides a calculation accuracy that is, in general, sufficient for the successful validation of candidate structures and, in particular, for the revision of structures that were originally determined incorrectly. QM chemical shift prediction is time consuming relative to empirical approaches, and calculation times can vary from several hours to tens of hours. It is therefore necessary to reduce the number of candidate structures to which QM calculations should be applied as much as possible before starting a series of calculations. It is natural to expect that prior to performing QM chemical shift calculations, a minimum set of candidate structures should be chosen on the basis of fast chemical shift prediction by empirical methods. As mentioned earlier (Section 9.4), structure generation and subsequent ranking of the candidate structures in descending order of their probability consumes only a few minutes on a modern PC using today's expert systems.[15] Obviously, the application of QM methods can play a decisive role in such cases when the structures to be analyzed contain "exotic" fragments that were absent from the training set.

The potential of QM methods was evaluated on molecules for which the number of heavy atoms was most frequently around 20 and rarely reached 30.[33] However, many natural product molecules contain 40–100 or more heavy atoms. For large molecules, we can only rely on empirical methods for chemical shift prediction.

In many publications, it has been suggested that the QM approach is a unique predictive method for proving or disproving a proposed structure. In our work,[99] we compared the capabilities of the newer empirical methods of chemical shift prediction with those of the QM approach when both approaches are used for identifying the most probable structure within a set of proposed structures. The efficiency of manual and automated methods of forming structural hypotheses was also compared.

For this purpose, we selected a series of articles in which the QM approach had been used successfully for the selection of the correct structure among a series of molecules suggested by a chemist or for revising the originally hypothesized chemical structure. For each case, if 2D-NMR data were available we made an attempt to solve the problem systematically using the StrucEluc expert system. We found that the correct structure was also assigned as the most probable one in the examples considered by both QM and fast empirical NMR chemical shift predictions,[48,64,66] while alternative and incorrect structures suggested by researchers were ranked lower. The examples studied enabled us to suggest a general approach in which the most probable structure is established as a result of the joint application of a CASE expert system in combination with both empirical and QM methods of chemical shift prediction. Let us consider an example showing the advantages of a systematic CASE-based approach over traditional methods.

9.7.2 Example

Balandina *et al.*[100] synthesized a novel quinoxaline and determined its molecular formula $C_{16}H_{10}N_2O_2$ from the MS data (*m*/*z* 262) combined with elemental analysis data. To elucidate the structure of this compound, they used ^{1}H, ^{13}C and ^{15}N NMR spectra. Assignment of the ^{1}H and ^{13}C NMR spectra was accomplished using data derived from DEPT, 2D-COSYGP, HSQC and HMBC experiments. Analysis of the NMR data provided two fragments containing H, C and N atoms with assigned chemical shifts. Three quaternary carbons (151.04, 138.29 and 134.68 ppm) without HMBC correlations, one hydrogen atom and two oxygen atoms were not assigned to either of the fragments. The initial data for forming structural hypotheses are presented in Figure 9.12.

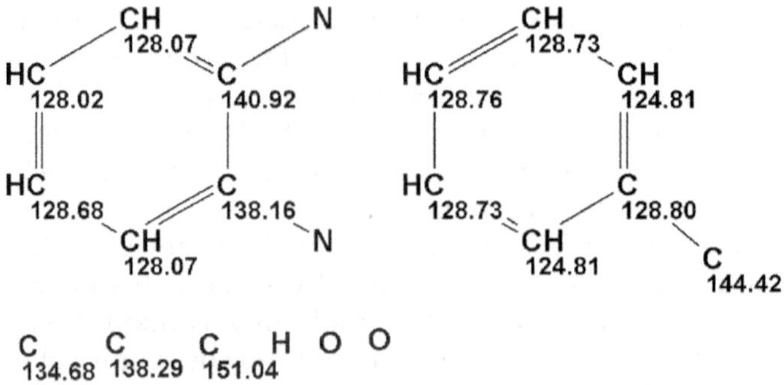

Figure 9.12 Initial structure information for the generation of structural hypotheses.

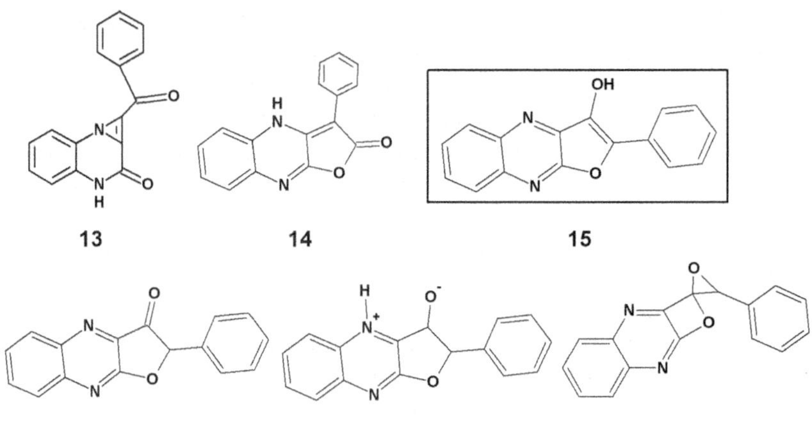

Figure 9.13 Six suggested structures derived from the experimental data.[100] Structure **15** corresponds to the correct structure.

Using these data and some additional chemical considerations, Balandina *et al.*[100] suggested six structures, presented in Figure 9.13.

To select the correct structure, ^{1}H, ^{13}C and ^{15}N chemical shifts were predicted for structures **13–18** using the DFT framework and using a hybrid exchange-correlation functional, GIAO B3LYP, at the 6–31G(d) level. Full geometry optimizations were performed under *ab initio* RHF/6–31G conditions. Linear correlation coefficients of the experimental *versus* calculated ^{13}C chemical shifts (R^2), root-mean-square errors (RMS), slope (a), standard deviations (SD) and mean absolute deviations [$\text{MAD} = \sum(|\delta_{\text{exp}} - \delta_{\text{calc}}|)/n$] for structures **13–18** were computed. As a result structure **9.15** was identified as the most probable ($R^2 = 0.9758$, RMS = 1.16 ppm, SD = 1.2 ppm, MAD = 7.03 ppm). Other proposed structures were rejected by the authors due to smaller R^2 values ($R^2 = 0.01$–0.57) and larger deviations. It should be noted that the R^2 values have a reasonable interpretation only in those cases when experimental chemical shifts are assigned to the atoms of competing structures. Otherwise, the preferred structure can be selected only by comparing the experimental with the calculated spectrum and by determining outliers. Applying an expert system for structure elucidation provides chemical shift assignments that agree with the 2D-NMR correlations and consequently the best structure is selected automatically (Section 9.5.3.2).

The spectral data reported by Balandina *et al.*[100] were entered into the StrucEluc system. Fragments and atoms shown in Figure 9.12 were eventually transformed into an MCD. The atom properties for three carbon atoms not included into the fragments were automatically set as sp^2/*not defined*. Structure generation was performed in the automatic mode and fuzzy structure generation was allowed. The following result was obtained: $k = 247 \rightarrow 16 \rightarrow 4$, $t_g = 1$ s.

Empirical chemical shift prediction was performed for ^{13}C, ^{1}H and ^{15}N nuclei. Subsequent structural ranking by $d_N(^{13}C)$ deviation resulted in the structure ordering shown in Figure 9.14.

Structure **15** is the best structure according to the shift predictions for all nuclei presented in Figure 9.14. Moreover, the deviations for structure **9.15** are dramatically smaller than those for the next ranked structure (No. 2) for all nuclei and suggest a high reliability for the solution.[15] Deviations were calculated for the chemical shift assignments performed for structure **9.15** by StrucEluc and deduced by the authors.[100]

All deviation values calculated for the automatic assignment [including the deviations $d(^{15}N)$] are markedly smaller than those found for the former assignment, which indicates the potential incorrectness of the initial assignment. It is also interesting that all suggested structures **13–18**, except **15**, were not generated by the program, being assessed as impossible.

Since the authors[100] did not report how the ^{13}C experimental chemical shifts were assigned to the carbons of all proposed structures **13–18** (except **15**), it was not possible to calculate the linear regressions for the ANN-calculated shifts for all supposed structures. Therefore, we predicted the ^{13}C NMR chemical shifts for structures **13–18** and then graphically

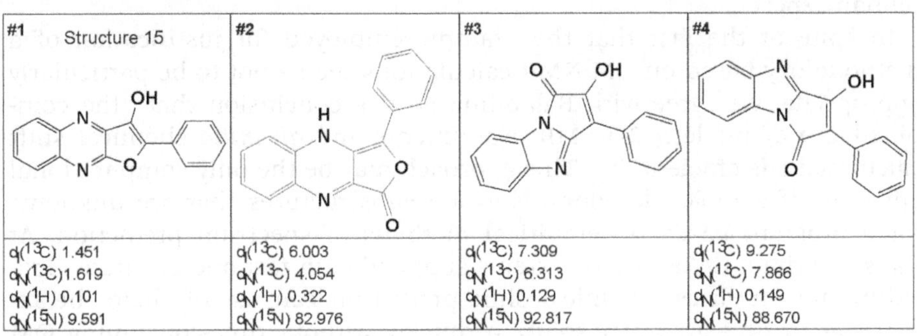

#1	#2	#3	#4
$d(^{13}C)$ 1.451	$d(^{13}C)$ 6.003	$d(^{13}C)$ 7.309	$d(^{13}C)$ 9.275
$d_N(^{13}C)$ 1.619	$d_N(^{13}C)$ 4.054	$d_N(^{13}C)$ 6.313	$d_N(^{13}C)$ 7.866
$d_N(^{1}H)$ 0.101	$d_N(^{1}H)$ 0.322	$d_N(^{1}H)$ 0.129	$d_N(^{1}H)$ 0.149
$d_N(^{15}N)$ 9.591	$d_N(^{15}N)$ 82.976	$d_N(^{15}N)$ 92.817	$d_N(^{15}N)$ 88.670

Figure 9.14 The output structural file ranked by $d_N(^{13}C)$ deviation.

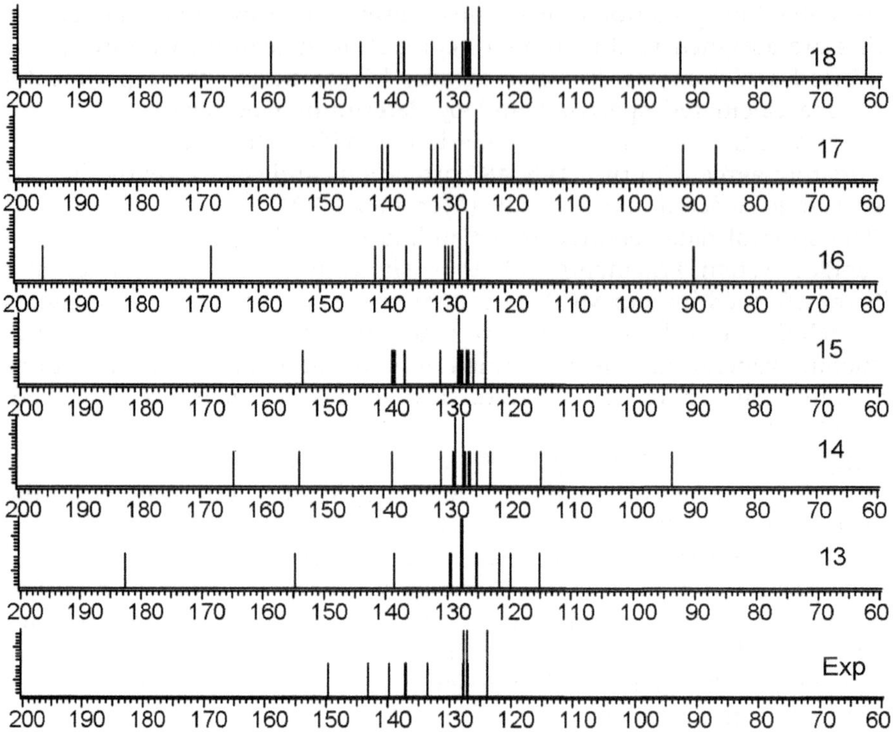

Figure 9.15 Experimental ^{13}C chemical shifts compared with the chemical shifts predicted by the neural net (NN) algorithm for proposed structures **13–18**.

compared the predicted spectra with the experimental one as shown in Figure 9.15.

The difference between the experimental and ANN-predicted spectra is dramatic for all structures except **15**. All incorrect structures could be immediately rejected before performing QM calculations and the correct structure would be quickly identified if hypotheses were offered by a human expert.

In spite of the fact that the example employed for justification of a methodology based on QM NMR calculations seems not to be particularly appropriate, we agree with Balandina *et al.*'s conclusion that "the combined use of modern 2D-NMR experiments and *ab initio* chemical shift calculations is efficient."[101] This approach may be the only computational approach if a molecule contains exotic substructures that are unknown for a program based on empirical methods of spectrum prediction. At the same time, CASE expert systems supplied with fast and accurate algorithms for empirical chemical shift prediction can usually help the researcher to find a correct solution quickly without time-consuming QM computations.

9.7.3 CASE as an Aid to Avoid Pitfalls During Structure Elucidation

In 2005, Nicolaou and Snyder[102] published a review entitled "Chasing molecules that were never there: misassigned natural products and the role of chemical synthesis in modern structure elucidation." The review posits that both imaginative detective work and chemical synthesis still have important roles to play in the process of solving Nature's most intriguing molecular puzzles.

According to Nicolaou and Snyder,[102] around 1000 articles were published between 1990 and 2004 where the originally determined structures needed to be revised. Figuratively, this means that 40–45 issues of the imaginary "*Journal of Erroneous Chemistry*" were published, where all articles contained only incorrectly elucidated structures and, consequently, at least the same number of articles were necessary to describe the revision of these structures. The associated labor costs necessary to correct structural misassignments and subsequent reassignments are very significant and, generally, are much higher than those associated with obtaining the initial solution. From these data, it is evident that the number of publications in which the structures of new natural products are incorrectly determined is fairly large and reducing this stream of errors is clearly a valid challenge. Nicolaou and Snyder[102] commented that "there is a long way to go before natural product characterization can be considered a process devoid of adventure, discovery, and, yes, even unavoidable pitfalls."

The Nicolaou and Snyder publication initiated our review[20] in which we tried to provide answers to the following important questions: (1) are the pitfalls that arise during the molecular structure elucidation unavoidable and (2) can modern CASE methods be used to minimize the probability of inferring incorrect structures from spectral data?

To investigate these questions, we analyzed ~20 examples for which the originally determined structures of novel natural products were revised in later publications. In all cases for which the 2D-NMR data were available, the expert system StrucEluc was used to determine whether the correct structure could be inferred from the experimental spectra and assumptions or "axioms" suggested by the researchers.

Our study showed that the application of modern CASE systems could indeed help the chemist avoid pitfalls or, in those cases when the researcher is challenged, the expert system could at least provide a cautionary warning. The various examples considered led us to conclude that the mistakenly identified chemical structure could be correctly elucidated if 2D-NMR data were available and the StrucEluc expert system was employed. If only 1D-NMR spectra were measured, then simply the empirical calculation of ^{13}C chemical shifts for the hypothetical structures most frequently enables a researcher to realize that their structural hypothesis is likely incorrect. We also tried to analyze how erroneous structural suggestions were made by highly qualified and skilled chemists. The investigation of these mistakes is very instructive and has facilitated a deeper understanding of the complicated logical-combinatorial process for deducing chemical structures.

It was shown that the CASE program can serve as a flexible scientific tool that assists chemists in avoiding pitfalls and obtaining the correct solution to a structural problem in an efficient manner. At the same time, chemical synthesis clearly still plays an important role in molecular structure elucidation. As multi-step synthesis requires the confirmation of the intermediate structures at each step, for which spectroscopic methods are commonly used, the application of a CASE system would be very helpful even in those cases when chemical synthesis is the crucial evidence to identify the correct structure. We also believe that the utilization of CASE systems will frequently reduce the number of compounds requiring synthesis.

Owing to space limits we will briefly describe only one example analyzed in detail in our review.[20] Sakuno *et al.*[103] isolated an aflatoxin biosynthesis enzyme inhibitor with molecular formula $C_{20}H_{18}O_6$. It was labeled as TAEMC161 and structure **19** was suggested for this alkaloid from the 1D-NMR, HMBC and NOE data (the chemical shift assignment suggested by authors is displayed):

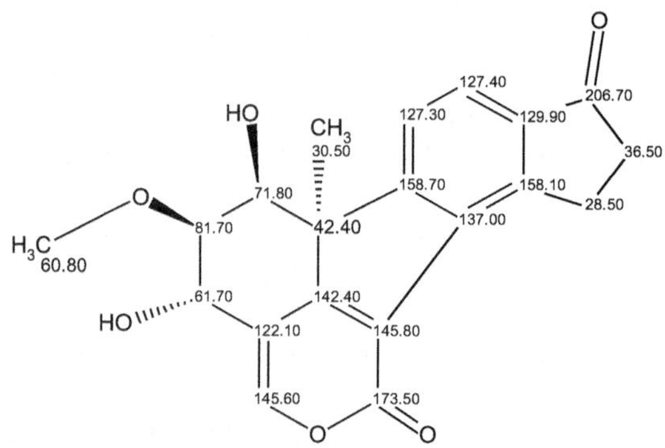

19

During the process of structure elucidation, Sakuno *et al.*[103] postulated that the ^{13}C chemical shift at 173.50 ppm was associated with the resonance of the ester group carbon. Assuming that this "axiom" is true, we obtained the following result: $k = 174 \rightarrow 80 \rightarrow 60$, $t_g = 30$ s. When the output file was ordered, structure **19** occupied the first position but with deviation values of about 4.5 ppm. Such large deviations suggest caution and warrant closer inspection of the data (the accuracy of chemical shift calculation was about 1.6–1.8 ppm).

Wipf and Kerekes[104] compared the NMR and IR spectra of TAEMC161 with a number of spectra of its structural relatives and found close similarity between the spectra of TAEMC161 and viridol (**20**). In this molecule, both carbonyl groups are ketones and the structure is in accord with the 2D-NMR data used for deducing structure **19**. Density functional theory calculations of ^{13}C chemical shifts were performed by the authors[104] for structures **19** and **20** using the GIAO approximation. It was proved that TAEMC161 is actually identical with **20**. We repeated structure generation from the 2D-NMR data without any constraints imposed on the carbonyl groups with the following result: $k = 494 \rightarrow 398 \rightarrow 272$, $t_g = 1$ min 40 s. Structure **20** was ranked first with $d_N = 2.14$ ppm and empirical prediction of the ^{13}C chemical shifts convincingly demonstrated the superiority of the revised structure **20** over the original structure **19** suggested for TAEMC161. With StrucEluc, the correct structure, supported with objective, minimized metrics, was obtained in just a few minutes.

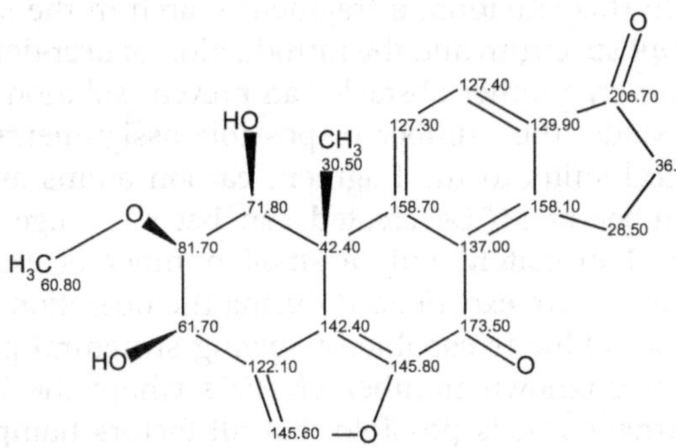

20

9.8 Performance and Limitations of StrucEluc

The different modes of operation of the StrucEluc system and the large number of examples of its application to the molecular structure elucidation of new complex organic compounds, mostly natural products, have been described in a series of articles.[13–15,70,72,83,105] These publications contain the detailed results of investigations of the system performance and the appropriate working parameters. The successful performance of the StrucEluc system and the efficiency of its application have been confirmed by elucidating more than 300 complex natural compounds. Among them, more than 100 molecules contained 30–106 skeletal atoms. Since the system is based on highly sophisticated, flexible and fast algorithms for structure generation, structure filtering and spectrum prediction, the total time for solving problems does not exceed 1 min for >80% of the problems solved. This time represents the time necessary to perform all calculations once all experimental data and all "axioms" formulated by the chemist have been entered into the program. The system provides an interface to programs developed for the calculation of physicochemical parameters of organic molecules such as $\log P$, pK_a, boiling point and many others. This allows for the prediction of many characteristics of new compounds elucidated using the system. It can also generate systematic names according to IUPAC recommendations.

It is appropriate to identify some limitations of StrucEluc. It is hoped that some of these will be removed in the process of further development of the system. They are as follows:

- The system is capable of generating only organic structures that obey classical valence theory including molecules containing formal charges. Metallo-organic compounds having non-classical structural units, for example ferrocene, cannot yet be elucidated using CASE methods.
- When there is a lack of connectivities in the 2D-NMR data, the number of structures generated and the calculation time required can become unmanageable. In this situation, a fragment search in the system database using a ^{13}C NMR spectrum and the introduction of user-defined fragments can help. However, another obstacle can prevent solution of the problem in Fragment Mode: the number of possible assignments of the experimental chemical shifts to the fragment carbon atoms and, correspondingly, the number of MCDs created can become huge and can cause system failure. Fortunately, only a small number of examples of such difficult problems were experienced during the operation of StrucEluc.
- In principle, StrucEluc is capable of solving structural problems in the presence of an unknown number of NSCs where the lengths are unknown. Nevertheless, it is possible that all factors hampering problem solving can act *simultaneously* – a lack of 2D-NMR correlations, the absence of appropriate fragments in the system DB, the number of NSCs is large and lengthening of them by more than one bond is required, *etc.* In such a situation, the program can fail and the acquisition of additional experimental data is necessary. In particular, it is expected that the combined application of both HMBC and 1,1-ADEQUATE data acquired using a CryoProbe will likely be very helpful.[25,26] If a single crystal of the unknown is available, then X-ray analysis is usually considered as a crucial experiment even though its results can also be ambiguous.[102]

It should be noted that the StrucEluc system is a commercial product of Advanced Chemistry Development (ACD/Labs) and is currently widely used in many pharmaceutical companies and universities worldwide for the identification of newly isolated natural products, synthetic impurities and degradants and to assist in the assignment of signals in 1D- and 2D-NMR spectra and the verification of structural hypotheses, *etc.*

9.9 Conclusion

CASE is an area of research that appeared at the interface frontier of spectroscopy, organic chemistry and analytical chemistry and has been developed and continually evolving over a period of more than 45 years. The development path to date has forced the developers of CASE systems to overcome many obstacles hindering the creation of a software application capable of drastically reducing the time and effort required to determine the structures of newly isolated organic compounds. Complex natural product molecules with up to 100 or more skeletal atoms can quickly (or in a reasonable time) be identified from MS and 2D-NMR data using modern CASE systems.

Among the modern CASE systems, Structure Elucidator (StrucEluc) is the most advanced at present. The system can be considered as an inference machine capable of deducing all logical consequences, without any exclusion, from the set of axioms and hypotheses that are automatically formed for each structural problem using the 1D- and 2D-NMR data and the knowledgebase of the system. As a result, the program produces a structural file containing all plausible structures and selects the most probable using NMR spectrum prediction. As the spectrum–structural information frequently may be fuzzy, inconsistent, incomplete and even false, the system provides the capability to adjust the structure elucidation process in an interactive manner. StrucEluc should therefore be considered as a potentially powerful "amplifier" of spectroscopists' intellect.

Automatic logical analysis of 2D-NMR data frequently allows the detection of the presence of COSY and HMBC correlations of "non-standard" length (those for which ${}^nJ_{HH}$, ${}^nJ_{CH}$, $n > 3$). Moreover, Fuzzy Structure Generation allows the identification of the correct structure even in those cases when an unknown number of non-standard correlations of unknown length are present in the spectra. Selecting a set of structures containing 1–3 of the most probable stereoisomers of an elucidated molecule is attained by generating all possible stereoisomers and then performing ^{13}C NMR spectrum prediction. The relative stereochemistry of large rigid molecules containing many stereocenters can be determined from NOESY/ROESY data using the StrucEluc system in semiautomatic mode.

StrucEluc is still being intensively developed in order to expand the general application of the system, to improve the workflows and usability of the system and to increase the reliability of the results. It is expected that expert systems similar to that described in this chapter will be increasingly accepted in the next decade and will ultimately be integrated directly into analytical instruments for the purpose of organic structure analysis. Efforts in this direction have already begun. Despite the many difficulties that have already been overcome to deliver on the spectroscopist's dream of "fully automated structure elucidation,"[16] there is still more work to do. Nevertheless, as the efficiency of expert systems is further enhanced, the solution of increasingly complex structural problems will be seen.

While we believe this chapter is a good representation of the state-of-the-art regarding computer-assisted structure elucidation the authors have also authored more extensive treatises regarding this work. Our recently issued books[106,107] offer much deeper examinations regarding the advantages of the CASE approach and readers are referenced to these works for more detail.

References

1. J. Lederberg, G. L. Sutherland, B. G. Buchanan, E. A. Feigenbaum, A. V. Robertson, A. M. Duffield and C. Djerassi, *J. Am. Chem. Soc.*, 1968, **91**, 2973.

2. D. B. Nelson, M. E. Munk, K. B. Gasli and D. L. Horald, *J. Org. Chem.*, 1969, **34**, 3800.
3. S. I. Sasaki, H. Abe, T. Ouki, M. Sakamoto and S. I. Ochia, *Anal. Chem.*, 1968, **40**, 2220.
4. M. E. Elyashberg and L. A. Gribov, *Zh. Prikl. Spectrosk.*, 1968, **8**, 296.
5. M. E. Elyashberg, L. A. Gribov and V. V. Serov, *Molecular Spectral Analysis and Computer*, Nauka, Moscow, 1980 (in Russian).
6. R. K. Lindsay, B. G. Buchanan, E. A. Feigenbaum and J. Lederberg, *Applications of Artificial Intelligence for Organic Chemistry: The DENDRAL Project*, McGraw-Hill, New York, 1980.
7. N. A. B. Gray, *Computer-Assisted Structure Elucidation*, Wiley, New York, 1986.
8. L. A. Gribov and M. E. Elyashberg, *Crit. Rev. Anal. Chem.*, 1979, **8**, 111.
9. N. A. B. Gray, *Anal. Chim. Acta.*, 1988, **9**, 210.
10. M. E. Elyashberg, *Russ. Chem. Revi.*, 1999, **68**, 525.
11. M. E. Munk, *J. Chem. Inf. Comput. Sci.*, 1998, **38**, 997.
12. M. E. Elyashberg, A. J. Williams and G. E. Martin, *Prog. NMR Spectrosc.*, 2008, **53**, 1.
13. M. E. Elyashberg, K. A. Blinov, S. G. Molodtsov, A. J. Williams and G. E. Martin, *J. Chem. Inf. Comput. Sci.*, 2004, **44**, 771.
14. K. A. Blinov, D. Carlson, M. E. Elyashberg, G. E. Martin, E. R. Martirosian, S. G. Molodtsov and A. J. Williams, *Magn. Reson. Chem.*, 2003, **41**, 359.
15. M. E. Elyashberg, K. A. Blinov, A. J. Williams, S. G. Molodtsov and G. E. Martin, *J. Chem. Inf. Model.*, 2006, **46**, 1643.
16. M. Elyashberg, K. Blinov, S. Molodtsov, Y. Smurnyy, A. J. Williams and T. Churanova, *J. Cheminform.*, 2009, http://www.jcheminf.com/content/1/1/3.
17. L. A. Gribov, M. E. Elyashberg and L. A. Moscovkina, *J. Mol. Struct.*, 1971, **9**, 357.
18. M. E. Elyashberg, in *Encyclopedia of Computational Chemistry*, ed. P. v. R. Schleyer, N. L. Allinger, T. Clark, J. Gasteiger, P. A. Kollman, H. F. Schaefer III and P. R. Schreiner, John Wiley & Sons Chichester, 1998, p. 1307.
19. M. E. Elyashberg, K. A. Blinov and A. J. Williams, *Magn. Reson. Chem.*, 2009, **47**, 371.
20. M. E. Elyashberg, A. J. Williams and K. A. Blinov, *Nat. Prod. Rep.*, 2010, **27**, 1296.
21. S. G. Molodtsov, M. E. Elyashberg, K. A. Blinov, A. J. Williams, E. E. Martirosian, G. E. Martin and B. Lefebvre, *J. Chem. Inf. Comput. Sci.*, 2004, **44**, 1737.
22. S. Berger and S. Braun, *200 and More NMR Experiments*, Wiley-VCH, Weinheim, 2004.
23. N. T. Nyberg, J. Ø. Duus and O. W. Sørensen, *J. Am. Chem. Soc.*, 2005, **127**, 6154.
24. N. T. Nyberg, J. Ø. Duus and O. W. Sørensen, *Magn. Reson. Chem.*, 2005, **43**, 971.
25. S. F. Cheatham, M. Kline, R. R. Sasaki, K. A. Blinov, M. E. Elyashberg and S. G. Molodtsov, *Magn. Reson. Chem.*, 2010, **48**, 571.
26. S. W. Meyer and M. Köck, *J. Nat. Prod.*, 2008, **71**, 1524.

27. A. W. T. Bristow, *Mass Spectrom. Rev.*, 2006, **25**, 99.
28. T. Kind and O. Fiehn, *BMC Bioinf.*, 2007, **8**, 105.
29. Y. Wang and M. Gu, *Anal. Chem.*, 2010, **82**, 7055.
30. K. A. Blinov, S. G. Molodtsov, M. E. Elyashberg, T. S. Churanova and A. J. Williams, presented in part at the SMASH-2010, Portland, Oregon, September 26th–29th, 2010.
31. L. A. Gribov, M. E. Elyashberg and V. V. Serov, *J. Mol. Struct.*, 1978, **50**, 371.
32. A. Tarantola, *Inverse Problem Theory and Methods for Model Parameter Estimation*, SIAM, Philadelphia, 2005.
33. M. E. Elyashberg, K. A. Blinov, Y. Smurnyy, T. Churanova and A. J. Williams, *Magn. Reson. Chem.*, 2010, **48**, 219.
34. L. Griffiths, *Magn. Reson. Chem.*, 2000, **38**, 444.
35. L. Griffiths, *Magn. Reson. Chem.*, 2000, **38**, 194.
36. L. Griffiths and J. D. Bright, *Magn. Reson. Chem.*, 2002, **40**, 623.
37. L. Griffiths and R. Horton, *Magn. Reson. Chem.*, 2004, **42**, 1012.
38. B. C. Hamper, D. M. Synderman, T. J. Owen, A. M. Scates, D. C. Owsley, A. S. Kesselring and R. C. Chott, *J. Comb. Chem.*, 1999, **1**, 140.
39. B. Lefebvre, March 3, 2005, NMR Discussion Group, available from http://www.acdlabs.com/publish/publ05/nmrdg_structure_identification.html.
40. D. M. Grant and E. G. Paul, *J. Am. Chem. Soc.*, 1964, **86**, 2984.
41. A. Fürst and E. Pretsch, *Anal. Chim. Acta*, 1990, **229**, 17.
42. J.-T. Clerc and H. A. Sommerauer, *Anal. Chim. Acta*, 1977, **95**, 33.
43. L. Chen and W. Robien, *Anal. Chem.*, 1993, **65**, 12282.
44. *Specinfo*, Chemical Concepts GmbH, Weinheim.
45. W. Bremser, *Magn. Reson. Chem.*, 1985, **23**, 271.
46. L. Chen and W. Robien, *Chemom. Intell. Lab. Syst.*, 1993, **19**, 217.
47. C. W. Crandall, N. A. B. Gray and D. H. Smith, *J. Chem. Inf. Comput. Sci.*, 1982, **22**, 48.
48. Advanced Chemistry Development, *ACD/NMR Predictors. Prediction suite includes ^{1}H, ^{13}C, ^{15}N, ^{19}F, ^{31}P NMR prediction*, 2010.
49. H. Kalchhauser and W. Robien, *J. Chem. Inf. Comput. Sci.*, 1985, **25**, 103.
50. Cambridge Soft Corporation, *CS Chem Draw PRO*.
51. Upstream Solutions, *NMR Prediction Products (SpecTool)*.
52. J. Zupan and J. Gasteiger, *Neural Networks for Chemists*, VCH, Weinheim, 1993.
53. J. Meiler, R. Meusinger and M. Will, *J. Chem. Inf. Comput. Sci.*, 2000, **40**, 1169.
54. J. Meiler, W. Maier, M. Will and R. Meusinger, *J. Magn. Reson.*, 2002, **157**, 242.
55. V. Kvasnicka, *J. Math. Chem.*, 1991, **6**, 63.
56. J. P. Doucet, A. Panaye, E. Feuilleaubois and P. J. Ladd, *J. Chem. Inf. Comput. Sci.*, 1993, **33**, 320.
57. Y. Miyashita, H. Yoshida, O. Yaegashi, T. Kimura, H. Nishiyama and S. Sasaki, *J. Mol. Struct.: THEOCHEM*, 1994, **311**, 241.
58. O. Ivanciuc, J.-P. Rabine, D. Cabrol-Bass, A. Panaye and J. P. Doucet, *J. Chem. Inf. Comput. Sci.*, 1996, **36**, 644.

59. W. Bremser, *Anal. Chim. Act. Comp. Techn. Optimiz.*, 1978, **2**, 355.
60. N. A. B. Gray, J. G. Nourse, C. W. Crandall, D. H. Smith and C. Djerassi, *Org. Magn. Res.*, 1981, **15**, 375.
61. V. Schutz, V. Purtuc, S. Felsinger and W. Robien, *Fresenius J. Anal. Chem*, 1997, **359**, 33.
62. W. Robien, *Nachr. Chem. Tech. Lab.*, 1998, **46**, 74.
63. W. Robien, CSEARCH; http://felix.orc.univie.ac.at/~wr/csearch_server_info.html.
64. Y. D. Smurnyy, K. A. Blinov, T. S. Churanova, M. E. Elyashberg and A. J. Williams, *J. Chem. Inf. Model.*, 2008, **48**, 128.
65. K. A. Blinov, Y. D. Smurnyy, M. E. Elyashberg, T. S. Churanova, M. Kvasha, C. Steinbeck, B. A. Lefebvre and A. J. Williams, *J. Chem. Inf. Model.*, 2008, **48**, 550.
66. K. A. Blinov, E. D. Smurnyy, T. S. Curanova, M. E. Elyashberg and A. J. Williams, *Chemom. Intell. Lab. Syst.*, 2009, **97**, 91.
67. M. E. Elyashberg, K. A. Blinov, A. J. Williams, S. G. Molodtsov and G. E. Martin, *J. Chem. Inf. Model.*, 2007, **47**, 1053.
68. D. Neuhaus and M. Williamson, *The Nuclear Overhauser Effect in Structural and Conformational Analysis*, Wiley, New York, 2000.
69. H. M. Ge, B. Huang, S. H. Tan, D. H. Shi, Y. C. Song and R. X. Tan, *J. Nat. Prod.*, 2006, **69**, 1800.
70. M. E. Elyashberg, K. A. Blinov, E. R. Martirosian, S. G. Molodtsov, A. J. Williams and G. E. Martin, *J. Heterocycl. Chem.*, 2003, **40**, 1017.
71. G. V. Subbaraju, M. Vanisree, C. V. Rao, C. Sivaramakrishna, P. Sridhar, B. Jayprakasam and M. G. Nair, *J. Nat. Prod.*, 2006, **69**, 1790.
72. G. E. Martin, B. D. Hadden, C. E. Russell, D. J. Kaluzny, J. E. Guido, W. K. Duholke, B. A. Stiemsma, T. J. Thamann, R. C. Crouch, K. A. Blinov, M. E. Elyashberg, E. R. Martirosian, S. G. Molodtsov, A. J. Williams and P. L. J. Schiff, *J. Heterocycl. Chem.*, 2002, **39**, 1241.
73. D. O. Collins, W. F. Reynolds and P. B. Reese, *J. Nat. Prod.*, 2004, **67**, 179.
74. J. M. Seco, E. Quiñoá and R. Riguera, *Chem. Rev.*, 2004, **104**, 17.
75. J. A. Dale and H. S. Mosher, *J. Am. Chem. Soc.*, 1973, **95**, 2543.
76. M. E. Elyashberg, K. A. Blinov and A. J. Williams, *Magn. Reson. Chem.*, 2009, **47**, 333.
77. C. Fattorusso, E. Stendardo, G. Appendino, E. Fattorusso, P. Luciano, A. Romano and O. Taglialatela-Scafati, *Org. Lett.*, 2007, **9**, 2377.
78. K. N. Maloney, M. Fujita, U. S. Eggert, F. C. Schroeder, C. M. Field, T. J. Mitchison and J. Clardy, *J. Nat. Prod.*, 2008, **71**, 1927.
79. Y. D. Smurnyy, M. E. Elyashberg, K. A. Blinov, B. A. Lefebvre, G. E. Martin and A. J. Williams, *Tetrahedron*, 2005, **61**, 9980.
80. M. Mitchell, *An Introduction to Genetic Algorithms*, MIT Press, Cambridge, MA, 1996.
81. Y.-Y. Lin, M. Risk, S. M. Ray, D. Van Engen, J. Clardy, J. Golik, J. C. James and K. Nakanishi, *J. Am. Chem. Soc.*, 1981, **103**, 6773.
82. M. S. Lee, D. J. Repeta, K. Nakanishi and M. G. Zagorksi, *J. Am. Chem. Soc.*, 1986, **108**, 7855.

83. K. A. Blinov, M. E. Elyashberg, E. R. Martirosian, S. G. Molodtsov, A. J. Williams, M. M. H. Sharaf, P. L. J. Schiff, R. C. Crouch, G. E. Martin, C. E. Hadden, J. E. Guido and K. A. Mills, *Magn. Reson. Chem.*, 2003, **41**, 577.
84. G. E. Martin, D. J. Russell, K. A. Blinov, M. E. Elyashberg and A. J. Williams, *Ann. Rep. NMR Spectrosc.*, 2003, **1**, 1.
85. B. D. Hilton and G. E. Martin, *J. Nat. Prod.*, 2010, **73**, 1465.
86. A. N. Tackie, G. L. Boye, M. H. M. Sharaf, P. L. J. Schiff, R. C. Crouch, T. D. Spitzer, R. L. Johnson, J. Dunn, D. Minick and G. E. Martin, *J. Nat. Prod.*, 1993, **56**, 653.
87. G. J. Sharman, I. C. Jones, M. P. Parnell, M. C. Willis, M. F. Mahon, D. V. Carlson, A. Williams, M. Elyashberg, K. Blinov and S. G. Molodtsov, *Magn. Reson. Chem.*, 2004, **42**, 567.
88. N. Lysek, E. Rachor and T. Lindel, *Z. Naturforsch.*, 2002, **57C**, 1056.
89. J.-P. Bouillon, B. Tinant, J.-M. Nuzillard and C. Portella, *Synthesis.*, 2004, 711.
90. G. N. Belofsky, M. Anguera, P. R. Jensen, W. Fenical and M. Köck, *Chem. Eur. J*, 2000, **6**, 1355.
91. C. Steinbeck, V. Spitzer, M. Starosta and G. von Poser, *J. Nat. Prod.*, 1997, **60**, 627.
92. A. Codina, R. W. Ryan, R. Joyce and D. S. Richards, *Anal. Chem.*, 2010, **82**, 9127.
93. A. Bagno, F. Rastrelli and G. Saielli, *Chemistry*, 2006, **12**, 5514.
94. A. Bagno and G. Saielli, *Theor. Chem. Acc.*, 2007, **117**, 603.
95. G. Barone, L. Gomez-Paloma, D. Duca, A. Silvestri, R. Riccio and G. Bifulco, *Chemistry*, 2002, **8**, 3233.
96. V. Barone, P. Cimino, O. Crescenzi and M. Pavone, *J. Mol. Struct.*, 2007, **811**, 323.
97. P. Cimino, L. Gomez-Paloma, D. Duca, R. Riccio and G. Bifulco, *Magn. Reson. Chem.*, 2004, **42**, S26.
98. S. D. Rychnovsky, *Org. Lett.*, 2006, **8**, 2895.
99. M. E. Elyashberg, K. A. Blinov and A. J. Williams, *Magn. Reson. Chem.*, 2009, **47**, 371.
100. A. Balandina, D. Saifina, V. Mamedov and S. Latypov, *J. Mol. Struct.*, 2006, **791**, 77.
101. A. A. Balandina, V. A. Mamedov, E. A. Khafizova and S. K. Latypov, *Russ. Chem. Bull.*, 2006, **55**, 2256.
102. K. C. Nicolaou and S. A. Snyder, *Angew. Chem., Int. Ed.*, 2005, **44**, 1012.
103. E. Sakuno, K. Yabe, T. Hamasaki and H. Nakajima, *J. Nat. Prod.*, 2000, **63**, 1677.
104. P. Wipf and A. D. Kerekes, *J. Nat. Prod.*, 2003, **66**, 716.
105. M. E. Elyashberg, K. A. Blinov, A. J. Williams, S. G. Molodtsow and E. R. Martirosian, *J. Nat. Prod.*, 2002, **65**, 693.
106. M. E. Elyashberg, A. J. Williams and K. A. Blinov, *Contemporary Computer-Assisted Approaches to Molecular Structure Elucidation*, RSC Publishing, Cambridge, 2012.
107. M. E. Elyashberg and A. J. Williams, *Computer-based Structure Elucidation from Spectral Data. The Art of Solving Problems*, Springer, Heidelberg, 2015.

CHAPTER 10

Multi-dimensional Spin Correlations by Covariance NMR

DAVID A. SNYDER*[a] AND RAFAEL BRÜSCHWEILER[b,c]

[a] Department of Chemistry, William Paterson University, Wayne, NJ 07470, USA; [b] Department of Chemistry and Biochemistry and Campus Chemical Instrument Center, The Ohio State University, Columbus, OH 43210, USA; [c] Chemical Sciences Laboratory, Department of Chemistry and Biochemistry and National High Magnetic Field Laboratory, Florida State University, Tallahassee, FL 32306, USA
*Email: snyderd@wpunj.edu

10.1 Introduction

Covariance nuclear magnetic resonance (NMR) spectroscopy encompasses methods that establish correlations between nuclear spins by means of statistical covariances.[1–3] The covariance transform serves as a complement to, or replacement for, the Fourier transform (FT) along indirect or direct dimensions in multi-dimensional NMR datasets. In its most basic form, the (direct) covariance transform applied to a homonuclear 2D-NMR data set, such as a 2D-TOCSY[4] or 2D-NOESY,[5] endows the indirect dimension with the same high resolution as the direct dimension, and thereby enhances the spectral resolution, reduces the experimental NMR time, or both.

Covariance of traces along the direct dimension of one or more proton-detected heteronuclear spectra yields a homonuclear spectrum correlating two relatively insensitive nuclei.[6] For example, indirect covariance of a ^{1}H–^{13}C HMBC spectrum[7] yields a spectrum that correlates carbon atoms separated by 1–6 bonds, but with a sensitivity characteristic of a

Modern NMR Approaches To The Structure Elucidation of Natural Products,
Volume 1: Instrumentation and Software
Edited by Antony J. Williams, Gary E. Martin and David Rovnyak

Published by the Royal Society of Chemistry, www.rsc.org

proton-detected spectrum rather than that of a carbon-detected homonuclear spectrum. Techniques such as unsymmetrical[8] and generalized indirect covariance[9] extend the covariance formalism to the reconstruction of non-symmetric NMR spectra. Generalized indirect covariance (GIC) of a ^{1}H–^{13}C HMBC spectrum with a ^{1}H–^{1}H TOCSY spectrum extends the reach of the HMBC spectrum to probe correlations between protons and carbons separated by more than four bonds,[9] whereas unsymmetrical covariance of a ^{1}H–^{13}C HSQC spectrum with a ^{1}H–^{13}C 1,1-ADEQUATE spectrum yields a dataset equivalent to a ^{13}C–^{13}C COSY spectrum.[10–12] Doubly indirect covariance can also provide ^{13}C–^{13}C COSY-type datasets with sensitivities characteristic of proton-detected spectra.[13]

The ability of covariance NMR to reconstruct homonuclear ^{13}C–^{13}C spectra with sensitivities characteristic of proton-detected spectra makes covariance NMR a valuable tool for the study of natural products, which may be present in small quantities and with ^{13}C at natural abundance. This chapter delineates the principles upon which covariance NMR rests, and highlights the benefits of covariance NMR for the reconstruction of homonuclear spectra, and also heteronuclear spectra correlating rare spins,[14,15] for which low experimental sensitivity impedes direct measurement. This chapter also describes how covariance NMR facilitates the elucidation of natural product structures.

10.2 Theory of Covariance NMR

The theoretical basis of covariance NMR rests upon three pillars: (1) 2D-NMR spectra can be treated as matrices and hence they are amenable to the operations of matrix algebra; (2) the experimental acquisition of multi-dimensional NMR spectra involves the acquisition of a set of 1D-NMR spectra in which statistical covariances between peak intensities correspond to physical correlations between spin-active nuclei; and (3) Parseval's theorem, which permits one to perform of covariance analysis in both the time and frequency domains.[1,16] Consider a 2D-NMR spectrum recorded with N_1 points in the indirect dimension and N_2 points in the direct dimension and subjected to Fourier transformation along the directly detected dimension but not the indirect dimension. The first pillar of covariance NMR conceptualizes this mixed time–frequency domain spectrum $\mathbf{M}$ as an $N_1 \times N_2$ matrix, subject to the operations of matrix algebra. The second pillar indicates that statistical covariances between column vectors of $\mathbf{M}$ correspond to physical correlations between spin-systems, thus the covariance matrix

$$\mathbf{C}^2 = \mathbf{M}^{\mathbf{T}} \cdot \mathbf{M}/\mathrm{N}_1 \tag{10.1}$$

is physically meaningful. We assume that the mean of the oscillating signals in the indirect time domain averages to zero, also known as "axial peak suppression," hence the matrix $\mathbf{C}^2$ is, indeed, the covariance matrix of $\mathbf{M}$, hence the name covariance NMR. We will drop the global scaling factor of $1/N_1$ from now on. Additional mathematical details can be found in Trbovic *et al.*[3]

The third pillar gives further meaning to the intuition captured by the second pillar. Consider the 2D Fourier transformed spectrum **S**, which is obtained from dataset **M** after Fourier transformation along the indirect dimension (columns), phase correction, and removal of the imaginary parts. By Parseval's theorem it follows that

$$\mathbf{S}^{\mathrm{T}} \cdot \mathbf{S} = \mathbf{M}^{\mathrm{T}} \cdot \mathbf{M} \tag{10.2}$$

If **S** is symmetric and positive semi-definite (*i.e.* has only non-negative eigenvalues), which is true for "symmetric" spectra with intense diagonal signals such as NOESY and TOCSY spectra with relatively short mixing times, $\mathbf{S} = (\mathbf{S}^{\mathrm{T}} \cdot \mathbf{S})^{1/2}$, then we may calculate the direct covariance spectrum as

$$\mathbf{C} = (\mathbf{S}^{\mathrm{T}} \cdot \mathbf{S})^{1/2} = (\mathbf{M}^{\mathrm{T}} \cdot \mathbf{M})^{1/2} = \mathbf{S} \tag{10.3}$$

obviating the need for Fourier transformation in the indirect dimension.[1,2] Eqn (10.3) shows that for the above-mentioned experiments, recorded with a sufficiently large number of N_1 increments, the covariance spectrum **C** is identical with the 2D FT spectrum **S**.

In practice, it is often still advisable to perform Fourier transformation along the indirect dimension to facilitate baseline correction in that dimension. The key feature of the direct covariance matrix is that for symmetric spectra recorded with values of $N_1 < N_2$, the direct covariance matrix, even as calculated in the frequency domain, $\mathbf{C} = (\mathbf{S}^{\mathrm{T}} \cdot \mathbf{S})^{1/2}$, is an $N_2 \times N_2$ matrix usually providing an excellent approximation to the full resolution spectrum recorded with N_2 points in the indirect dimension (Figure 10.1). For typical natural product samples, direct covariance accurately yields a high-resolution spectrum from data recorded with as few as 48 t_1 increments, *i.e.* in a small fraction of the time required to obtain data with the same high resolution by 2D FT processing.[17] Therefore, direct covariance NMR, which is applicable to any inherently symmetric 2D-NMR experiment, enables high-resolution NMR data to be obtained with relatively short measurement times.

Since the covariance spectrum is symmetric, asymmetric artifacts are either suppressed or propagated. In the case of strong t_1-noise ridges, direct covariance may display additional signals. On the other hand, computation of the *indirect* covariance matrix defined by

$$\mathbf{C}_{\mathbf{ind}} = (\mathbf{S} \cdot \mathbf{S}^{\mathrm{T}})^{1/2} = (\mathbf{M} \cdot \mathbf{M}^{\mathrm{T}})^{1/2} \tag{10.4}$$

in which the covariance is calculated between rows instead of columns, helps to suppress ridge artifacts parallel to the indirect dimension, such as solvent artifacts.[18] In studies of natural products dissolved in multi-solvent systems, indirect covariance NMR may play an important role in solvent suppression as other processing (*e.g.* time-domain baseline correction) and even experimental (*e.g.* gradient-based) techniques are often only suitable for eliminating the signal arising from a single solvent. In contrast to a direct covariance spectrum, the resolution of an indirect covariance spectrum, which is an $N_1 \times N_1$ matrix, is limited by the resolution of the indirect dimension.

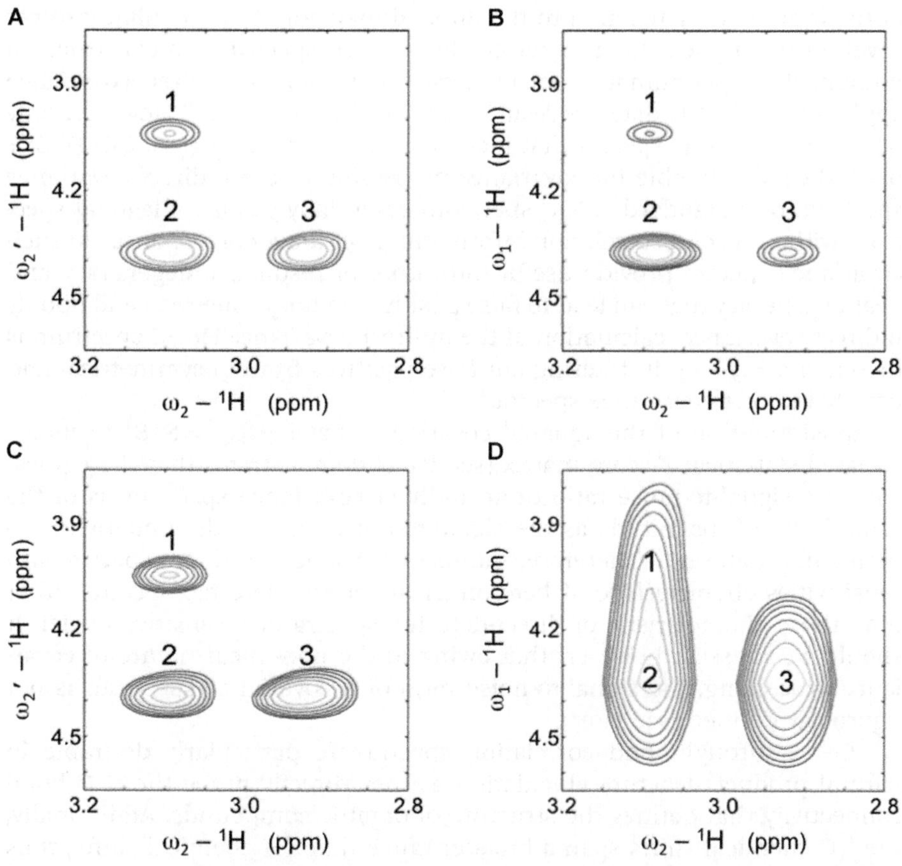

Figure 10.1 (A, B) Covariance *versus* (C, D) 2D Fourier transform (FT) TOCSY spectra of the protease inhibiting peptide antipain collected with different numbers of points along the indirect dimension. (A, C) TOCSY spectrum collected with 256 complex points along the indirect dimension. (B, D) TOCSY spectrum truncated to have only 64 complex points along the indirect dimension. Note that the covariance spectra possess the same resolution along the indirect dimension and look mostly identical, thus demonstrating the resolution enhancement provided by the direct covariance transform. However, the 2D FT TOCSY spectrum (D) with only 64 complex points along the indirect dimension fails to resolve one of the phenylalanine Hβ–Hα cross peaks (1; the other such cross peak is peak 3) from the arginine Hδ–Hα (2) cross peak, whereas in the corresponding covariance spectrum (C) the peaks are well resolved.

10.3 Homonuclear NMR *via* Indirect and Doubly Indirect Covariance

In general, direct covariance maps the high resolution of the direct dimension onto the indirect dimension. On the other hand, indirect covariance maps the resolution, sweep width, nucleus probed, and other characteristics of the indirect dimension onto the direct dimension. In particular, indirect covariance, applied to a heteronuclear X–^{1}H spectrum, reconstructs a homonuclear spectrum for nuclei of type X. For example, indirect covariance applied to a ^{1}H–^{13}C data set results in a ^{13}C–^{13}C spectrum.[6] However, only heteronuclear X–^{1}H spectra that include a "relay" effect (*e.g.* HSQC–TOCSY or HMBC) are suitable for covariance processing as the indirect covariance spectrum of a standard HSQC spectrum essentially yields a diagonal spectrum without useful correlation information. Off-diagonal responses in such covariance spectra provide useful indicators of resonance degeneracy and near degeneracy that can lead to false positives in unsymmetrical and doubly indirect covariance; calculation of the indirect covariance HSQC spectrum is therefore a key step in filtering out false positives from unsymmetrical and doubly indirect covariance spectra.[13,19]

Transformation of the squared covariance matrix $\mathbf{C_{ind}}^2 = \mathbf{S} \cdot \mathbf{S}^\mathrm{T}$ to its associated statistical *Z*-score matrix (see later) demonstrates that the equivalent of a signal-to-noise ratio of an indirect covariance spectrum is of the same order of magnitude as the signal-to-noise ratio of the underlying experimental dataset.[20] Therefore, indirect covariance produces spectra with sensitivities characteristic of heteronuclear, proton-detected spectra rather than those characteristic of direct-detected spectra of insensitive nuclei. It should be stressed, however, that owing to the non-linear nature of covariance processing, the signal-to-noise ratio of a covariance spectrum is not rigorously defined (see later).

^{13}C–^{13}C through-bond correlation spectra are particularly desirable in natural product structure elucidation as they directly probe the C–C bond connectivity that defines the structure of organic compounds. Additionally, the ^{13}C chemical shifts span a broader range than ^{1}H chemical shifts, thus reducing the likelihood of chemical shift degeneracy. As described above, indirect covariance allows for the reconstruction of ^{13}C–^{13}C correlations at sensitivities comparable to those of the experimentally obtained ^{13}C–^{1}H correlation spectra used in the covariance calculation. For example, indirect covariance of a ^{1}H–^{13}C HMBC spectrum probes medium-range [1–6 bond (*i.e.* between carbons directly bonded to each other or connected by a series of up to six bonds)] carbon–carbon correlations (Figure 10.2C) whereas indirect covariance processing of a heteronuclear ^{1}H–^{13}C HSQC–TOCSY spectrum yields a homonuclear ^{13}C–^{13}C TOCSY spectrum,[6] each with the sensitivity of a proton detected experiment.

The *doubly indirect covariance* (DIC) method converts homonuclear proton correlation spectra to homonuclear ^{13}C correlation spectra. Specifically, doubly indirect covariance can generate ^{13}C–^{13}C COSY-type spectra for which the correspondence to molecular structure is self-evident.[13] Unsymmetrical or, alternatively, generalized indirect (see later) covariance (GIC) of a ^{1}H–^{13}C HSQC spectrum with a ^{1}H–^{13}C 1,1-ADEQUATE spectrum also reconstructs ^{13}C–^{13}C COSY-type spectra.[10–12] Figure 10.3 shows how doubly indirect covariance yields a COSY-type spectrum isomorphic, in terms of graph theory, to the structure of isoleucine.[13] Figure 10.4 shows

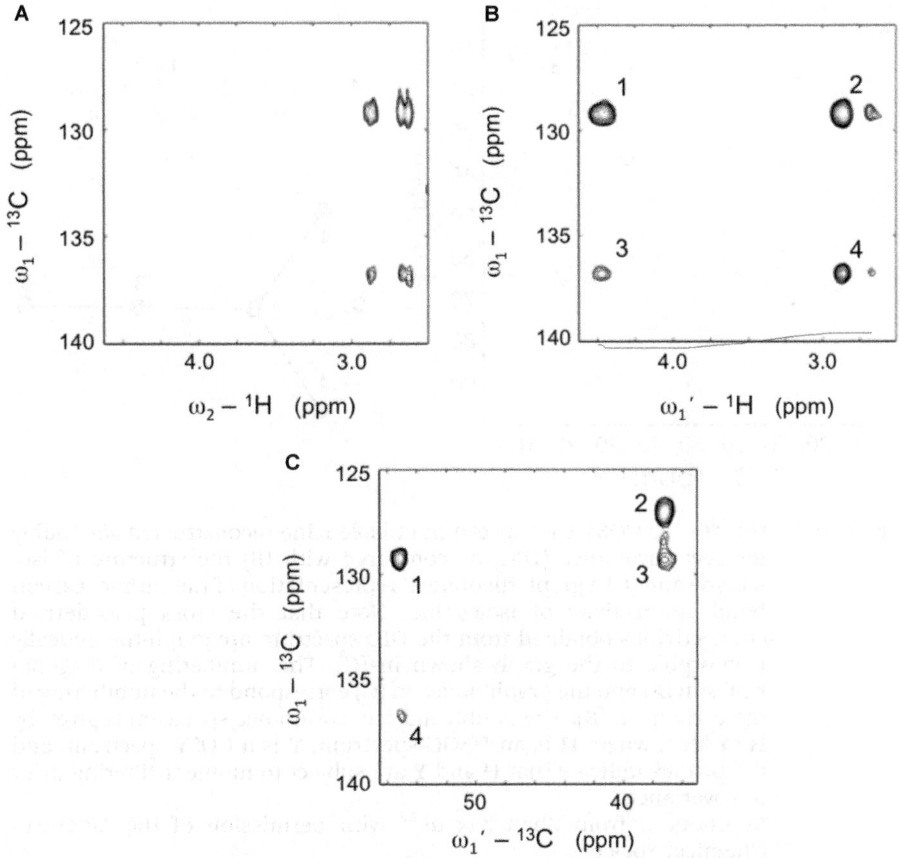

Figure 10.2 (A) 2D ^{1}H–^{13}C HMBC spectrum, (B) 2D GIC $[HMBC*TOCSY]^{1/2}$ (for a detailed discussion of the $[X*Y]^{\lambda}$ notation, see ref. 9), and (C) indirect covariance spectrum calculated from the ^{1}H–^{13}C HMBC spectrum of the protease inhibiting peptide antipain. The displayed portions of the spectrum contain peaks arising from the phenylalanine residue. Peaks in the $[HMBC*TOCSY]^{1/2}$ spectrum include (1) Hα–Cδ, (2) Hβ–Cδ, (3) Hβ–Cγ and (4) Hα–Cγ. The corresponding region of the HMBC spectrum lacks cross peaks between the Hα and aromatic carbons and TOCSY transfer is generally inefficient between aliphatic and aromatic protons. However, the combination of TOCSY and HMBC information *via* GIC is capable of recovering longer range, through-bond connectivities. Indirect covariance of the HMBC spectrum also yields correlations between aliphatic and aromatic carbons that are difficult to obtain directly from Fourier transform NMR including (1) Cα–Cδ, (2) Cβ–Cη, (3) Cβ–Cδ (with the satellite peak belonging to Cβ–Cε) and (4) Cα–Cγ. Note that in antipain the two Cδ and Cε carbons in the aromatic ring have degenerate chemical shifts.

the GIC ^{13}C–^{13}C COSY (HSQC–1,1-ADEQUATE) spectrum of the drug candidate Dinaciclib, the structure of which has features typical of many natural product structures.[12]

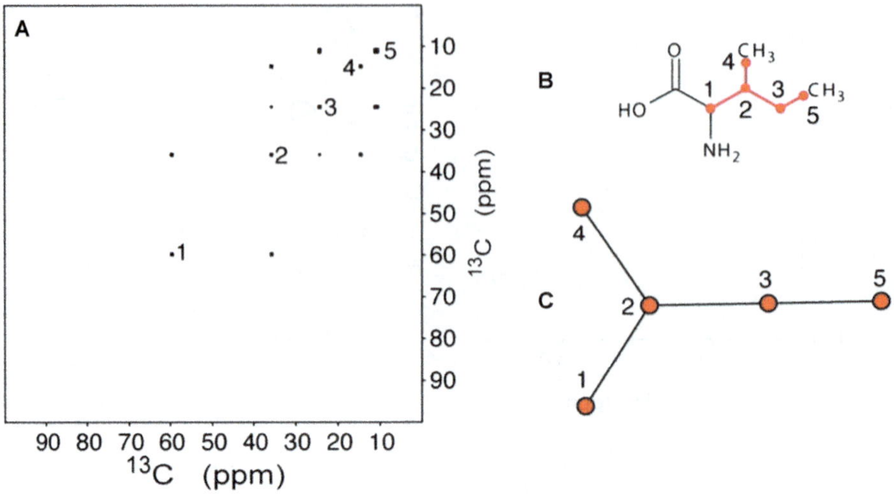

Figure 10.3 (A) ^{13}C–^{13}C COSY-type spectrum of isoleucine reconstructed *via* doubly indirect covariance (DIC) as compared with (B) the structure of isoleucine and (C) a graph theoretical representation of the carbon–carbon bond connectivity of isoleucine. Note that the cross peak-derived connectivities obtained from the DIC spectrum are graph-theoretically isomorphic to the graph shown in (C). The numbering of diagonal peaks in (A) and the graph nodes in (C) correspond to the numbering of the carbons in (B). The doubly indirect covariance spectrum is given by $\mathbf{H}'*\mathbf{Y}'*\mathbf{H}'^{T}$, where **H** is an HSQC spectrum, **Y** is a COSY spectrum, and the primes indicate that **H** and **Y** are subject to moment filtering prior to covariance.
Reproduced from Zhang, *et al.*[13] with permission of the American Chemical Society.

The combination of several factors leads to the higher sensitivity of covariance-reconstructed ^{13}C–^{13}C spectra compared with their direct-detected analogs. Covariance reconstruction of a ^{13}C–^{13}C spectrum from a proton-detected spectrum maintains the $(\gamma_H/\gamma_C)^{3/2} \approx 8$-fold increase of proton detection sensitivity over carbon detection.[21] Moreover, since the natural abundance of ^{13}C is only 1.1%, an experiment correlating two ^{13}C atoms in a sample without isotope enrichment has 0.012% of the signal of the same experiment performed on a sample with 100% isotope enrichment. On the other hand, a heteronuclear experiment on such a non-enriched sample has 1.1% of the signal of the corresponding isotopically pure sample and thus is over 90 times more sensitive than the homonuclear experiment.

The combined effect of the higher sensitivity of proton detection and the smaller "penalty" due to low natural abundance of ^{13}C can in principle yield up to a 700-fold increase in sensitivity for ^{1}H–^{13}C and hence for covariance-reconstructed ^{13}C–^{13}C spectra over the sensitivity available by direct acquisition of ^{13}C–^{13}C spectra. In practice, however, larger proton linewidths and the presence of proton–proton *J*-couplings decrease peak intensities in

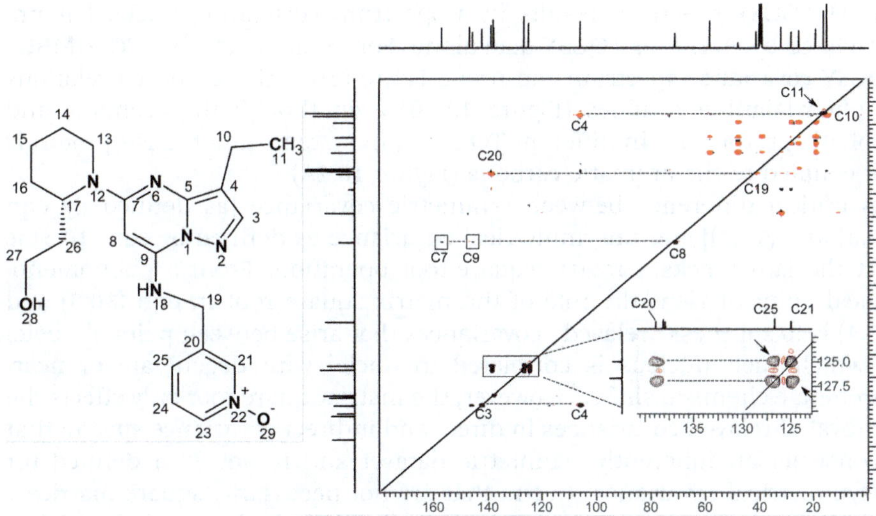

Figure 10.4 $^{13}C-^{13}C$ COSY spectrum of Dinaciclib, a compound with a molecular mass and functional groups typical of many secondary metabolites as in the structure shown (with carbon numbering). The spectrum was obtained by covariance of a $^{1}H-^{13}C$ multiplicity-edited gHSQC and $^{1}H-^{13}C$ 1,1-ADEQUATE spectrum, where negative peaks (red) indicate correlations to methylene carbons.[10,12] Lines demonstrate "steps" in a "COSY walk" used in chemical shift assignment and structure elucidation. The expanded region shows how peak assignment reaches into the pyridine ring.
Reproduced from Martin and Sunseri[12] with permission of Elsevier.

proton-detected spectra and reduce the potential sensitivity advantage of covariance NMR spectra relative to their directly detected analogs. Nevertheless, the sensitivity advantage of covariance NMR is still significant and covariance techniques greatly expand the arsenal of datasets applicable in the characterization of natural products and natural product mixtures.

10.4 Unsymmetrical and Generalized Indirect Covariance

The concept of indirect covariance NMR has been extended by Blinov, Martin, and co-workers and Kupče and Freeman to the reconstruction of non-symmetric spectra from pairs of spectra **F** and **G**:

$$\mathbf{C} = \mathbf{F} \cdot \mathbf{G}^{\mathrm{T}} \tag{10.5}$$

termed unsymmetrical covariance or hyperdimensional NMR, respectively.[8,22,23] Unsymmetrical covariance NMR concatenates spectra by calculation of covariances along their common, direct-detected dimension. For instance, unsymmetrical covariance of a $^{1}H-^{13}C$-HMBC spectrum with a

^{1}H–^{1}H TOCSY spectrum results in a spectrum correlating nuclei left uncorrelated by HMBC or TOCSY data alone. For example, the ^{1}H–^{13}C–HMBC–TOCSY covariance spectrum can probe Hα to aromatic carbon correlations in phenylalanine residues (Figure 10.2B) even though the aromatic and aliphatic protons are in different TOCSY spin systems and the Hα proton is not coupled to the aromatic carbons (Figure 10.2A).

A critical difference between symmetric covariance [as defined by eqn (10.3) and (10.4)] and unsymmetrical covariance as defined by eqn (10.5) is that the latter lacks a matrix square-root operation. From a phenomenological point of view, the role of the matrix square root in eqn (10.3) and (10.4) is to suppress "relayed" covariances that arise between pairs of nuclei in which each nucleus is correlated to nuclei with degenerate or near-degenerate chemical shifts.[9] However, the matrix square root only effects the removal of relayed covariances in direct and indirect covariance spectra that reconstruct an inherently symmetric dataset and is not even defined for unsymmetrical covariance spectra that are not necessarily square matrices.

In doubly indirect covariance, "moment filtering" is also applied, which is a masking procedure to eliminate automatically regions of the spectra that would lead to false peaks.[13] Moment filtering pursues similar goals to other filtering procedures previously applied in the context of unsymmetrical covariance.[19,24] Generalized Indirect Covariance (GIC) involves a simple extension of the unsymmetrical covariance procedure that embeds the unsymmetrical covariance spectrum as a sub-matrix of a larger symmetric matrix, which is then subjected to a matrix square root in order to suppress relayed covariance signals.[9]

In FT-NMR, signal-to-noise ratios (S/Ns) provide a convenient statistic for evaluating the sensitivity of a spectrum by comparison of signal intensities with a (mostly) uniform noise floor. Unlike FT, the covariance transform of eqn (10.5) constitutes a non-linear operation, which non-linearly scales not only the signal but also the noise. This non-linear scaling renders the noise floor of a spectrum non-uniform, *i.e.* potentially induces a chemical shift dependence in the noise level of covariance spectra. In this case, sensitivity estimates (*e.g.* from S/Ns comparing signal intensities with signal-free baseline regions) either under- or overestimate the sensitivity of the non-linearly processed spectrum.[20,25] The *Z*-matrix formalism[20] converts the indirect covariance spectrum **C** of eqn (10.5) into one that has a uniform noise floor that lends itself to the same type of sensitivity analysis as a 2D FT spectrum. Calculation of such *Z*-matrices for unsymmetrical covariance spectra confirms the general observation that unsymmetrical covariance spectra preserve the sensitivity of their underlying datasets.[20]

10.5 Computational Aspects

Practical implementations of covariance NMR must take into account several computational concerns. Perhaps most important is the computationally efficient implementation of the matrix square root as it is critical for many practical applications of the covariance method.[26] The singular value decomposition (SVD)[3] is the method used in implementations of covariance NMR such as the Covariance NMR Toolbox for MATLAB and OCTAVE.[27]

Additionally, covariance NMR assumes that the spectrum it reconstructs, when represented as a matrix, is positive semi-definite or, in the case of GIC, that it is a sub-matrix of a positive semi-definite matrix. This applies to the reconstruction of NOESY-type spectra but not necessarily for the reconstruction of TOCSY-type spectra, which potentially lack certain diagonal peaks at longer mixing times. To remedy this problem, the 2D FT spectrum **F** can be regularized by adding a properly scaled unit matrix to **F** prior to the covariance transform, which is followed by subtraction of the same unit matrix from the output covariance matrix **C**.[28]

Programs implementing covariance NMR include the ACD/NMR Processor package[29] and also stand-alone programs and Bruker AU programs implementing particular covariance techniques such as covNMR/covNMR2.0.au,[30] which implement both direct and indirect covariance. The recently released Covariance NMR Toolbox[27] bundles together many covariance methods (including direct, indirect, unsymmetrical, generalized indirect, and 4D covariance[31]) into a single, easy-to-use software package compatible with both the MATLAB and OCTAVE computing environments. This toolbox has facilities for viewing 2D-NMR spectra in MATLAB and also for storing and manipulating 2D-NMR spectra as MATLAB/OCTAVE arrays, thus allowing users to explore novel heuristics, such as the application of non-negative matrix factorization,[32] to take advantage of the matrix representation of NMR spectra. As MATLAB/OCTAVE toolboxes are collections of functions written in the high-level MATLAB/OCTAVE programming language, they are readily modified and users can easily use this toolbox as a starting point to develop further extensions to the palette of methods that comprise covariance NMR.

10.6 Applications of Covariance NMR to Natural Product Structure Elucidation

Figure 10.2 illustrates the role that indirect covariance and GIC (or similar methods) play in the elucidation of natural product structures as exemplified using data recorded on an unlabeled sample (50 mM in D_2O) of the protease inhibitor antipain (obtained from Sigma Chemicals), a bacterial tripeptide with an additional phenylalanine residue attached at the N-terminus *via* a carbamoyl linkage.[33] Without covariance, a ^1H–^{13}C HMBC spectrum (Figure 10.2A) cannot directly correlate the phenylalanine Hα to the aromatic carbons. Incorporation of TOCSY data *via* GIC connects the Hα chemical shifts to the aromatic carbons (Figure 10.2B), and indirect covariance of the HMBC spectrum itself establishes aliphatic to aromatic carbon–carbon correlations (Figure 10.2C).

As three-bond ^{1}H–^{13}C correlations are strongest in the HMBC, the strongest cross-peaks in the indirect covariance of an HMBC spectrum are typically those between carbon pairs that are three bonds away from the same proton, which correspond to carbons that are separated by four consecutive carbon–carbon bonds. Thus, the Cβ carbon has the strongest correlation to the Cη carbon (Figure 10.2C, peak 2). The peaks associated with the Cα carbon (Figure 10.2C, peaks 1 and 4) illustrate an exception to this rule: since Cγ of phenylalanine is unprotonated, there is no three-bond correlation peak from Cα, hence the strongest correlation is not to the Cε carbons as expected but rather to the Cδ carbons. Covariance-derived constraints, such as these, can therefore be used to elucidate natural product structures by "hopping" through them three or four carbons at a time.

Covariance techniques, such as doubly indirect covariance and unsymmetric covariance processing of a ^{1}H–^{13}C HSQC with a ^{1}H–^{13}C 1,1-ADEQUATE spectrum, can also yield data equivalent to ^{13}C–^{13}C COSY data, using just a double resonance probe. In spectrometers with multiple receivers, which can operate in parallel, the PANACEA approach can produce similar data.[34,35] The one-bond carbon–carbon constraints obtained from ^{13}C–^{13}C COSY (HSQC–1,1-ADEQUATE) data, obtained *via* either PANACEA or covariance methods, are graph-theoretically isomorphic to the carbon bonding network of the organic compound under study (Figure 10.3A). The "2D walk" through a COSY spectrum is already a standard technique for proton chemical shift assignment and the analogous process applied to ^{13}C–^{13}C covariance COSY spectra provides a powerful method for the identification of natural products and the assignment of carbon chemical shifts (Figure 10.3B). The high sensitivity of covariance spectra relative to the equivalent Fourier transform experiments and the large nominal S/N values for covariance datasets allow for rapid acquisition of sufficient data for structure elucidation. Rapidly obtaining high-sensitivity spectra, *e.g. via* covariance NMR, is particularly critical for natural products, many of which are typically scarce, may not be stable, and typically have only natural abundance ^{13}C. For example, a few hours of measurement time on an unlabeled, sub-milligram sample of an alkaloid such as strychnine provides sufficient data to reconstruct a high-quality ^{13}C–^{13}C COSY-type spectrum from which the structure of the interrogated compound is derived with few remaining ambiguities.[11]

A key advantage of using doubly indirect covariance NMR to reconstruct ^{13}C–^{13}C COSY data is the high sensitivity of the ^{1}H–^{1}H COSY experiment relative to that of the ^{1}H–^{13}C 1,1-ADEQUATE experiment (especially for samples with natural abundance ^{13}C). However, doubly indirect covariance requires an additional step of moment filtering prior to covariance processing. An advantage of covariance between a ^{1}H–^{13}C HSQC and a ^{1}H–^{13}C 1,1-ADEQUATE experiment is that a multiplicity-edited HSQC may be used so that the phase of peaks in the resulting covariance spectrum yields information about the number of protons attached to the "donor" carbon probed.[10]

10.7 NMR Analysis of Mixtures of Natural Products

The ability of covariance techniques to extract high-resolution and high-sensitivity spectra from rapidly collected NMR data makes covariance NMR a natural approach for the analysis of biochemical mixtures, for example, in the context of metabolomics studies. The analysis of the resulting covariance spectra can be considerably facilitated by matrix factorization methods. For example, since TOCSY peaks are generally positive, linear algebraic non-negative matrix factorization (NMF) applied to 2D FT or covariance TOCSY spectra allows for the rather robust deconvolution of TOCSY spectra of complex mixtures into the 1D spectra of each mixture component, provided that one has a good estimate of the number of compounds of the mixture.[32]

NMF and principle component analysis (PCA), which is equivalent to SVD used to implement many covariance NMR techniques, each perform unsupervised clustering of cross-peaks into groups that belong to individual components. Another clustering method, termed DemixC, has shown promise in the deconvolution of TOCSY spectra of mixtures that exhibit a moderate amount of peak overlap.[36,37] For more severely overlapped spectra, the related DeCoDeC method, which uses consensus trace clustering for the identification of clean and unique TOCSY traces, can be applied instead.[38] DemixC, DeCoDeC, and NMF can be applied not only to homonuclear TOCSY but also to ^{1}H–^{13}C-HSQC–TOCSY spectra, allowing the identification of compounds in a natural product mixture *via* the extraction of both the ^{1}H and ^{13}C 1D-NMR spectra of the spin systems present,[39] as demonstrated, for example, in an NMR study of a cancer cell extract.[40] Other spectroscopic approaches based on statistical correlations of 1D spectra, such as STOCSY[41], which uses 1D spectra of different samples as input, and GEN2D,[42] are also amenable to DemixC, DeCoDeC, and NMF analysis. These approaches applied to TOCSY, STOCSY, covariance, GEN2D, or similar data may prove useful in identifying active components and drug candidates found in crude biochemical extracts and also in identifying impurities in drug or cosmetic preparations.

Doubly indirect covariance NMR is also particularly suited for the analysis of mixtures of natural products. Since correlation information obtained from such a spectrum is isomorphic to the carbon connectivity graph for the compounds in the mixture, identification of connected components in the connectivity graph can be used to isolate and assign the spectra of individual components in a given sample. Doubly indirect covariance analysis of an extract obtained from the prostate cancer cell line DU145 demonstrates that this procedure correctly traces out the structures of key components in a natural product mixture.[13]

10.8 Conclusion and Outlook

Because covariance NMR allows spin correlations to be probed at spectral resolutions or sensitivities that are often not achievable *via* direct experimental measurements, it affords a substantial gain in the resolution and/or sensitivity obtainable within a fixed amount of measurement time. The gain in available resolution and sensitivity is particularly important for the rapid assessment, identification, and structural analysis of natural products, which are often present in impure form or in dilute solution with ^{13}C (or other NMR-active heteronuclei) at their relatively low natural abundance.

Recent advances in covariance NMR include doubly indirect covariance NMR and the application of unsymmetrical and generalized indirect covariance NMR to reconstruct ^{13}C–^{13}C COSY-type spectra in which carbon–carbon bond connectivity is self-evident. Covariance NMR has also found entrance in other NMR subfields such as solid-state NMR[43–45] with non-uniform sampling (NUS) applications, which can be easily handled by covariance processing.[46] The recently released Covariance NMR Toolbox uses MATLAB/OCTAVE scripts to implement many covariance techniques in a user-friendly and highly extensible fashion. Current work on this toolbox includes implementation of doubly indirect covariance NMR and the associated moment filtering approach for spectrum editing.

Acknowledgements

We thank Ama Berko, Gary Martin, Timothy Short, and Fengli Zhang for helpful discussions. This work was supported by NIH grant GM 066041 (to R.B.) and with assigned release time for research and start-up funds (to D.A.S.) from the Office of the Provost, William Paterson University of New Jersey. The antipain sample used to generate examples for this chapter was obtained with funds from a College Cottrell Grant from the Research Corporation for Science Advancement.

References

1. R. Brüschweiler, *J. Chem. Phys.*, 2004, **121**, 409.
2. R. Brüschweiler and F. Zhang, *J. Chem. Phys.*, 2004, **120**, 5253.
3. N. Trbovic, S. Smirnov, F. Zhang and R. Brüschweiler, *J. Magn. Reson.*, 2004, **171**, 277.
4. L. Braunschweiler and R. R. Ernst, *J. Magn. Reson.*, 1983, **53**, 521.
5. J. Jeener, B. H. Meier, P. Bachmann and R. R. Ernst, *J. Chem. Phys.*, 1979, **71**, 4546.
6. F. Zhang and R. Brüschweiler, *J. Am. Chem. Soc.*, 2004, **126**, 13180.
7. A. Bax and M. F. Summers, *J. Am. Chem. Soc.*, 1986, **108**, 2093.
8. K. A. Blinov, N. I. Larin, A. J. Williams, K. A. Mills and G. E. Martin, *J. Heterocycl. Chem.*, 2006, **43**, 163.
9. D. A. Snyder and R. Brüschweiler, *J. Phys. Chem. A*, 2009, **113**, 12898.
10. G. E. Martin, B. D. Hilton and K. A. Blinov, *Magn. Reson. Chem.*, 2011, **49**, 248.

11. G. E. Martin, B. D. Hilton, M. R. Willcott and K. A. Blinov, *Magn. Reson. Chem.*, 2011, **49**, 350.
12. G. E. Martin and D. Sunseri, *J. Pharm. Biomed. Anal.*, 2011, **55**, 895.
13. F. Zhang, L. Bruschweiler-Li and R. Brüschweiler, *J. Am. Chem. Soc.*, 2010, **132**, 16922.
14. G. E. Martin, P. A. Irish, B. D. Hilton, K. A. Blinov and A. J. Williams, *Magn. Reson. Chem.*, 2007, **45**, 624.
15. E. Kupče and R. Freeman, *Magn. Reson. Chem.*, 2007, **45**, 103.
16. T. F. Havel, I. Najfeld and J. X. Yang, *Proc. Natl. Acad. Sci. U. S. A.*, 1994, **91**, 7962.
17. Y. Chen, F. Zhang, W. Bermel and R. Brüschweiler, *J. Am. Chem. Soc.*, 2006, **128**, 15564.
18. Y. Chen, F. Zhang and R. Brüschweiler, *Magn. Reson. Chem.*, 2007, **45**, 925.
19. K. A. Blinov, N. I. Larin, M. P. Kvasha, A. Moser, A. J. Williams and G. E. Martin, *Magn. Reson. Chem.*, 2005, **43**, 999.
20. D. A. Snyder, A. Ghosh, F. Zhang, T. Szyperski and R. Brüschweiler, *J. Chem. Phys.*, 2008, **129**, 104511.
21. R. K. Harris, E. D. Becker, S. M. Cabral de Menezes, R. Goodfellow and P. Granger, *Pure Appl. Chem.*, 2001, **73**, 1795.
22. K. A. Blinov, A. J. Williams, B. D. Hilton, P. A. Irish and G. E. Martin, *Magn. Reson. Chem.*, 2007, **45**, 544.
23. E. Kupče and R. Freeman, *Prog. Nucl. Magn. Reson. Spectrosc.*, 2008, **52**, 22.
24. G. E. Martin, B. D. Hilton, K. A. Blinov and A. J. Williams, *Magn. Reson. Chem.*, 2008, **46**, 138.
25. D. L. Donoho, I. M. Johnstone, A. S. Stern and J. C. Hoch, *Proc. Natl. Acad. Sci. U. S. A.*, 1990, **87**, 5066.
26. D. A. Snyder and R. Brüschweiler, in *Encyclopedia of Magnetic Resonance*, ed. R. K. Harris and R. E. Wasylishen, John Wiley, Chichester, 2009.
27. T. Short, L. Alzapiedi, R. Brüschweiler and D. A. Snyder, *J. Magn. Reson.*, 2011, **209**, 75.
28. Y. Chen, F. Zhang, D. A. Snyder, Z. Gan, L. Bruschweiler-Li and R. Brüschweiler, *J. Biomol. NMR*, 2007, **38**, 73.
29. ACD/NMR Processor (v.12), Advanced Chemistry Development, Inc. Toronto, Ont., Canada, 2011.
30. F. Zhang and R. Brüschweiler, CovNMR and CovNMR2.0.au.
31. D. A. Snyder, F. Zhang and R. Brüschweiler, *J. Biomol. NMR*, 2007, **39**, 165.
32. D. A. Snyder, F. Zhang, S. L. Robinette, L. Bruschweiler-Li and R. Brüschweiler, *J. Chem. Phys.*, 2008, **128**, 052313.
33. H. Umezawa, *Methods Enzymol.*, 1976, **45**, 68.
34. E. Kupče and R. Freeman, *J. Magn. Reson.*, 2010, **206**, 147.
35. E. Kupče and R. Freeman, *Magn. Reson. Chem.*, 2010, **48**, 333.
36. F. Zhang and R. Brüschweiler, *Angew. Chem., Int. Ed.*, 2007, **64**, 2639.
37. F. Zhang, A. T. Dossey, C. Zachariah, A. S. Edison and R. Brüschweiler, *Anal. Chem.*, 2007, **79**, 7748.

38. K. Bingol and R. Brüschweiler, *Anal. Chem.*, 2011, **83**, 7412.
39. F. Zhang, S. L. Robinette, L. Bruschweiler-Li and R. Brüschweiler, *Magn. Reson. Chem.*, 2009, **47**, S118.
40. F. Zhang, L. Bruschweiler-Li, S. L. Robinette and R. Brüschweiler, *Anal. Chem.*, 2008, **80**, 7549.
41. A. C. Alves, M. Rantalainen, E. Holmes, J. K. Nicholson and T. M. D. Ebbels, *Anal. Chem.*, 2009, **81**, 2075.
42. B. W. Hu, P. Zhou, I. Noda and G. Z. Zhao, *Anal. Chem.*, 2005, 77, 7534.
43. C. Kaiser, J. J. Lopez, W. Bermel and C. Glaubitz, *Biochim. Biophys. Acta*, 2007, **1768**, 3107.
44. B. Hu, J. P. Amoureux, J. Trebosc, M. Deschamps and G. Tricot, *J. Chem. Phys.*, 2008, **128**, 134502.
45. M. Weingarth, P. Tekely, R. Brüschweiler and G. Bodenhausen, *Chem. Commun.*, 2010, **46**, 952.
46. Y. Li, B. Hu, Q. Chen, Q. Wang, Z. Zhang, J. Yang, I. Noda, J. Trébosc, O. Lafon, J. P. Amoureux and F. Deng, *Analyst*, **138**, 2411.

CHAPTER 11

Future Approaches for Data Processing

KIRILL BLINOV*[a] AND ANTONY J. WILLIAMS[b]

[a] Molecule Apps, LLC, Wilmington, DE 19808, USA; [b] ChemConnector Inc., Wake Forest, NC 27587, USA
*Email: kirill.blinov@gmail.com

11.1 General Description of the Structure Elucidation Process

The process of structure elucidation can be separated into a series of steps: the acquisition of the spectra, data processing, extracting information from the spectra (especially the process of peak picking), combining different source data, finding all possible structures corresponding to the data, and then ranking the structures. These stages can differ in detail between manual and automated approaches to structure elucidation, but all of the individual stages are nevertheless present in both methods.

The different stages are illustrated schematically in Figure 11.1. In this chapter, we deal only with the processing of data and the subsequent extraction and combination of information from different spectra. All of these stages are combined to define "data preparation".

The structure elucidation process can be very sensitive to the quality of the data. Errors made in the early stages of data preparation can, and in most cases do, produce incorrect results. Most errors made in the earlier stages of data preparation are very difficult to correct during the later stages of analysis and generation of hypothetical structures. Data preparation is, in

Modern NMR Approaches To The Structure Elucidation of Natural Products,
Volume 1: Instrumentation and Software
Edited by Antony J. Williams, Gary E. Martin and David Rovnyak

Published by the Royal Society of Chemistry, www.rsc.org

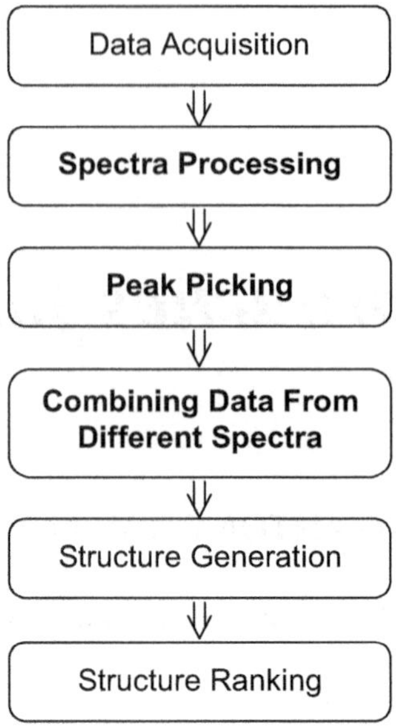

Figure 11.1 The main steps in structure elucidation. The stages related to data preparation are in bold font.

many ways, the most important task in the structure elucidation process, whether manual or automated. The criticality of data processing can be illustrated using a very simple example. If the resonance associated with a particular carbon atom is missed in the initial stages of analysis (due to incorrect peak picking, low signal-to-noise ratio, *etc.*), then the correct structure can never be identified because all hypothetical structures will have an incorrect number of carbon atoms.

As indicated in Figure 11.1, the above data preparation can be divided into three stages:

1. *Processing:* Manipulation of the data matrix using Fourier transformation or any other procedure that converts the time domain data to the frequency domain, including weighting, linear prediction, zero-filling, removing noise, phase correction, *etc.* This stage can be automatic but generally experts prefer an interactive approach to processing.
2. *Peak picking:* The main goal of peak picking is to determine the positions of signals. Currently this stage is generally manual and only a few examples of efficient automated peak picking have been described in the literature.[1]
3. *Combining information from different spectra:* This is required to transfer C–H and H–H (and other X–H) connectivity information to C–C connectivities that are necessary to determine the molecular scaffold. This stage is generally fairly straightforward and can be easily automated; manual elucidation at this stage can be tedious because it requires the

manual replacement of proton chemical shifts with carbon chemical shifts. Interestingly, this procedure can be applied before peak picking using Unsymmetrical Indirect Covariance (UIC) processing as described earlier in this volume in Chapter 10. In this case, peak picking applied to the resultant spectrum produces information about the C–C connectivities. The HSQC–1,1-ADEQUATE spectrum[2] obtained by the UIC processing of the HSQC and 1,1-ADEQUATE experiment is very useful for manual structure elucidation because it provides information about such direct C–C connectivities.

There are two primary approaches to increasing quality and robustness during the data preparation stage. First, if possible, the quality of the spectral data should be improved by acquiring the data at the highest resolution and signal-to-noise ratio and to remove artifacts, *etc.* Second, the data should be prepared so that peak picking is as optimal as possible: the number and positions of peaks should be determined correctly, overlapped peaks should be resolved, solvent and impurity peaks should be identified and removed from final peak list, *etc.* Obviously, these approaches are not independent, and improvements in spectral quality help substantially in peak determination (*e.g.* increased spectral resolution simplifies the separation of overlapped peaks and an increased signal-to-noise ratio allows small peaks to be found).

11.2 General Features of Natural Product Spectra

The standard set of spectra used for the structure elucidation of natural products generally includes a 1D-^{1}H-NMR spectrum (sometimes in multiple solvents) and several 2D spectra: HSQC (or preferably multiplicity-edited HSQC), HMBC, and COSY. NOESY (more generally ROESY) and TOCSY can be used in addition. A 1D-^{13}C NMR spectrum is also very helpful but owing to the amount of material available it may be almost impossible, in many cases, to acquire a carbon spectrum. With small amounts of material, only those investigators with high-field magnets and a small-volume cryoprobe can generate a ^{13}C spectrum.

As discussed in Chapter 4, cryoprobe technology is now widespread. The sensitivity of these probes allows for the acquisition of ^{13}C spectra and, more importantly, the acquisition of 1,1-ADEQUATE or even INADEQUATE spectra. As described in Chapter 4, a 1,1-ADEQUATE spectrum contains information regarding the connectivity between adjacent carbon atoms (except for pairs of quaternary carbons) and makes the structure elucidation process significantly easier and faster.[3] A comprehensive review of the application of ADEQUATE spectra is available.[4] Low sensitivity is the main disadvantage of this method and any processing techniques that can reduce the acquisition time are therefore very useful. It should be noted that low sensitivity is certainly a relative term, and small-volume cryoprobes do allow for the analysis of sub-milligram samples.[5]

Since 2D-NMR spectra are the main source of data for performing structure elucidation, most spectrometer time is spent acquiring 2D data and, as a result, most modern processing techniques are focused on enhancing and improving 2D spectra. Almost all algorithms described in this chapter are applications to 2D spectra.

11.3 Common Problems with Spectral Data

Most challenges associated with the analysis of spectra during the elucidation process are *missing signals*, *signal overlap*, and the *presence of extra signals* (artifacts). The most problematic issue is the presence of extra signals in the spectra that can imply extra atoms that are in fact absent in the structure. This finally produces an incorrect structure. Extra signals can be mistakenly processed as connectivities between atoms, which may produce some contradictions in the initial data. A simple example is when the number of COSY correlations associated with a single atom is usually treated as indications of chemical bonds and therefore exceeds the atom's valency. Missing signals and peak overlaps make the initial information more ambiguous and can lead to significant increases in the time associated with structure elucidation. Additionally, ambiguous initial data inputs can correspond to a larger number of structures that are consistent with the data, and the structure elucidation process (especially an automated process) can produce thousands or even millions of structures. This makes the whole elucidation process almost useless. Brief examples for each of the cases discussed are given below.

11.3.1 Missing Signals

This particular situation does not happen very often but is more common when the signal-to-noise ratio is low. This occurs when the amount of substance is small and/or the acquisition time is too short. Signals can also be absent because of too narrow a spectral window and misinterpretation of folded signals as noise spikes, *etc.* Additionally, some peaks produced by constants with "unusual" values can be missed in spectra. Figure 11.2 presents two 1,1-ADEQUATE spectra of strychnine (**1**) optimized to slightly different J values: 55 and 60 Hz. The carbon signals at 168 ppm are almost gone in the spectrum that was optimized to a 60 Hz coupling constant.

11.3.2 Signal Overlap

Signal overlap usually occurs along the F_1 axis where resolution is low. Typically, resolution along the directly acquired t_2 axis is well digitized at >1024 points, whereas the typical number of indirect t_1 increments is 160–384 points, corresponding to a digital resolution of >0.5 ppm in the frequency domain for carbon. In many cases, this resolution is not sufficient to separate unambiguously carbon resonances with similar chemical shifts or even to assign the 2D peak to the corresponding carbon. Figure 11.3 contains

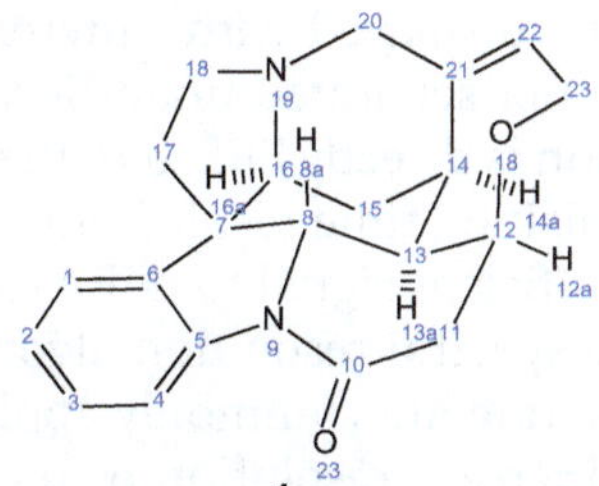

1

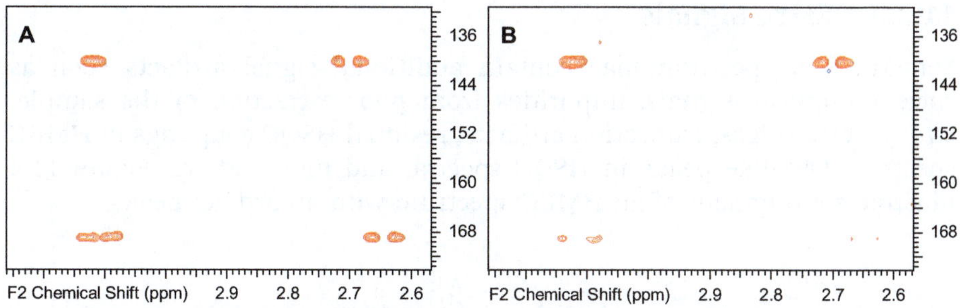

Figure 11.2 Spectra with a different number of signals. The same expansion of two 1,1-ADEQUATE spectra of strychnine (**1**) is shown. The first is optimized for 55 Hz (A) and the second for 60 Hz (B). The signals from the carbon at 168 ppm are negligible in the second spectrum.

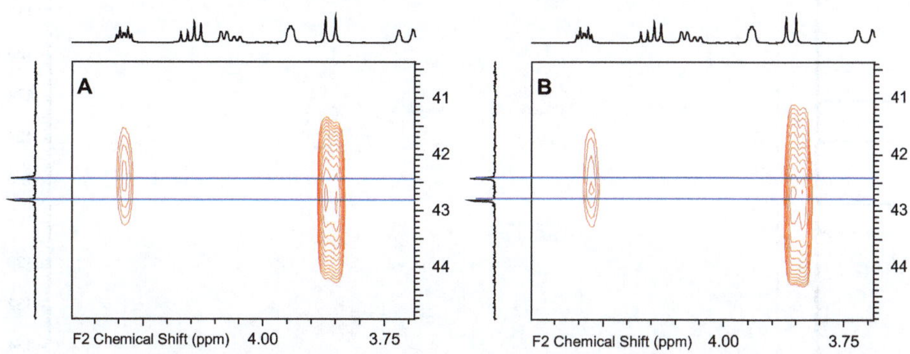

Figure 11.3 Spectra with different resolution: the same expansions of two HMBC spectra of strychnine (**1**). Spectrum A was acquired using 1024 points along t_1 while spectrum B was acquired using only 256 points along t_1 (both spectra have a spectral window of 27 905 Hz acquired at a frequency of 125.8 MHz and digitized to 1024 points along t_1. Linear prediction was not used). In spectrum A, peaks very clearly can be assigned to the corresponding carbon atoms (the blue lines correspond to the positions of the carbon peaks in the 1D spectrum). In spectrum B, the assignment of the left peak is not clear and it can be assigned to both of the carbon atoms at 42.4 or 42.8 ppm. In general, when carbon resonances are <1 ppm apart in a 2D spectrum, it is not feasible to make assignments. This depends a great deal, however, on the spectral window employed.

the same spectral subsections displayed with different resolutions and it is obvious that some peaks are not separated under low-resolution conditions.

Generally, spectral resolution is directly linked to the time it takes to acquire a spectrum. Increases in acquisition time result from a greater number of data increments when there is a sufficient signal-to-noise ratio. There are, however, alternatives to improving the spectral resolution algorithmically. Well-known methods such as linear prediction are commonly applied to enhance spectral resolution but really cannot increase resolution by more than a factor of two.

11.3.3 Extra Signals

Sometimes a spectrum may contain additional signal artifacts such as solvent impurity signals, impurities from poor extraction of the sample, strong noise ridges, truncation artifacts, residual HSQC couplings in HMBC spectra, COSY-like peaks in HSQC spectra, and many others. Figure 11.4 illustrates a fragment of an HMBC spectrum with an artifact peak.

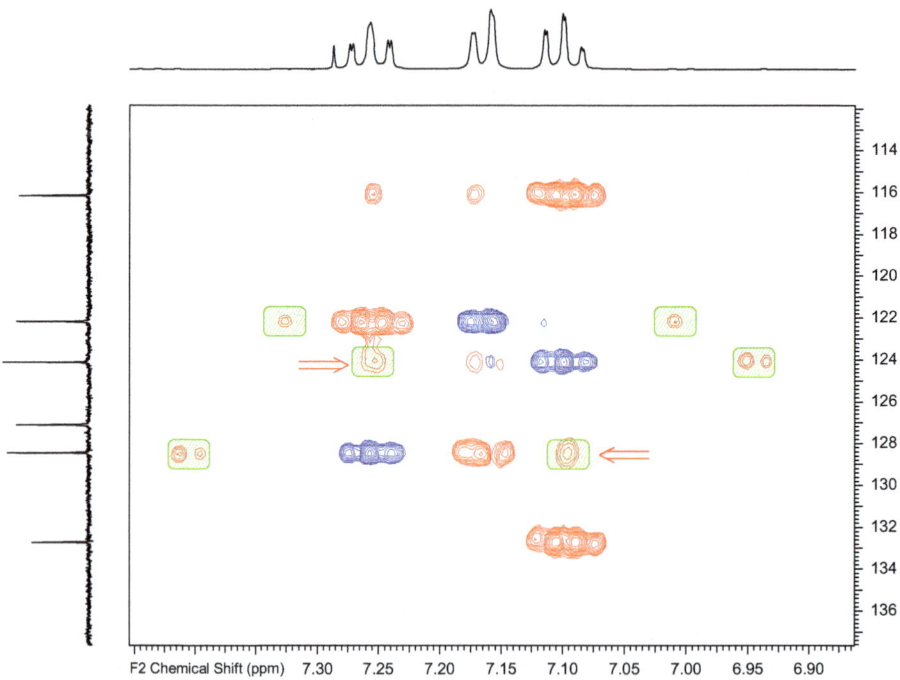

Figure 11.4 A fragment of an HMBC spectrum containing residual ^{1}J couplings which can mistakenly be interpreted as valid HMBC correlations. The figure displays an expansion of the superimposed HMQC (blue) and HMBC (red) spectra of strychnine (**1**). Residual ^{1}J coupling is marked by green squares. Most ^{1}J peaks can be filtered by the position along the ^{1}H axis because there are no protons in these positions. Two ^{1}J residual peaks, indicated by the red arrows, have positions along ^{1}H axis that correspond to protons and therefore cannot easily be filtered. These peaks can therefore be mistakenly processed as real HMBC peaks.

Certain types of signals can be removed by modifying the data matrix. Such signals include truncation artifacts or noise ridges. Other artifacts can only be removed by analysis of the signals using peak-picking procedures for detection prior to analysis. For example, the artifact signal displayed in Figure 11.4 can be removed during the analysis of connectivities produced from a peak list by examining for the presence of contradictions or searching for corresponding symmetrical peaks along the positions of the corresponding peak in an HSQC spectrum.

11.4 Main Approaches for Improved Processing

As already discussed, the two main directions for improved processing are to increase the quality of the spectra (improved resolution/sensitivity) and to increase the quality of the peak picking. Obviously these can be mutually beneficial in that high-quality spectra simplify the process of peak picking significantly.

11.4.1 Improving Spectral Quality or Reducing the Acquisition Time

A large number of methods to improve spectrum resolution have been suggested over the last 30 years. Some, such as linear prediction, are now routine in NMR processing packages. A comprehensive review of different processing methods has been published.[6] This section focuses only on promising new methods.

11.4.1.1 Non-uniform Sampling

The concept of non-uniform sampling (NUS) was originally suggested nearly 30 years ago.[7] However, until recently, NUS had only been used very marginally. In the context of this book, its applications to enhance the signal-to-noise ratio in natural product NMR is explored in Chapter 6. The main idea of NUS is that n-dimensional NMR spectra are relatively sparse and it is not necessary to acquire all points to obtain spectra with high quality. In 2006 Candes and Tao[8] demonstrated that a set of signals can almost always be reconstructed from an incomplete set of data. This work initiated investigations into a method called Compressing Sensing or Compressive Sampling (CS)[9,10] which is presently growing very quickly with the first application in magnetic resonance imaging reported in 2007,[11] and multiple applications in classical NMR[12–14] following quickly.

The most useful areas for the application of NUS is in 3D and higher dimensional NMR spectroscopy of proteins.[13] Even 2D-NMR acquisition times can be reduced significantly with this approach. It has been demonstrated[12] using the HSQC spectrum of the globular protein azurin that random sampling of a spectrum with only 18.3% of the points resulted in a spectrum of almost equal quality to a conventionally acquired spectrum.

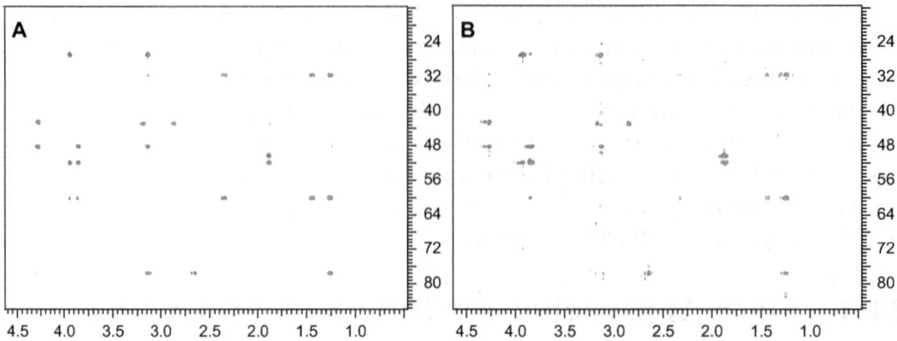

Figure 11.5 An expansion of an HSQC spectrum of strychnine (**1**) obtained with different percentages of randomly sampled points. Both spectra have a t_1 spectral window of 27 268 Hz acquired using 32 transients per increment. The number of increments along t_1 is 384 for spectrum A and 126 for B. Both spectra were digitized to 1024 points along t_1. Spectrum A is the conventionally sampled spectrum whereas spectrum B is a non-uniformly sampled spectrum with 33% randomly sampled points. Spectrum B contains more noise than A but all resonances are present.

Now algorithms for the processing of NUS data are increasingly available within processing software. Figure 11.5 displays the HSQC spectrum of strychnine (**1**) acquired with different numbers of points (expressed as percentages) and processed using Bruker's Topspin software.[15] It should be obvious that 33% of the number of points is sufficient to obtain a spectrum with quality equal to that of a fully sampled spectrum.

Walker reported[16] a comparison of conventional and NUS COSY, HSQC, and HMBC spectra. Spectra equivalent to those digitized with 1024 points along t_1 were achieved using NUS spectra acquired at only 25%. This significant time saving may certainly be very useful for the analysis of natural products as many spectra display signal overlap along t_1. It is worth noting that the compressed sensing algorithm can be used as an alternative to conventional linear prediction, *i.e.* the algorithm can be applied to conventional uniformly sampled spectra as a more powerful alternative to general linear prediction. Currently, compressed sensing has only one clear disadvantage – the processing time is minutes instead of the seconds associated with conventional processing.

11.4.1.2 *Increasing* t_1 *Resolution by Spectral Aliasing*

Spectra aliasing, a technique commonly applied to the acquisition of INADEQUATE spectra, is a well-known technique that allows the spectral resolution to be enhanced since a spectrum is acquired using a relatively small spectral window. It allows close signals to be resolved because the number of acquired points is applied to a small spectral window. The main disadvantage of this technique is that all signals that do not fit into a chosen spectral window are still present in a spectrum but are not shown in the correct positions. The correct positions can be restored by using prior knowledge about the real chemical shifts for the signals. In addition, aliasing may produce overlaps in the spectrum, which can add significant confusion.

The spectral aliasing approach is especially important for HMBC spectra. The method termed computer-optimized spectral aliasing (COSA), introduced by Jeannerat,[17] can resolve most of the problems associated with spectral aliasing. The correct positions of resonances along the carbon axis are restored using the knowledge of the positions of the carbon resonances obtained from a previously recorded carbon spectrum. The key part of the algorithm is optimization of the spectral window along t_1 based on the knowledge of the position of the carbon signals obtained from the carbon spectrum and, optionally, the algorithm can take into account the position of resonances in low-resolution HSQC and/or HMBC spectra.

The application of the method to acquire an HSQC spectrum of cyclosporin A has been described.[1a] A fully resolved aliased spectrum was acquired 126 times faster than for a conventional spectrum with the same resolution. The main disadvantage of this method, which significantly restricts it application in natural product structure elucidation, is the requirement for a carbon spectrum, which is certainly not always accessible for natural products owing to sample limitations.

11.4.2 Peak Picking

The development of an accurate peak picking procedure would seem very simple, as in theory it should simply identify the extrema in a data matrix. However, in reality the following problems occur:

1. More than one extrema can correspond to one atom, especially along the proton axis. In this case, several extrema (peaks) need to be combined into one "peak" or multiplet, which is commonly not a simple task.
2. Peaks may often overlap, and in these cases one peak really corresponds to two or more nuclei and algorithmic analysis of the peak picking result is required to resolve this problem.
3. Peaks may have non-ideal forms. Examples include small satellite peaks that can be removed by proper weighting of the spectrum, some phase distortion that may be removed by appropriate phasing, or some other issue contributing to the non-ideal peak form.
4. The number of atoms and, therefore, the number of expected resonances are not always known in real-world structure elucidation. In addition, a structure may be symmetric, which can also influence the number of observed signals. This makes peak picking significantly more difficult in many cases.
5. The intensity of peaks may vary depending on the atom type. For example, a CH_3 group may produce a very intense, single peak whereas a

CH group may produce a very broad, coupled peak with low height. Sometimes the t_1 ridges associated with methyl groups are very intense and mask the peaks from the CH groups.
6. A spectrum can also contain "false" peaks. These may be artifacts, solvent peaks, impurities resulting from poor isolation, residual HSQC peaks in an HMBC spectrum, COSY-like peaks in a HSQC spectrum, *etc.*

Some of these problems, such as incorrect peak shape and peak overlap, can, in many cases, be resolved before peak picking during the processing of the data matrix. Some "false" peaks can be identified after peak picking using additional procedures. In any case, automated procedures should be able to detect at least some cases that require reprocessing of the spectrum. Figure 11.6 shows an expansion of the relatively complex HMBC spectrum of brevetoxin B (**2**), demonstrating the issues that have been described. The structure of brevetoxin B (**2**) is complex (50 carbon atoms) and the spectrum contains a large number of signals that can be significantly overlapped in many cases. The structure also contains seven methyl groups that have very intense peaks and mask some of the signals from the CH and CH_2 groups.

Generally all problems described may be solved, and an experienced spectroscopist can resolve all of these issues and perform manual peak picking relatively quickly. However, the authors are not aware of any ideal automatic algorithms that can outperform manual peak picking to provide an ideal data set that can be used for the purpose of structure elucidation.

Automated peak picking for the purpose of structure elucidation should be able to

1. correctly identify multiplets and combine them as appropriate into one signal corresponding to one atom;
2. find and resolve overlapped signals from several atoms;
3. identify peaks with low intensity;
4. cope with different noise levels in different parts of a spectrum;
5. identify those cases when a spectrum needs to be reprocessed before peak picking;
6. identify and ignore erroneous peaks.

No specific structure elucidation-oriented peak picking algorithms have been described in the literature. A few articles related to the peak picking of complex protein spectra (2D to *n*D) have been published in recent years. It should be emphasized that the peak picking of protein spectra is different from peak picking for structure elucidation. The main difference is that natural product structure elucidation is much more sensitive to absent and erroneous peaks than protein structure determination. Nevertheless, some parts of the algorithms are common for both types of peak picking. Two detailed algorithms for the peak picking of 2D-NMR spectra have been published in recent decades, namely AUTOPSY[1b] and PICKY.[1a] Both were

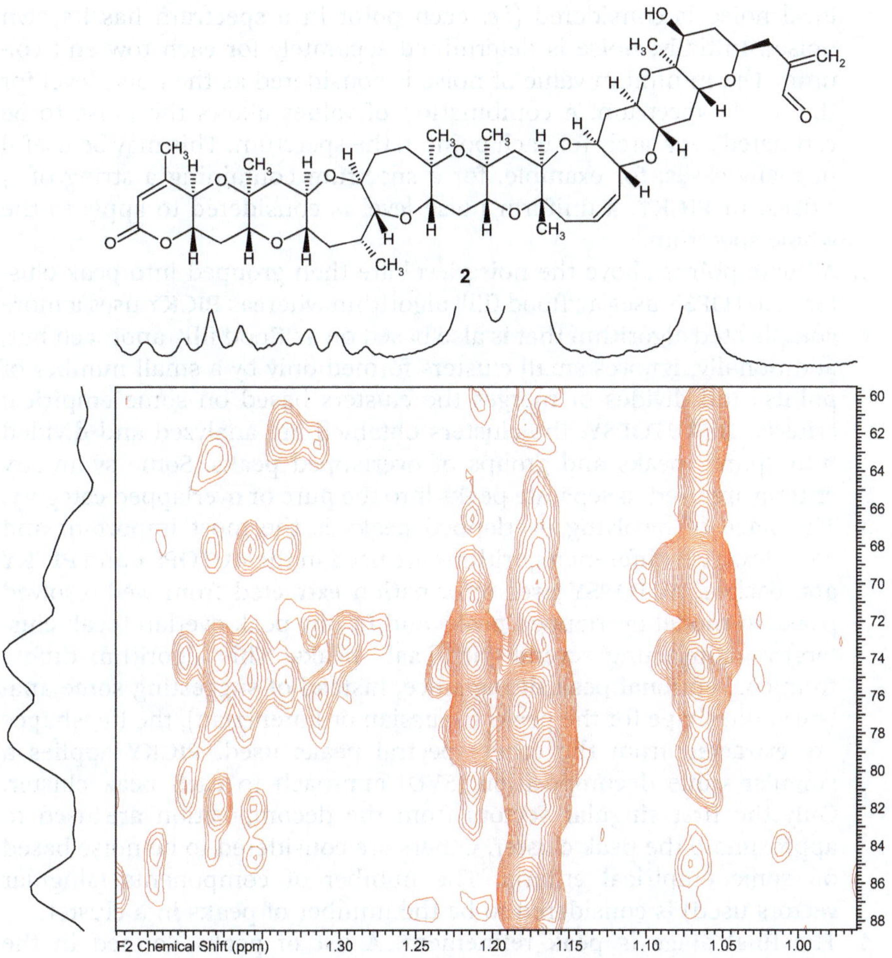

Figure 11.6 An example of the most common peak picking problems, showing an expansion of the HMBC spectrum of brevetoxin B (2) containing both resonances from CH_3 groups (right side) and CH or CH_2 groups (left side). The t_1 spectrum window is 30 166 Hz with 128 original points and a final count of 512. The number of transients per increment is 256. The resonances of the CH_3 groups are very intense and even the shoulders of the peaks are more intense than the resonances of the CH/CH_2 groups, and can be mistakenly processed as real peaks. Additionally, owing to strong peak overlap, the shape of some of the peaks is distorted and it is difficult to pick some peaks.

designed for the peak picking of 2D (or *n*D) spectra of proteins and use the following staged approaches:

1. The first stage is noise determination. Both algorithms initiate the process by determining the noise level. In AUTOPSY, the presence of

local noise is considered (*i.e.* each point in a spectrum has its own noise). Initially, noise is determined separately for each row and column. The minimum value of noise is considered as the noise level for the whole spectrum. A combination of values allows the noise to be estimated separately for each point in the spectrum. This may be useful in many cases, for example, for a spectrum containing a string of t_1 ridges. In PICKY, a uniform noise level is considered to apply to the whole spectrum.

2. All data points above the noise level are then grouped into peak clusters. AUTOPSY uses a "flood fill" algorithm whereas PICKY uses a more complicated algorithm that is also based on a "flood fill" approach but, additionally, ignores small clusters formed only by a small number of points, and divides or merges the clusters based on some empirical criteria. In AUTOPSY, the clusters obtained are analyzed and divided into "pure" peaks and groups of overlapped peaks. Some symmetry criteria are used to separate peaks into the pure or overlapped category.
3. The stage of resolving overlapped peaks is the most important and complex. Two different algorithms are used in the AUTOPSY and PICKY approaches. AUTOPSY uses information extracted from well-resolved peaks to model overlapped peaks and fit any peak overlap (peak clusters) by combining several "artificial" peaks. This algorithm differs from conventional peak fitting since, instead of suggesting some analytical lineshape for the peaks (Gaussian or Lorentzian), the lineshapes are extracted from the other spectral peaks used. PICKY applies a singular value decomposition (SVD) approach to each peak cluster. Only the first singular vectors from the decomposition are used to approximate the peak cluster. Others are considered to be noise based on some empirical criteria. The number of components (singular vectors used) is considered to be the number of peaks in a cluster.
4. The final stage is peak refinement. A list of peaks defined in the previous stages is filtered using criteria such as peak intensity and the correspondence to peaks in other spectra.

Both of the algorithms described have been compared with manual peak picking and are claimed to be at least as efficient for one example. AUTOPSY automated peak picking was tested on the 2D-NOESY spectrum of yeast killer toxin WmKT protein.[1b] A total of 2761 peaks were selected (compared with 1698 selected by manual peak picking). The protein structure obtained using automated peak picking had a comparable RMSD to that of the structure obtained using manual peak picking. PICKY was used for the structure determination of the TM1112 protein.[18] Automatic peak picking found 94% of the peaks (averaged over several spectra) and the correct protein structure was identified on the basis of the peaks found.

The algorithms described solve some aspects of the problem of peak picking but other parts of the algorithm need to be enhanced further for optimal use in structure elucidation. For example, the noise determination algorithms appear to be good enough to use with real-world examples. Resolution of overlapping peaks and peak refinement needs to be very accurate for the purposes of structure elucidation as the process is very sensitive to false or missing peaks in HSQC spectra. The approach therefore needs to produce a 100% accurate peak list because absences in the carbon spectrum or signal overlap in the proton spectrum cause issues with the refinement of peaks in HSQC spectra based on 1D data. The most realistic scenario for the future is likely to test the PICKY algorithm on natural product HSQC spectra and, based on the test results, adjust some empirical parameters or replace some part(s) of the algorithm. For example, SVD can be replaced by independent component analysis (ICA), which is another method of matrix factorization that produces more meaningful results in many cases.

11.5 Combining Information from Different Spectra. Unsymmetrical Indirect Covariance

Initially, indirect covariance (IC) (see Chapter 10 for more detail) was suggested as a method to convert an HSQC-TOCSY spectrum from a C–H to C–C representation.[19] This method produced a diagonally symmetrical C–C correlation spectrum that is more convenient for manual assembly of a structure than a conventional C–H correlation plot. Unfortunately, this method can produce artifact peaks. The sources of the artifacts and approaches to avoid them were described by Blinov *et al.*[20]

The same group subsequently described approaches that allow IC to be applied to any pair of spectra whose nuclei are equivalent along the F_2 axis. This method, called unsymmetrical indirect covariance (UIC), allows the combination of, for instance, HSQC and HMBC to produce a C–C spectrum,[21] as shown in Figure 11.7. This is very useful because this pair of experiments is routinely used in structure elucidation and a C–C combination spectrum can provide significant parts of a molecular skeleton. Various combinations of spectra have been described following the initial work.[19–23] As commented earlier, the HSQC–1,1-ADEQUATE[2] UIC spectrum is very useful for structure elucidation because it contains direct C–C connectivity information and therefore can be used to assemble the molecular skeleton. The HSQC–1,1-ADEQUATE UIC spectrum of strychnine (**1**) is displayed in Figure 11.8.

Technically, UIC is equal to matrix multiplication of the data matrix from the first spectrum and the transposed matrix of the second spectrum. In conventional IC an additional procedure of calculating the square root of the matrix is performed. This is generally impossible in UIC because the resultant matrix is not always a square matrix, which is a condition for calculating the square root. Another method, called generalized indirect covariance (GiC), has also been suggested to overcome this restriction.[24]

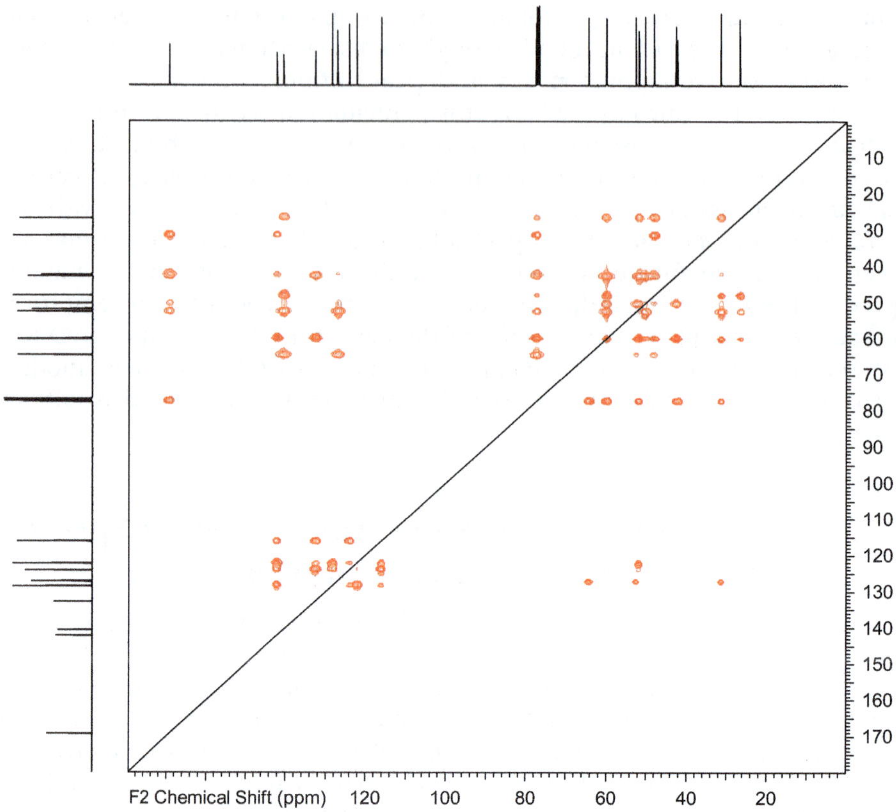

Figure 11.7 The UIC spectrum obtained from combining the HSQC and HMBC spectra of strychnine (**1**).[19] The resulting spectrum has a diagonal form but not all responses appear on both sides of the diagonal. Responses from quaternary carbons, which are absent in the HSQC spectrum, appear only in the upper left part of spectrum. Additionally, some peaks may not be a diagonally symmetric pair because peaks from different sides of the diagonal are formed by different pairs of HSQC–HMBC responses that may have different relative intensities.

This method also can reduce artifacts in some cases in the resultant spectra. The presence of artifacts in IC spectra may be the main reason why IC processing is not yet widely used. Artifacts in IC spectra appear as a result of partial overlap of proton peaks (or, more precisely, projections of 2D proton peaks onto the proton axis) in different spectra. The UIC spectrum, with artifacts highlighted, is displayed in Figure 11.8.

Several attempts have been made in recent years to remove or reduce artifacts in IC spectra. Generally, the problem is unsolvable in those cases when there are two equal (equal position and shape) proton signals. In those cases of partial overlap, the problem can be solved in theory, and partially in practice, using various methods.[20,22–24] In practice, complete overlap is very rare and can be ignored, but the partial overlap of proton peaks appears often

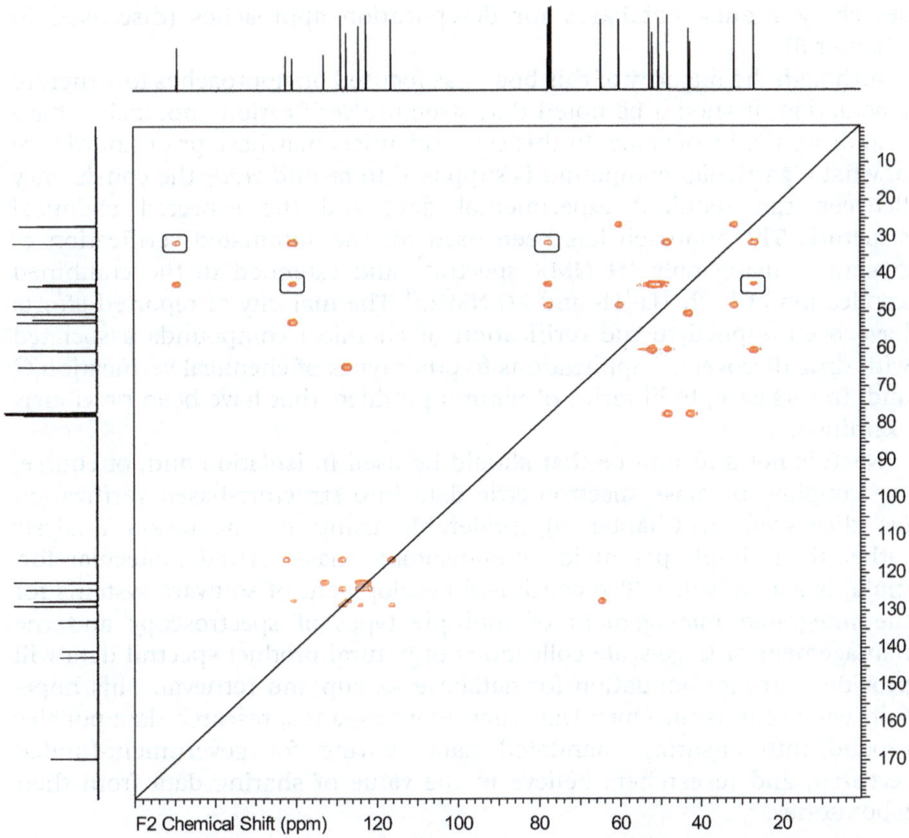

Figure 11.8 The UIC spectrum obtained from combining the HSQC and HMBC spectra of strychnine (**1**). Each peak in the spectrum corresponds to a C–C bond in the structure. The spectrum also contains several artifact peaks marked with squares. The artifact peaks are produced by partial peak overlap of peaks along the proton axis in the range 3.13–3.15 ppm. The source of artifacts has been described in detail elsewhere.[2]

enough that this problem cannot be ignored. A robust solution for the removal of artifacts resulting from partial peak overlap is required to make UIC a routine processing procedure that can contribute to structure elucidation.

11.6 Automated Data Processing for Structure Identification

As discussed earlier, the primary reason to improve data processing techniques is to provide the highest quality data possible in as short a time as possible so that the data can be used for the purpose of manual structure elucidation or computer-assisted structure elucidation, or as the basis for searching against databases for dereplication approaches (discussed in Chapter 8).

Although the majority of this book has focused on approaches to structure elucidation, it should be noted that structure verification approaches have also proven to be of value. In this case, scientists may have prior knowledge of what a particular compound is supposed to be and *verify* the consistency between the acquired experimental data and the expected chemical structure. The approach has been used for the automated verification of structures using only ^{1}H NMR spectra[25] and extended to the combined application of both 1D-^{1}H- and 2D-NMR.[26] The majority of reported efforts have been applied to the verification of chemical compounds associated with drug discovery,[27] applications to other types of chemical verification,[28] and, in this case, to libraries of natural products that have been previously examined.

NMR is not a technique that should be used in isolation and, of course, the coupling of mass spectrometric data into structure-based verification (as discussed in Chapter 9), preferably using fragmentation analysis rather than simply parent ion monoisotopic mass-derived molecular formula, is also of value. The continued development of software systems for the integrated management of multiple types of spectroscopy and the management of large-scale collections of natural product spectral data will provide a strong foundation for database lookup and retrieval. This hopefully will occur as the Open Data movement expands, research data policies expand into ensuring mandated data sharing for government-funded research, and researchers believe in the value of sharing data from their laboratories.

11.7 Conclusion

Two of the most important directions for the future development of NMR data processing as applied to structure elucidation, especially for natural products, have been discussed. First, methods that allow for a reduction in spectral acquisition time will be very important. These techniques include non-linear (non-uniform) sampling and, in theory, others that will be elaborated in the future. Second, automated peak-picking procedures, which are really the last barrier to the general application of automated structure elucidation, need to be developed and applied as a standard procedure in the elucidation process. Ultimately, an array of advanced processing algorithms will be developed that will be able to provide a complete and accurate dataset extracted from the experimental data. These algorithms will account for signal overlap, for experimental artifacts, and for issues associated with low signal-to-noise ratios. The resulting data set provided will be ideal not only as data feeds for CASE systems but also as the basis of improved dereplication procedures and searching across spectral databases that will, undoubtedly, continue to grow in size and scope.

References

1. (a) R. Koradi, M. Billeter, M. Engeli, P. Güntert and K. Wüthrich, *J. Magn. Reson.*, 1998, **135**, 288; (b) B. Alipanahi, X. Gao, E. Karakoc, L. Donaldson and M. Li, *Bioinformatics*, 2009, **25**, 268.
2. G. E. Martin, B. D. Hilton and K. A. Blinov, *Magn. Reson. Chem.*, 2011, **49**, 248.
3. S. F. Cheatham, M. Kline, R. R. Sasaki, K. A. Blinov, M. E. Elyashberg and S. G. Molodtsov, *Magn. Reson. Chem.*, 2010, **48**, 571.
4. G. E. Martin, *Annu. Rep. NMR Spectrosc.*, 2011, **74**, 215.
5. B. D. Hilton and G. E. Martin, *J. Nat. Prod.*, 2010, **73**, 1465.
6. D. Jeannerat, *Annu. Rep. NMR Spectrosc.*, 2002, **46**, 151.
7. (a) J. C. J. Barna, E. D. Laue, M. R. Mayger, J. Skilling and S. J. P. Worrall, *J. Magn. Reson.*, 1987, **73**, 69; (b) J. C. J. Barna and E. D. Laue, *J. Magn. Reson.*, 1987, **75**, 384.
8. E. J. Candès, J. Romberg and T. Tao, *IEEE Trans. Inf. Theory*, 2006, **52**, 489.
9. D. Donoho, *IEEE Trans. Inf. Theory*, 2006, **52**, 1289.
10. E. J. Candès and M. B. Wakin, *IEEE Signal Processing Magazine*, 2008, **25**, 21.
11. M. Lustig, D. Donoho and J. M. Pauly, *Magn. Reson. Med.*, 2007, **58**, 1182.
12. K. Kazimierczuk and V. Y. Orekhov, *Angew. Chem., Int. Ed.*, 2011, **50**, 5556.
13. D. J. Holland, M. J. Bostock, L. F. Gladden and D. Nietlispach, *Angew. Chem., Int. Ed.*, 2011, **50**, 6548.
14. Y. Shrot and L. Frydman, *J. Magn. Reson.*, 2011, **209**, 352.
15. *TopSpin*, 3.1; Bruker.
16. G. S. Walker, in *The Utility of Non-Uniform Sampling in 2D NMR Analysis of Small Molecules*, SMASH Small Molecule NMR Conference, Chamonix, France, 2011.
17. D. Jeannerat, *J. Magn. Reson.*, 2007, **186**, 112.
18. (a) P. Comon, *Signal Process.*, 1994, **36**, 287; (b) A. Hyvärinen and E. Oja, *Neural Networks*, 2000, **13**, 411.
19. F. Zhang and R. Brüschweiler, *J. Am. Chem. Soc.*, 2004, **126**, 13180.
20. K. A. Blinov, N. I. Larin, M. P. Kvasha, A. Moser, A. J. Williams and G. E. Martin, *Magn. Reson. Chem.*, 2005, **43**, 999.
21. K. A. Blinov, N. I. Larin, A. J. Williams, M. Zell and G. E. Martin, *Magn. Reson. Chem.*, 2006, **44**, 107.
22. G. E. Martin, B. D. Hilton, K. A. Blinov and A. J. Williams, *Magn. Reson. Chem.*, 2008, **46**, 138.
23. D. A. Snyder and R. Brüschweiler, *J. Phys. Chem. A*, 2009, **113**, 12898.
24. K. Bingol, R. K. Salinas and R. Brüschweiler, *J. Phys. Chem. Lett.*, 2010, **1**, 1086.
25. S. S. Golotvin, E. Vodopianov, B. A. Lefebvre, A. J. Williams and T. D. Spitzer, *Magn. Reson. Chem.*, 2006, **44**, 52.

26. S. S. Golotvin, E. Vodopianov, R. Pol, B. A. Lefebvre, A. J. Williams, R. D. Rutkowske and T. D. Spitzer, *Magn. Reson. Chem.*, 2007, **45**, 803.
27. Automated Structure Verification by NMR, Part 2: Return on Investment. http://www.americanlaboratory.com/913-Technical-Articles/37957-Automated-Structure-Verification-by-NMR-Part-2-Return-on-Investment/.
28. Automated Structure Verification by NMR, Part 1: Lead Optimization Support in Drug Discovery. http://www.americanlaboratory.com/913-Technical-Articles/37311-Automated-Structure-Verification-by-NMR-Part-1-Lead-Optimization-Support-in-Drug-Discovery/.

CHAPTER 12

NMR: The Emerging New Analytical Tool for Nutraceutical Analysis

KIMBERLY L. COLSON,*[a] JIMMY YUK[a] AND CHRISTIAN FISCHER[b]

[a] Bruker BioSpin Corporation, Billerica, MA 01821, USA; [b] Bruker BioSpin GmbH, 76287 Rheinstetten, Germany
*Email: kim.colson@bruker.com

12.1 Introduction

12.1.1 Nutraceuticals

Nutraceuticals constitute a wide range of products that include dietary supplements, herbal products, functional foods and beverages, and isolated nutrients. These products are utilized for a wide range of health benefits from general wellness to cures for specific diseases. The reliance on nutraceuticals is long standing, with aboriginal populations relying on traditional herbal products for many thousands of years. These populations were dependent on local suppliers that provided quality, properly identified products, and instructions of the material's use. The promise of beneficial effects raised the curiosity and demand for these materials in other populations. With expanded use and cultivation of these materials outside their original location, confusion has resulted as to the product identity, proper material collection, material preparation, and use. With this confusion came mistrust of the nutraceuticals that are highly regarded in the original

Modern NMR Approaches To The Structure Elucidation of Natural Products,
Volume 1: Instrumentation and Software
Edited by Antony J. Williams, Gary E. Martin and David Rovnyak

Published by the Royal Society of Chemistry, www.rsc.org

population. Fortunately, passage of key US regulation in 1994, the Dietary Supplement Health and Education Act (DSHEA), "gave birth to a regulatory environment that provided a legal definition and therefore credibility, to the fledgling, fragmented, but growing dietary supplement industry, driven by a passionate belief in the health values of its products," according to Kathie Wrick in an excellent review about the impact of regulations in the book *Regulation of Functional Foods and Nutraceuticals*.[1] With the regulations set in place, markets for nutraceuticals expanded first to multinational markets and then to global markets. The global market for nutraceutical products is significant at US$142.1 billion in 2011.[1] With the further expanded trade, issues related to identity and quality have grown as a concern and enhanced regulation of nutraceutical products has resulted. The Dietary Supplement Current Good Manufacturing Practices Rule of 2007[2] requires that herbal products meet safety standards and control systems be put in place to ensure that dietary supplements meet identity, purity, strength, and composition specifications. Such regulations enhance the product safety and require that suppliers demonstrate compliance with the regulations in order for them to participate in commercial trade to the USA. Many other countries are adopting similar regulations. In the USA, non-compliance results first in warnings issued through the Food and Drug Administration (FDA) and, if not corrected, legal action may result. Another concern noted by Kathie Wrick in the same review stated the urgency for the development of enhanced analytical methods for the evaluation of nutraceuticals, and wrote, "a very real problem is the absence of validated analytical methods for use in manufacturing controls and finished product testing to assure that label claims, and therefore customer expectations, are met. ... Natural products are some of the most complex matrices found in the world of analytical chemistry. Sometimes they contain thousands of phytochemicals in one plant, which may vary in composition and quantity depending on season and soil."[3]

Botanical identification and composition analysis are challenging tasks. Typically, identity is established in the field when plant material is harvested, and is accurately performed with the use of voucher specimens and trained experts. Identifying the material taxonomically and avoiding confusion that often results from the use of common names and similar looking plants is essential. It is especially important as the wrong identification of a botanical material can be harmful for human consumption. There have been many cases where toxic plants have been substituted for non-toxic species. For example, the substitution of *Teucrium* genus (germander) in *Scutellaria lateriflora* L. (skullcap) products had major repercussions owing to its hepatotoxic properties.[4] Once plant material including leaves, roots, stems, or fruit is pressed, ground, or extracted, the product identity relies on analytical techniques that make comparisons with standards. DNA fingerprinting has commonly been used in botanical analysis for species identification through genetic composition and is not influenced by age or physiological or environmental conditions.[5] However, DNA analysis can only verify species when DNA is present and not physicochemical content such as metabolites and other potential chemicals that could be present, such as adulterants. In addition, if the sample integrity is compromised (degradation, harsh solvent extraction processes, *etc.*), species verification by DNA becomes very difficult. Direct comparisons with botanical standards are challenged by the variability of material that results from different growing conditions, different local cultivars and landraces, and different harvest and processing techniques. Additionally, the high cost, and often instability, of purified botanical metabolites as standards makes the comparison of metabolites present in botanical material non-trivial. Similar challenges in analysis are also observed with other dietary supplements, functional foods, and beverages, such as energy drinks, which contain a wide variety of chemical components and, in some cases, botanical materials.

12.1.2 Unique Strengths of NMR

Traditional analytical techniques for nutraceuticals include HPLC, TLC, Raman spectroscopy, GC-MS and near-infrared spectroscopy.[5] Only recently has NMR emerged as an important technology in the nutraceutical analysis toolbox, resulting from the need for high reproducibility, precision quantitation, and high compound specificity to meet the demands of the new regulations.[6,7] Reproducibility of NMR data assures consistency in measurements between different laboratories and permits the exchange of NMR spectral databases and statistical models between laboratories.[8,9] These abilities empower the analyst to investigate the variances in the commercial product or raw materials for material assessment and not have doubts resulting from analytical techniques having a low capacity to obtain reproducible results. For example, techniques that utilize chromatographic methods, such as HPLC and GC, requires more preparation time and frequent calibrations for each sample or experimental run to obtain reproducible results. Each chromatographic instrument can be subject to wide variations in the sample profile, as factors such as mobile phase selection and volume and column type, length, and conditions can cause shifts of the retention times of materials.[10] The age and history of the chromatographic column can have major effects on retention times and performance.[10] In a study by Pham-Tuan *et al.*,[11] differences in a urine profile measured by HPLC were observed when different column lengths (30–100 mm) were used. In addition, the column degradation was evident during a high-throughput analysis after 50–100 sample runs. Mass spectrometry (MS) is commonly used for the analysis of nutraceuticals owing to its high sensitivity in detecting materials at very low concentrations (10^{-12} mol). However, MS is a destructive technique, which means each sample injected is considered a new sample acquired, and therefore careful preparation such as calibrations must be performed, especially to account for the degradation of the source.[12] According to Bristow *et al.*,[12] the design of the ion source may contribute significantly to reproducibility and, therefore, different instruments could perform differently and therefore comparisons must be made with attention to this detail.

Two key principles of NMR make it an inherently reproducible and quantitative technique: (a) the NMR frequency of a material is directly proportional to the strength of the magnetic field and (b) the intensity of the signal is proportional to the number of atoms giving rise to the signal.[13] An example of the level of reproducibility achievable on different instruments, and with different people preparing the samples, was demonstrated by Spraul and co-workers as shown in Figure 12.1. Achieving this high level of reproducibility made possible by NMR requires attention to standard operating procedures (SOPs) for sample preparation, instrument optimization, data acquisition, and data processing. Additionally, proper instrument maintenance and well-designed sampling conditions are essential to realizing the desired reproducibility results.

Distinctive to NMR is the ease of obtaining highly quantitative results. The non-destructive nature of NMR spectroscopy, along with the principles noted above, makes it a highly quantitative method provided that the material is soluble in the NMR solvent. Because these materials are used for wellness improvement and/or general nutrition, quantitation of key components in nutraceuticals is essential to the evaluation of product quality and potential efficacy. Relative to other analytical techniques, quantitation by NMR is fast; data can be acquired in minutes and it is capable of quantifying the material without the need to obtain the actual material standard to be measured at the same time. This saves both time and money for the analyst. Two standard approaches are used for conducting quantitative NMR measurements, namely the use of (1) an internal reference material or (2) an externally calibrated spectrometer. These approaches were reviewed by Wider and Dreier[14] and Burton *et al.*[15] The externally calibrated NMR approach has

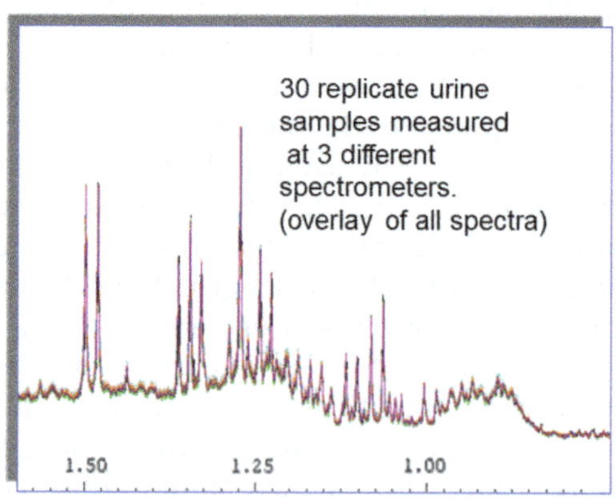

Figure 12.1 Reproducibility of NMR as shown by an overlay of 30 replicate NMR spectra of urine including sample preparation acquired on three different Bruker AVIII 400 MHz spectrometers. Samples were prepared by six different people. Courtesy of Manfred Spraul, Bruker BioSpin.

gained strength from recent advances in NMR spectrometer technology that have improved the linearity. Further, the improved linearity allowed the use of single-point calibrations to external quantification standards as shown by Hicks *et al.*[9] (Figure 12.2) in a single laboratory validation of NMR for lowbush blueberry leaf material (*Vaccinium angustifolium*) that is used as a traditional natural health product for the treatment of diabetic symptoms.[16] Especially in the case of nutraceuticals, where complex mixtures are often involved (Figure 12.3), externally calibrated spectrometers are beneficial owing to resonance overlap and sample stability issues. Mixtures having many metabolites are more likely to experience sample stability issues than

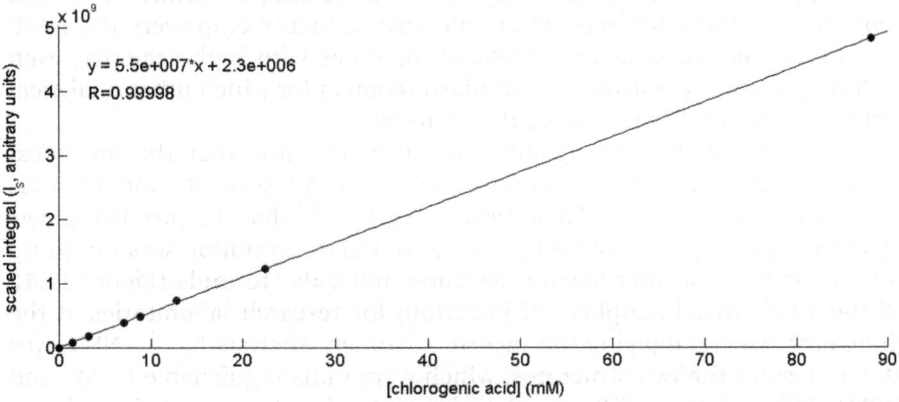

Figure 12.2 Linearity of NMR as demonstrated with a calibration curve of known chlorogenic acid standards (mM) measured on a five decimal place gravimetric scale, prepared in DMSO-d_6 against the scaled integral (IS) in arbitrary units of the NMR signal at 6.8 ppm.

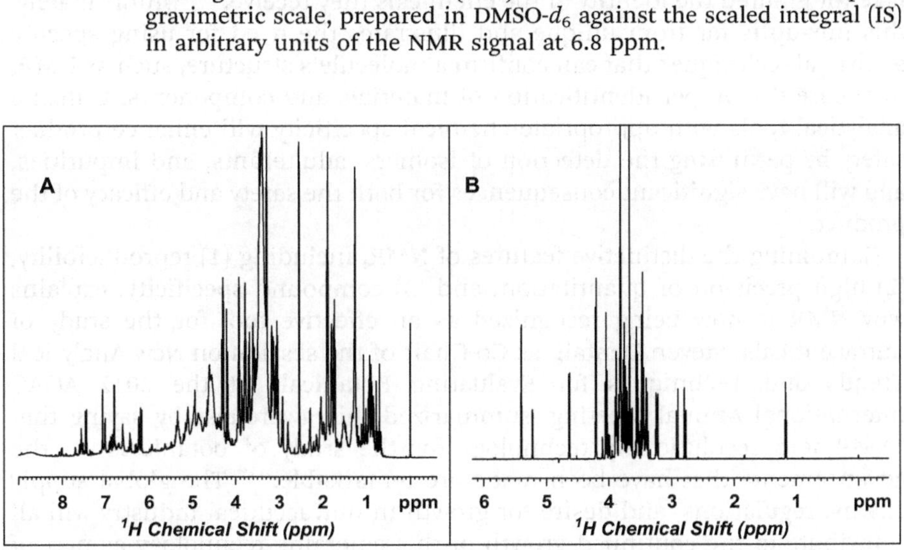

Figure 12.3 NMR spectrum of two nutraceuticals: (A) *Vaccinium angustifolium* and (B) Red Bull energy drink.

NMR samples having a single component. Further factors that may contribute to interference in obtaining quantitative results on a botanical product were evaluated by Hicks *et al.*,[9] where it was determined that, within the approaches tested, there was no significant interference from different field strengths, extraction method, mesh size, gravimetric scale precision, NMR spectroscopy tube type, pulse program, amount of starting dry material, or day-to-day operation. Attention to instrument optimization and experiment choice additionally plays a significant role in the level of quantification accuracy that is achievable.[15,17]

Also important to the uniqueness of NMR for nutraceutical analysis is its inherently high compound specificity, enabling the user to identify the actual components with high accuracy. Utilizing the unique chemical shift and coupling constants for specific atomic environments empowers the NMR user to distinguish structurally related materials with high accuracy, even including many stereoisomers and diastereomers for which other analytical techniques would fail to resolve the materials.

A general principle for quality assurance requires that the analytical technique being used be "fit for purpose" for the question at hand. In 2012, it was reported in *Chemical and Engineering News*[18] that the promising new pharmaceutical agent bosutinib, a selective kinase inhibitor, was mixed up with a structural isomer having the same molecular formula (Figure 12.4). Of the 10 chemical suppliers of bosutinib for research laboratories at the time, eight were supplying the incorrect isomer. Analysis by ^{13}C NMR was able to resolve the two structures, which were indistinguishable by MS and HPLC. Steven Boxer of Stanford University, who discovered the mix-up, stated: "The whole bosutinib saga illustrates that researchers should never take for granted the identity of the chemicals they receive."[18] Unfortunately, this mix-up is far from unique and illustrates the need for using specific analytical techniques that can confirm a molecule's structure, such as NMR, to ensure the proper identification of materials and components. Utilizing analytical tools with appropriate chemical specificity will enhance product safety by permitting the detection of isomers, adulterants, and impurities, and will have significant consequences for both the safety and efficacy of the product.

Combining the distinctive features of NMR, including (1) reproducibility, (2) high precision of quantitation, and (3) compound specificity, explains why NMR is now being recognized as an effective tool for the study of nutraceuticals. Steven Dentali, as Co-Chair of the session on New Analytical Trends and Techniques for Evaluating Botanicals at the 2012 AOAC International Annual Meeting, summarized this awareness by saying that "NMR is a revolutionary technology for the study of botanicals ... the advancements that have been made are remarkable."[19] The global supply chains, regulations, and desire for growth in nutraceutical industry will all contribute to the continued growth of this emerging technology as part of the analytical toolbox for the determination of the identity, purity, strength, and composition of nutraceuticals.

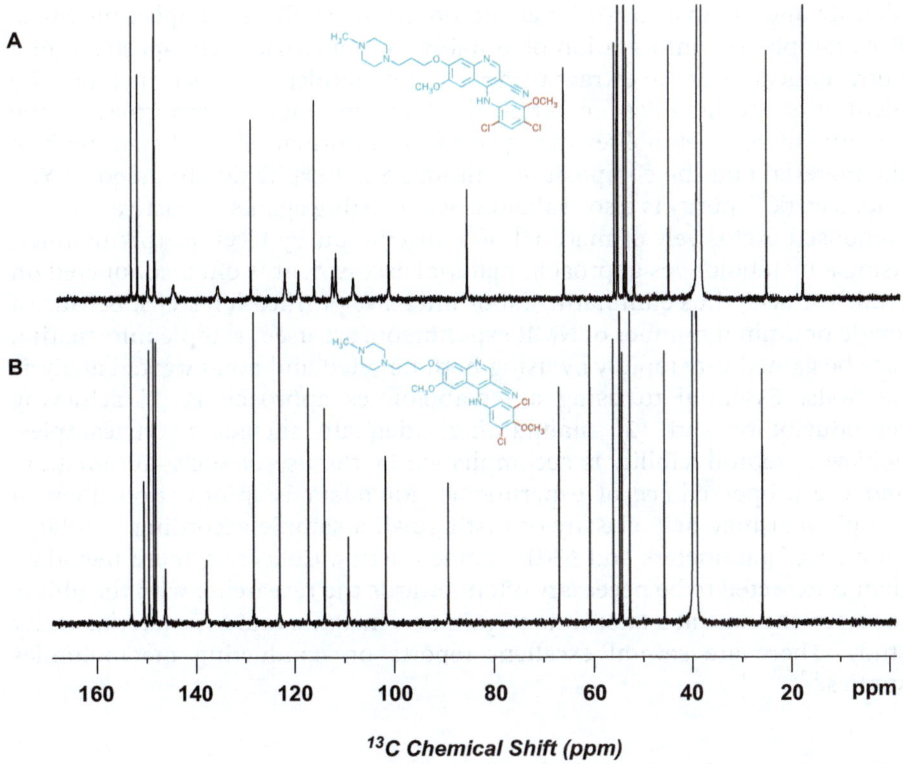

Figure 12.4 ^{13}C NMR spectra of (A) bosutinib and (B) bosutinib isomer acquired at 600 MHz in DMSO-d_6 at 298 K. These two materials are clearly distinguished using ^{13}C NMR.

12.1.3 Highly Complex Mixtures and the Metabolomics Approach

NMR has traditionally been applied to highly purified samples to study the molecular structure in detail (primary, secondary, and tertiary) and molecular dynamics. These studies have used a large number of different NMR experiments to obtain data for sophisticated interpretation. In the early 1990s, studies on human body fluids became common and this began a new era of looking at highly complex mixtures. Metabolomics, the study of small-molecule endogenous metabolites produced by an organism, is now well established, and the application of this approach to natural product material is emerging as not only a research tool but also a quality control tool.[20–22] NMR fingerprinting and NMR profiling of botanicals often includes the study of hundreds of components to characterize the sample.[23–26] In contrast to NMR studies on pure materials, experimentation utilizing a metabolomics approach typically involves a limited number of routine NMR experiments. From these experiments, a wealth of information, including identity and sample history, may be obtained on these complex mixtures. For example, the information obtainable from a single NMR spectrum of a nutraceutical or crude extract using a metabolomics approach includes the identity of the botanical material, the location where it was grown,[26] the quantity of key metabolites that give rise to information on the strength of the material and the composition. Although not explicitly discussed in Yuk *et al.*'s work,[26] purity is also evaluated when testing against statistical models composed exclusively of material of a specific purity level. In this manner, using a metabolomics approach, material assessment is often conducted on crude extracts, thus eliminating labor-intensive purification steps. Because a single or limited number of NMR experiments are used, sample information may be gained very rapidly by using both targeted and non-targeted analysis methods. Essential to using a metabolomics approach is (1) achieving reproducibility and (2) maintaining adequate metadata on samples. Achieving reproducibility is accomplished by the use of SOPs, automation, and the proper choice of experiments. Metadata is information about a sample that may help classify or distinguish a sample according to a large number of parameters and NMR characteristics. Collecting more metadata than is expected to be necessary often rewards the researcher with the ability to review the data at a later time to gain new insights into the material under study. There are several excellent reports on conducting metabolomics studies.[21,27–29]

12.2 Sample Evaluation Procedures

Prior to the analysis of nutraceuticals by NMR, it is helpful to develop a set of SOPs that can be easily followed and used for a large and diverse set of samples. Because using a metabolomics approach typically results in either a consistent means of analysis from sample to sample or a very large number of samples being studied, the SOPs can ensure that the analyst has an easy path for data analysis and the ability to harvest significant details from a relatively small number of NMR experiments. In the following, examples are given of SOPs used by Bruker BioSpin for the analysis of botanical materials from various botanical research laboratories.

12.2.1 Example Bruker SOP Considerations Used for Nutraceutical Analysis

Nutraceuticals from various sources and at different stages of processing are analyzed. In some cases, the material will be not be processed at all, and in other cases the material will be highly processed and in a complex matrix. SOPs for these different processing levels of materials need to be developed for each facility. The example used here as demonstration for considerations in the development of SOPs is leaf materials from *Vaccinium angustifolium* (lowbush blueberry)[8]. Through this example, a process of NMR analysis of blueberry leaf material is brought from the natural state to the end NMR result to demonstrate all possible stages where SOPs may be required.

12.2.1.1 SOP for Sample Collection and Processing

Specifics on how the material is collected and processed to the point prior to NMR sample preparation are included in the SOP. Example: blueberry leaves[8] are harvested and dried overnight in a dehydrator. Dried leaves are ground in a Wiley Mill through a size 40 mesh and extracted with 95% ethanol (10 mL per gram of leaf material) with shaking at room temperature for 24 h. After 24 h, the solvent is decanted (phase 1) and the ground material is extracted again using 95% ethanol (5 mL per gram of leaf material) and shaken for a further 24 h. Subsequently, the solvent is decanted (phase 2) and phase 1 and 2materials are pooled and centrifuged at 3000 rpm for 5 min at room temperature. The solvent is decanted and all alcohol is removed in a Speed Vac at 37 °C. To remove water, the samples are lyophilized overnight. All extracts are stored at −20 °C.

12.2.1.2 SOP for Metadata

Metadata is information on the sample that can empower the analyst to harvest extensive information on the sample from the NMR data. For the blueberry leaf sample, information on the taxonomic name, location of collection, time and date of collection, plant health status, name of collector and processor, and processing method is the minimum that is collected. This information, in addition to other information shown in Figure 12.5, positions the NMR spectroscopist to be capable of obtaining significant information such as establishing correlations between the metadata and statistically significant trends in the NMR data during the current analysis and at any time in the future.

Name Information
- Common
- Taxonomic Name
- Name of person making identification
- Method used for identification
- Cultivar
- Other

Collection Site Information
- Collector Name
- Date of Collection
- Time of Day of Collection
- Site of Collection
- GPS location

Harvest information
- Seasonal weather information
- Current farm practice (fertilizers and/or pesticides)
- Storage procedure
- Specimen Information
- Plant part (s)
- Age of plant
- Growing conditions
- Health status

Processing information
- Processed by
- Processing date
- Processing procedure
- Storage details

Figure 12.5 Example MetaData collected on a plant sample.

12.2.1.3 SOP for Preparation of NMR Samples

Considerations for SOPs for sample preparation are typically (1) the choice of solvent and (2) the concentration needed to utilize a constant receiver gain setting for the sample set to be compared. In the case of blueberry leaf material, the study's aim was to develop a screening tool to establish the species of the material in the *Vaccinium* genus and determine the amount of chlorogenic acid and hyperoside present in the sample. Chlorogenic acid and hyperoside are thought be some of the metabolites that contribute to the health benefits of this material when used as a traditional medicine.[16,30] Initial solubility experiments were conducted across a few species of *Vaccinium,* and the data were acquired with the most sensitive NMR probe that would be utilized in the study, which in our case was a 5 mm TCI *z*-gradient CryoProbe. For *Vaccinium* analysis, DMSO-d_6 was the solvent of choice, offering complete solubility of the test material. Nutraceutical NMR analysis often utilizes DMSO-d_6 or D_2O as solvent to achieve high solubility of the test material.[26,31,32] However, it should be noted that if D_2O is used, it is recommended to utilize a buffered solution such as sodium or potassium phosphate (NaH_2PO_4 or KH_2PO_4) solution to control pH to minimize chemical shifts in the sample.

An example of NMR sample preparations is as follows:

- *For samples soluble in DMSO:* A dimethyl sulfoxide–4,4-dimethyl-4-silapentane-1-sulfonic acid-d_6 (DMSO–DSS) solvent solution is prepared by adding 157 μL of a 300 mM solution of DSS-d_6 (in DMSO-d_6) to 100 mL of DMSO-d_6. A 25 mg amount of crude botanical extract is dissolved in 1.0 mL of DMSO–DSS solvent solution. Samples are vortexed for 10 s and subsequently centrifuged for 10 s at 6400 rpm. 600 μL of the resulting supernatant is transferred to a 5 mm NMR tube for spectroscopy.
- *For samples soluble in D_2O:* A buffer solution of 150 mM KH_2PO_4 (pH 7.4), 200 μM NaN_3, 0.01% sodium -3-trimethylsilylpropionate (TMSP-2,2-3,3-d_4) (or DSS) is prepared. A 25 mg amount of crude botanical extract is dissolved in 1 mL of buffer. Samples are vortexed for 1 min and subsequently centrifuged for 10 s at 6400 rpm. A 600 μL volume of the resulting supernatant is transferred to a 5 mm NMR tube for spectroscopy.

12.2.1.4 SOP for NMR Instrument Optimization

Evaluating the NMR spectrometer's readiness to acquire reproducible data is an essential step in analyzing nutraceuticals. This includes two levels of optimization of NMR instrumentation. The first level is performed after any hardware or software changes have been made to the instrument to be fully automated for data acquisition and data processing. On a Bruker spectrometer, this may include the lock parameters (edlock), the probe ringdown delay (DE), pulse power for each nucleus (edprosol), temperature control (edte), and the starting shim set for the probe (setshim). In most cases, each vendor will have a set of optimization conditions for their NMR spectrometers. The second level of NMR optimization is performed to address daily magnet drifts and unexpected instrument issues. Such optimization is performed on a regular basis, often daily, and typically includes instrument shim optimization, temperature optimization, a ^{1}H lineshape test, and sensitivity tests for each nucleus used in the evaluation of the nutraceutical. This second level of optimization additionally validates that the NMR instrument is performing to expected specifications and as required for facilities operating under good laboratory practices (GLP). With the current state of NMR instrumentation, the instrument validation and daily optimization may be performed with automation as shown in Figure 12.6 to simplify the workflow for a research laboratory. The NMR optimization steps described above are generally applied and are not specific to any given material.

12.2.1.5 SOP for Data Acquisition and Processing

SOPS for NMR acquisition and processing typically involve standard parameter sets and a set routine for exponential multiplication, Fourier transformation, spectral phasing, baseline correction, and referencing. For blueberry leaf material, all samples are run at 298 K on a temperature-calibrated NMR spectrometer. A 90° pulse width calibration is performed

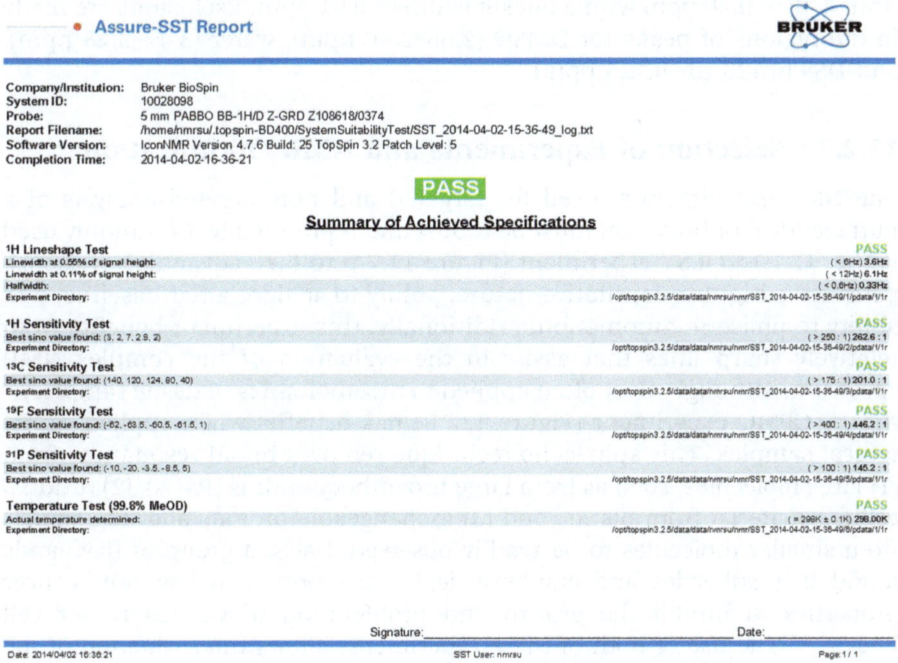

Assure-SST Report

BRUKER

Company/Institution: Bruker BioSpin
System ID: 10028098
Probe: 5 mm PABBO BB-1H/D Z-GRD Z108618/0374
Report Filename: /home/nmrsu/.topspin-BD400/SystemSuitabilityTest/SST_2014-04-02-15-36-49_log.txt
Software Version: IconNMR Version 4.7.6 Build: 25 TopSpin 3.2 Patch Level: 5
Completion Time: 2014-04-02-16-36-21

PASS

Summary of Achieved Specifications

^{1}H Lineshape Test PASS
Linewidth at 0.55% of signal height: (< 6Hz) 3.6Hz
Linewidth at 0.11% of signal height: (< 12Hz) 6.1Hz
Halfwidth: (< 0.6Hz) 0.33Hz
Experiment Directory: /opt/topspin3.2.5/data/data/nmrsu/nmr/SST_2014-04-02-15-36-49/1/pdata/1/1r

^{1}H Sensitivity Test PASS
Best sino value found: (3, 2, 7, 2.8, 2) (> 250 : 1) 262.6 : 1
Experiment Directory: /opt/topspin3.2.5/data/data/nmrsu/nmr/SST_2014-04-02-15-36-49/2/pdata/1/1r

^{13}C Sensitivity Test PASS
Best sino value found: (140, 120, 124, 80, 40) (> 175 : 1) 201.0 : 1
Experiment Directory: /opt/topspin3.2.5/data/data/nmrsu/nmr/SST_2014-04-02-15-36-49/3/pdata/1/1r

^{19}F Sensitivity Test PASS
Best sino value found: (-62, -63.5, -60.5, -61.5, 1) (> 400 : 1) 446.2 : 1
Experiment Directory: /opt/topspin3.2.5/data/data/nmrsu/nmr/SST_2014-04-02-15-36-49/4/pdata/1/1r

^{31}P Sensitivity Test PASS
Best sino value found: (-10, -20, -3.5, -8.5, 5) (> 100 : 1) 145.2 : 1
Experiment Directory: /opt/topspin3.2.5/data/data/nmrsu/nmr/SST_2014-04-02-15-36-49/5/pdata/1/1r

Temperature Test (99.8% MeOD) PASS
Actual temperature determined: (= 298K ± 0.1K) 298.00K
Experiment Directory: /opt/topspin3.2.5/data/data/nmrsu/nmr/SST_2014-04-02-15-36-49/6/pdata/1/1r

Signature:__________ Date:__________

Date: 2014/04/02 16:36:21 SST User: nmrsu Page: 1 / 1

Figure 12.6 Daily instrument optimization and validation may be performed with complete automation, as demonstrated by Bruker BioSpin's Assure-SST product. The automatically generated report indicates the current instrument performance compared with required performance specifications.

with automation on each sample prior to acquiring NMR data. Proton NMR spectra acquired include a one-dimensional proton nuclear Overhauser effect spectroscopy 1D-NOESY pulse sequence experiment utilizing ^{13}C decoupling[8] and a 1D-Carr–Purcell-Meiboom-Gill experiment (1D-CPMG).[26] The 1D-NOESY sequence is performed using hard pulses (non-selective mode) to achieve highly quantitative data across the entire spectrum. For optimal baseline, the Bruker digital filter "baseopt" is used. Spectra are processed using exponential line broadening (0.3 Hz) and phased with zero-order phase corrections. Spectra are referenced to DSS at 0.0 ppm.

12.2.1.6 SOP for Data Analysis

Constructing SOPs for data analysis is needed when sample-to-sample comparisons are made. This may include the method used for (1) identification of the signal (spectral database comparison or knowledge based), (2) signal integration (region, linefit, or other means), (3) generation of a bucket table, and (4) testing against models or spectra of known material. For blueberry leaf material, chlorogenic acid and hyperoside are identified using a comparison with an NMR spectral database. Quantification is performed using a linefit integration of the signals at 6.8 and 7.7 ppm. Species differentiation is performed by principal component analysis (PCA) and using inclusivity panels. Bucket table parameters include a chemical shift region from 8.40 to 0.40 ppm with a bucket width of 0.01 ppm. Exclusions are made in the regions of peaks for DMSO (2.66–2.40 ppm), water (3.42–3.23 ppm), and DSS (0.125 to −0.125 ppm).

12.2.2 Selection of Experiments and NMR Optimization

The NMR experiment(s) used for targeted and non-targeted analysis of a nutraceutical or botanical must be robust and reproducible. Commonly used is the ^{1}H 1D-NOESY experiment (Figure 12.7A) to take advantage of the experiment's highly quantitative nature, ability to achieve a flat baseline, and ability to phase in automation. Additionally, this experiment benefits from relatively sharp lines that assist in the evaluation of the complex NMR spectrum. Although most often applied to metabolomics of tissue samples,[33] the 1D-CPMG experiment (Figure 12.7B) has benefits when applied to botanical samples. This spin–echo technique removes broad resonances from (1) large molecules, such as from large proanthocyanidins (PACs), (2) residual cellular material from plants, and (3) exchangeable protons allowing signals from smaller molecules to be readily observed. PACs, a group of flavonoids found in cranberries and cranberry leaf, are reported to have anti-cancer properties to inhibit the growth and proliferation of various tumor cell lines.[34–36] The use of both of these experiments allows the evaluation of the PACs and also the smaller metabolites that resonate in the same chemical shift region as the PACs, as seen in Figure 12.7. Both of these experiments are highly reproducible and may be used for targeted and non-targeted analysis.

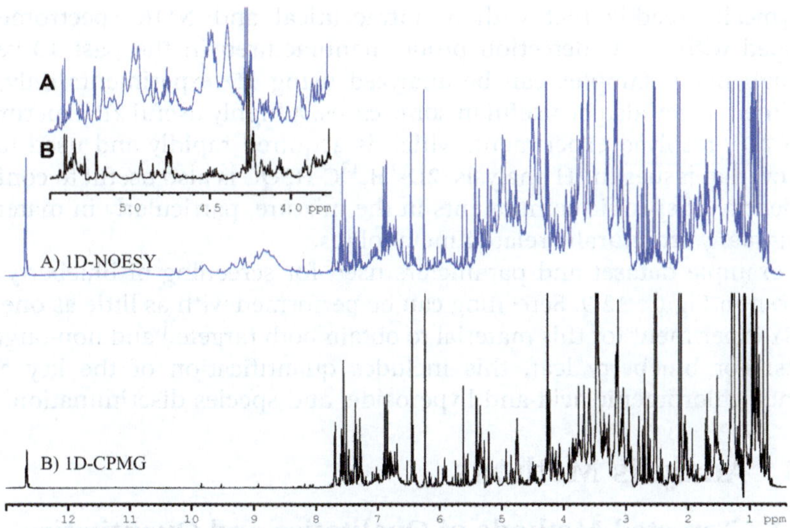

Figure 12.7 NMR spectra of cranberry leaf (*Vaccinium macrocarpon*) extract including (A) 1D-NOESY with a spoil gradient and (B) 1D-CPMG that filters out the broad resonances from the large molecules such as polycyclic aromatic compounds.

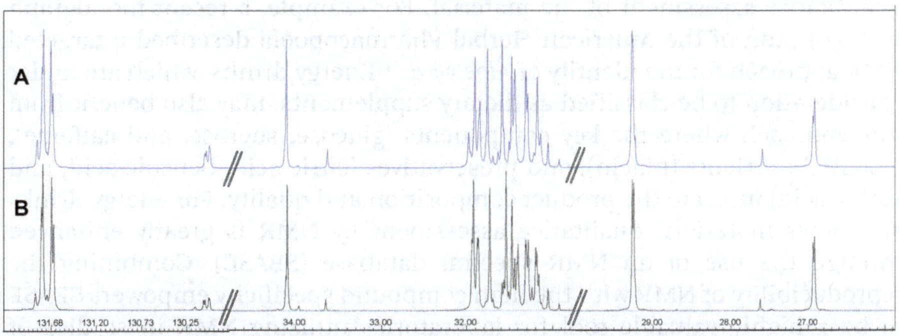

Figure 12.8 Select regions of representative ^{13}C NMR spectra showing the distinction between (A) olive oil and (B) hazelnut oil. Data were acquired at 500 MHz with 48 scans using a direct detect 5 mm DCH CryoProbe using the zgpg30 pulse program.

Decoupling ^{13}C to eliminate the observation of the ^{1}H–^{13}C couplings simplifies the data analysis when using these experiments.

Decoupled ^{13}C experiments can afford highly quantitative results and the analyst benefits from the large ^{13}C chemical shift range that often clearly distinguish closely similar materials. For example, olive oil may be distinguished from hazelnut and other seed oils by using this experiment (Figure 12.8). Utilizing this experiment often requires a significant amount of sample or a probe designed for high ^{13}C sensitivity. Both of these criteria are typically readily met with a nutraceutical and NMR spectrometers equipped with a ^{13}C detection probe manufactured in the past 10 years. Although many samples can be analyzed using 1D experiments only, 2D experiments may also be useful in some cases. A highly useful 2D experiment is the 2D *J*-resolved experiment, which is acquired rapidly and used to resolve overlap issues in ^{1}H analysis. 2D-^{1}H,^{13}C-HSQC is also useful to confirm the identity of specific components in the mixture, particularly in materials having many structurally related metabolites.

An example dataset and parameters used for screening of blueberry leaf is shown in Figure 12.9. Screening can be performed with as little as one 1D-NOESY experiment for this material to obtain both targeted and non-targeted results. For blueberry leaf, this includes quantification of the key components, chlororgenic acid and hyperoside, and species discrimination.[8,9]

12.3 Analysis Methods

12.3.1 Targeted Methods of Qualitative and Quantitative Assessment: Identity, Purity Strength, and Composition

Targeted approaches involve evaluating a sample for specific metabolites, components, impurities, or adulterants to provide a qualitative and often a quantitative assessment of the material. For example, a recent monograph by Roy Upton of the American Herbal Pharmacopoeia described a targeted NMR approach for the identity of *Aloe vera.*[37] Energy drinks, which are under consideration to be classified as dietary supplements, may also benefit from this approach where the key components (glucose, sucrose, and caffeine), essential nutrients (niacin), and preservatives (citric acid, benzoic acid, and sorbic acid) indicate the product composition and quality. For energy drinks and other materials, qualitative assessment by NMR is greatly enhanced through the use of an NMR spectral database (SBASE). Combining the reproducibility of NMR with the high compound specificity empowers SBASE to be a highly valuable tool for laboratories utilizing NMR in studies of nutraceuticals. An NMR spectral database of a specific metabolite typically contains a "cleaned" NMR spectrum devoid of signals from solvent, reference standards, impurities, and noise. The SBASE entry may be of a pure material or a commonly used mixture. Utilizing appropriate SOPs, SBASEs can be developed to be used for the identification of specific components with relative ease. Comparison of line positions, *J*-couplings, and the line-shape of the individual resonances from the nutraceutical with the pure component enables the analyst to match the resonances and determine the presence of components, as shown in Figure 12.10 for Monster Energy Drink. Spectral complexity of a molecule benefits the analysis because it brings assurance that the identity of the component is properly assigned, as seen for glucose in Figure 12.11. Materials with simple spectra, such as acetic acid that produces only a singlet peak, are much more difficult to

Blueberry Leaf NMR Dataset Parameters

Minimal Data Acquired

- 1D-NOESY with inverse gated decoupling:
  - o Bruker pulse program = noesyig1d
  - o Automation scripts: au_assure (AUNM) which includes a pulse calibration and acquisition without automated receiver gain adjustment, proc_assureshim (AUNMP) which includes a first-order automated phase adjustment
  - o Delays: D1 = 10 s, D8 = 0.0 s
  - o Scans: DS = 4, NS = 64
  - o Other: SW = 20ppm, TD = 64K, RG = 32, DIGMOD = baseopt, LB = 0.3Hz

Optional Experiments

- 1D-CPMG: T2 filter using Carr–Purcell–Meiboom–Gill sequence
  - o Bruker pulse program = cpmg1d.baseopt
  - o Automation scripts: au_assure (AUNM) which includes a pulse calibration and acquisition without automated receiver gain adjustment, proc_assureshim (AUNMP) which includes a first-order automated phase adjustment
  - o Delays: D1 = 4 s, T2 filter time = 0.126 s
  - o Scans: DS = 4, NS = 32
  - o Other: SW = 20 ppm, TD = 64K, RG = 32, DIGMOD = baseopt, LB = 0.3Hz

- 2D *J*-resolved: homonuclear *J*-resolved 2D correlation
  - o Bruker pulse program = jresqf
  - o Automation scripts: au_assure (AUNM) which includes a pulse calibration and acquisition without automated receiver gain adjustment, proc_assureshim (AUNMP) which includes a first-order automated phase adjustment
  - o Delays: D1 = 2 s
  - o Scans: DS = 16, NS = 16
  - o Other: SW = 20 ppm, RG = 32, 8K data points in F_2, 40 data points in F_1, LB = 0.3Hz, window function = SINE, sine bell shift = 0

- 2D-^{1}H, ^{13}C-HSQC : 2D, H/C correlation via double inept transfer
  - o Bruker pulse program = hsqcedetgpsisp2.3
  - o Automation scripts: au_assure (AUNM) which includes a pulse calibration and acquisition without automated receiver gain adjustment
  - o Delays: D1 = 1.5 s, optimized for a ^1JCH = 145 Hz
  - o Scans: DS = 16, NS = 8
  - o Other: SW2 = 20 ppm, SW1 240 ppm, 4K data points in F_2, 256 data points in F_1, RG = 32

Figure 12.9 Parameters used for a complete NMR screening run for blueberry leaf. Parameter sets are available on request from the corresponding author.

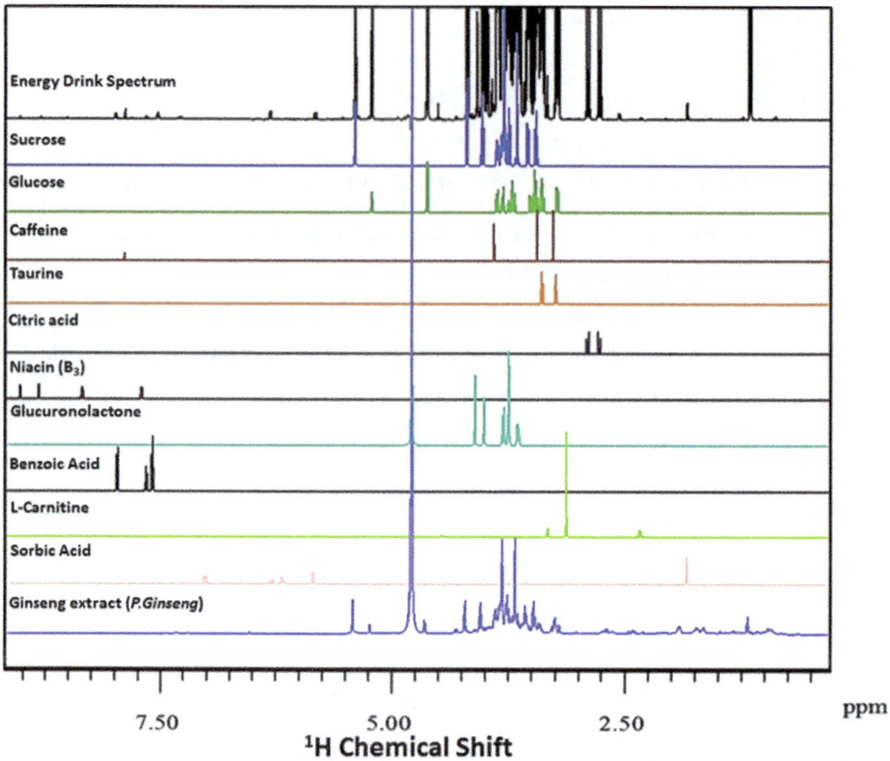

Figure 12.10 NMR spectrum of Monster Energy Drink compared with NMR SBASE entries of common components, degradation products, and NMR reference standards. Data for all the samples were acquired at 600 MHz in 150 mM phosphate buffer at pH 7.4.

assign with confidence. With careful attention to experimental conditions, the NMR user may reduce the chemical shift differences between the metabolite in a pure material, and hence the SBASE, as compared with the chemical shift within the nutraceutical to make identification through spectral matching of single peaks easier.

To aid in the interpretation of highly overlapped regions of the NMR spectrum of a nutraceutical, two-dimensional NMR may be employed. A 2D *J*-resolved experiment (Figure 12.12) can be acquired in a few minutes and benefits the analyst by resolving protons having the same chemical shift by utilizing differences in the coupling constants. In molecules with a small number of protons or a small number of proton–proton couplings, the identification of the component is more complex and often requires experiments utilizing heteronuclear information. The 2D-^{1}H,^{13}C-HSQC experiment is commonly employed to permit a clear distinction between even closely related structures such as various sugars, *e.g.* sucrose and glucose, in a complex mixture such as energy drinks, as seen in Figure 12.13.

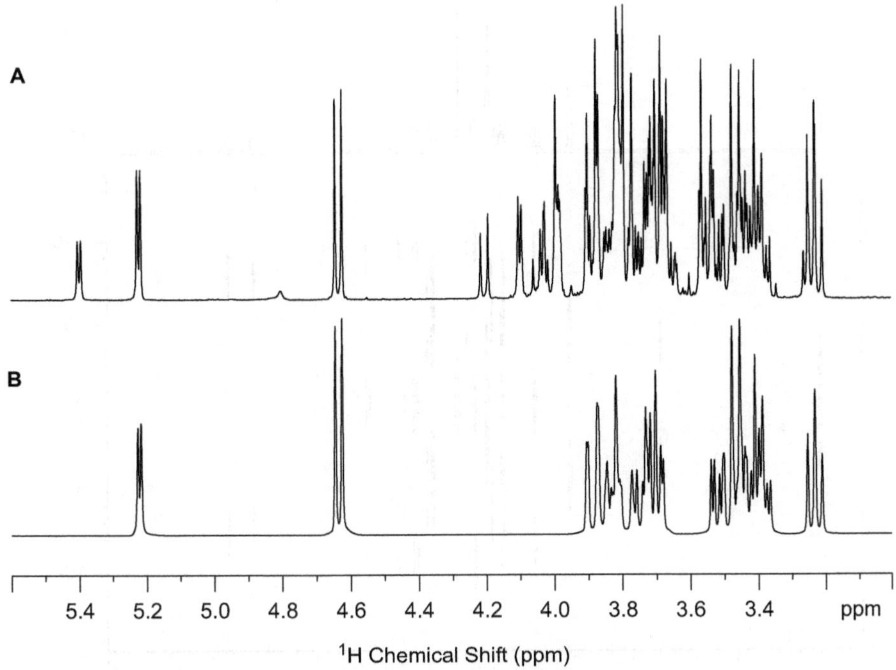

Figure 12.11 NMR spectrum of (A) Red Bull energy drink compared with (B) a glucose NMR SBASE entry. The complexity of the glucose spectrum assists the identification of this material in this complex region of the spectrum. Data for all the samples were acquired at 600 MHz in 0.15 M phosphate buffer at pH 7.4.

The targeted approach of nutraceutical evaluation by NMR provides identity and composition analysis of the material through the identities of specific components. Information on the nutraceutical's strength may also be ascertained from the crude material through quantification, if the active component has been identified as the active component and has resolved signals in the NMR spectrum. The metabolites must also be present in adequate quantity in the mixture. For example, eugenol, a key component in holy basil (*Ocimum tenuiflorum*), may be identified in an extract of holy basil by spectral comparison, as shown in Figure 12.14. The eugenol concentration may be determined by integration of the peak at 6.74 ppm. Integration approaches generally utilize either region integration or linefit integration, as shown in Figure 12.15. Linefit integration typically affords greater accuracy in complex spectra where spectral overlap and baseline distortions from broad signals are observed.

In addition to the means of integration, other factors influence the accuracy of quantitation from an NMR spectrum and are well documented in papers by Burton *et al.*,[15] Pauli *et al.*,[38] and Saude *et al.*[17] Attention to several influences, such as flat baselines, pulse calibration, adequate relaxation delay, and signal-to-noise ratios, are needed to obtain reliable results.

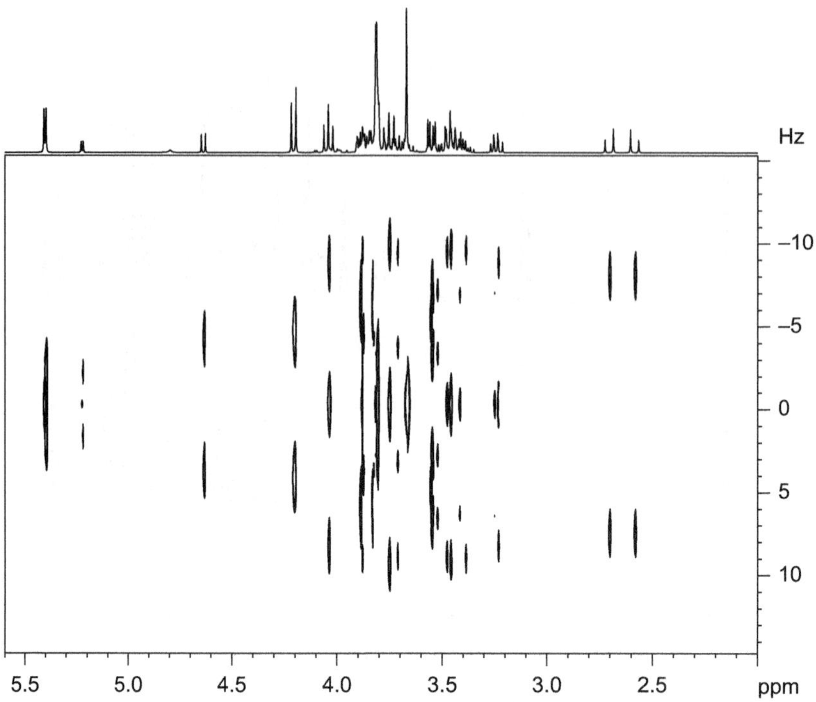

Figure 12.12 Expansion of the 2D J-resolved spectrum Red Bull energy drink showing the heavily overlapped sugar region. The 2D *J*-resolved spectrum shows a specific proton signal as a function of chemical shift (*x*-axis) and coupling constant (*y*-axis). The pattern for each proton in this spectrum may be used for identification and quantification of signals that are difficult to evaluate in the 1D ^{1}H spectrum.

Additionally, quantitative comparisons for multiple spectra are enhanced by utilizing SOPs for sample preparation, NMR acquisition, data processing, and analysis.

12.3.2 Non-targeted NMR Approaches of Qualitative Assessment

Non-targeted NMR approaches to nutraceutical analysis allow a rapid assessment of material and can provide extensive information on the sample because of the large numbers of components that are simultaneously evaluated. Information pertaining to material identification, purity, adulterants, plant health state, growing location and/or processing location, processing method, and product conformance are some of the information accessible from a non-targeted assessment of a nutraceutical.[16,39] Accordingly, botanical origin, identity, purity, strength, and composition of a nutraceutical product may be evaluated with the combined use of statistical methods (non-targeted) and targeted analysis as previously discussed.

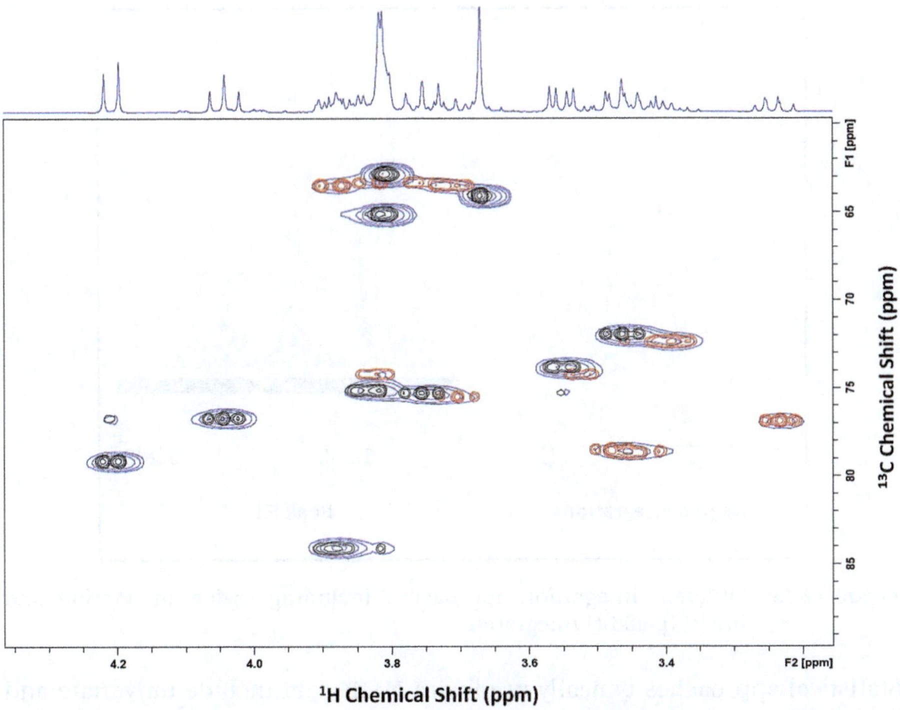

Figure 12.13 ^{1}H,^{13}C-HSQC spectrum of the Red Bull energy drink showing the heavily overlapped sugar region. The 2D-HSQC spectrum shows the connectivity of a specific proton signal to carbon as a function of chemical shift of proton (*x*-axis) and carbon (*y*-axis). The blue spectrum is the Red Bull energy drink, the red spectrum is the glucose signals, and the black spectrum is the sucrose signals.

Figure 12.14 600 MHz NMR spectra of (A) eugenol and (B) extract of holy basil (*Ocimum tenuiflorum*). The resonance at 6.74 ppm in the holy basil sample was integrated to determine the concentration of eugenol as 2.70 mM in the extract by linefit (peakfit) integration.

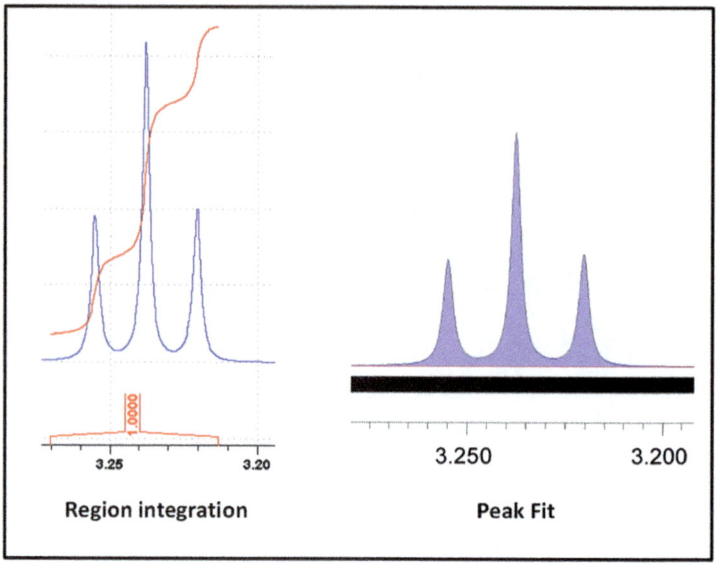

Figure 12.15 Different integration approaches including region integration and linefit (peakfit) integration.

Statistical approaches typically used with NMR data include univariate and multivariate analysis. Any statistical approach relies on the evaluation of a set of spectra rather than individual spectra. For this reason, SOPs are rigorously adhered to and the scope of the analysis is evaluated thoroughly prior to commencement of the analysis. Methodologies utilizing NMR-based metabolomics were well summarized in an excellent review by Cox *et al.*[28] For a general understanding of statistical approaches, there are many good sources.[40–43]

12.3.2.1 Number of Samples Needed for a Statistical Analysis

Before undertaking a statistical evaluation of a nutraceutical, the expected variation in the material is established. Knowing this expected variation allows the analyst to define a set of reference spectra that are used to compare to a new sample to be analyzed. In some cases, as in the case of an energy drink of a specific formula, there will be minimal variation. Other cases, such as in the case of wild-collected *Vaccinium angustifolium* leaf, the variation from sample to sample is high. Determining the number of samples needed for statistical analysis is therefore variable and is dependent on the material tested and the analysis desired. The model spectra need to describe the natural variation.

With vitamin B tablets, where the product may be produced with high conformity, 10 samples define the variation in composition of the product, whereas in *Vaccinium angustifolium*, where hybridization is common, a larger sample set of 43 was needed, as shown in Figure 12.16.

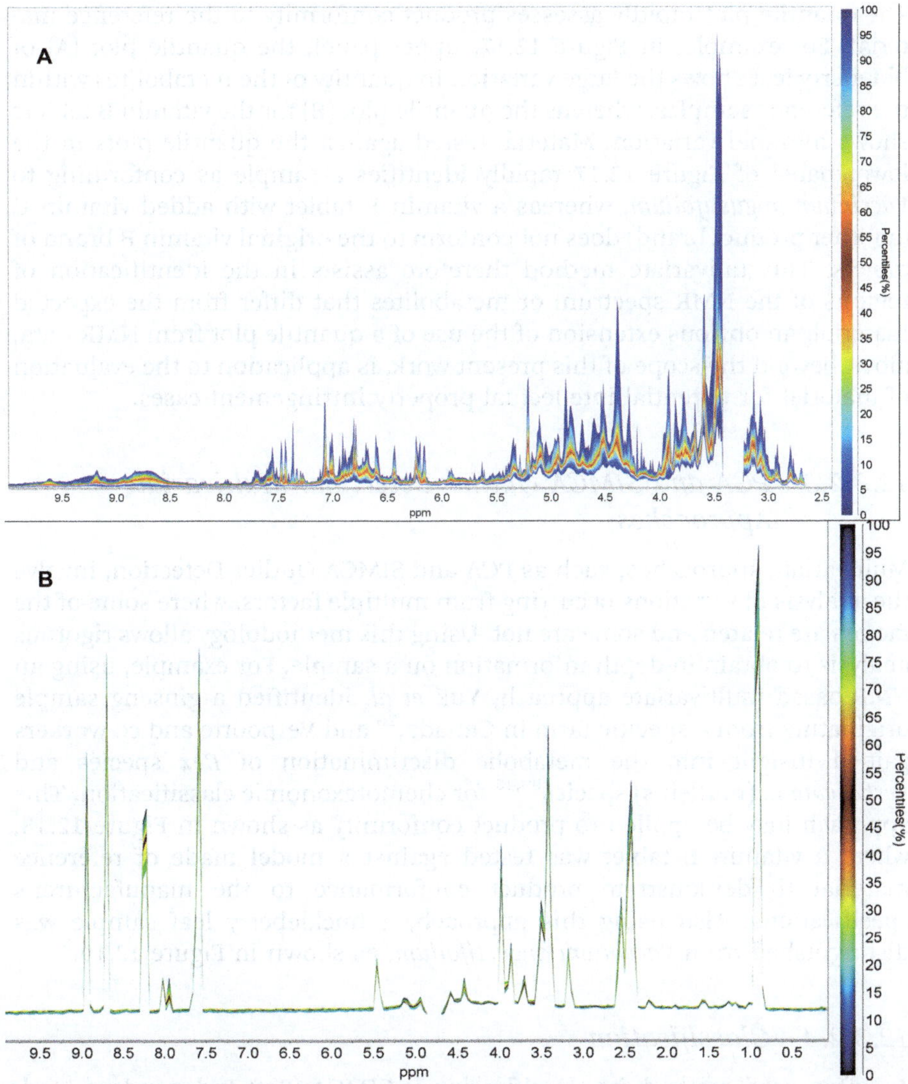

Figure 12.16 Quantile plots of: (A) *Vaccinium angustifolium* showing large variation over the sample set and (B) vitamin B tablets showing little product variation from sample to sample.

12.3.2.2 Quantile Plot: a Univariate Method

A valuable tool for establishing the conformance of a sample to the reference samples is the quantile plot.[44] This plot calculates the intensity distribution (percentiles) for the group of spectra that one is analyzing and is an excellent method for understanding the natural variation within the group and if a spectrum fits the group. For NMR, the distributions are chemical shift and intensity, where intensity reflects the quantity of the component present.

The quantile plot rapidly assesses product conformity to the reference material. For example, in Figure 12.17, upper panel, the quantile plot (A) of blueberry leaf shows the large variation in quantity of the metabolites within the reference samples, whereas the quantile plot (B) for the vitamin B tablets shows minimal variation. Material tested against the quantile plots in the lower panel of Figure 12.17 rapidly identifies a sample as conforming to *Vaccinium angustifolium*, whereas a vitamin B tablet with added vitamin C (another product brand) does not conform to the original vitamin B brand of tablets. This univariate method therefore assists in the identification of regions of the NMR spectrum or metabolites that differ from the expected material. An obvious extension of the use of a quantile plot from NMR data, albeit beyond the scope of this present work, is application to the evaluation of material for potential intellectual property infringement cases.

12.3.2.3 PCA and SIMCA Outlier Detection: Multivariate Approaches

Multivariate approaches, such as PCA and SIMCA Outlier Detection, involve the analysis of variations occurring from multiple factors, where some of the factors are related and some are not. Using this methodology allows rigorous analysis to obtain in-depth information on a sample. For example, using an NMR-based multivariate approach, Yuk *et al.* identified a ginseng sample originating from a specific farm in Canada,[26] and Verpoorte and co-workers gained insight into the metabolic discrimination of *Ilex* species and *Verbascum* L. (mulleins) species[39,45] for chemotaxonomic classification. This approach may be applied to product conformity as shown in Figure 12.18, where a vitamin B tablet was tested against a model made of reference material to demonstrate product conformance to the manufacturer's specifications. Also using this approach, a huckleberry leaf sample was distinguished from *Vaccinium angustifolium*, as shown in Figure 12.19.[8]

12.3.2.4 Classification

An often used method for classification is SIMCA (Soft Independent Modeling of Class Analogies). Classification of a sample requires an analysis of the natural variance within the group to compare first. This is done by a principal component analysis (PCA), which projects the high-dimensional data on to lower dimensional space while retaining as much as possible of the variance of the data set.[46,47] PCA transforms the data set from one coordinate system (*e.g.* position/intensity) into a new coordinate system. The new principal components axes (PCs) are uncorrelated and are sorted by variance: the first PC has the highest variance, and higher PCs explain only small variances. To reduce the dimensionality, the first few PCs are typically interpreted because these describe the natural variance within the group. The remaining dimensions are typically regarded as "noise" and ignored. This is an unsupervised method because the membership of individual

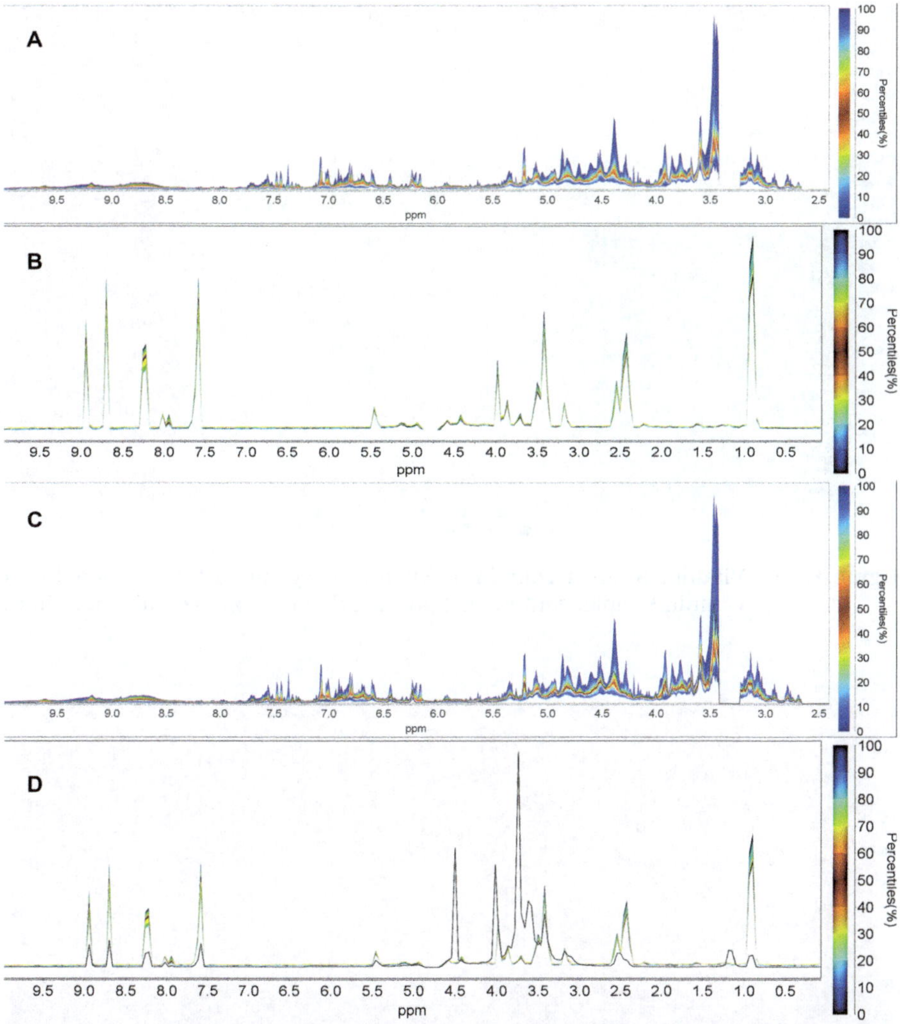

Figure 12.17 In the upper panels, the quantile plots of (A) *Vaccinium angustifolium* and (B) vitamin B tablets generated from 43 and 10 samples, respectively, show the distribution of reference samples for each of these materials. The lower panels show a black line that represents material tested against the models where (C) shows a new *Vaccinium angustifolium* sample conforming to the product and (D) shows a vitamin B tablet containing vitamin C (another product brand) not conforming to the original product distributed as vitamin B tablets. The plot in (D) additionally shows the outliers from the expected product and may be used to identify the differences between the two vitamin B brands.

samples of a data set is not known to the PCA algorithm. Prior to the statistical analysis, the data are centered and scaled. An overview of standard scaling methods can be found in the book by Axelson.[48]

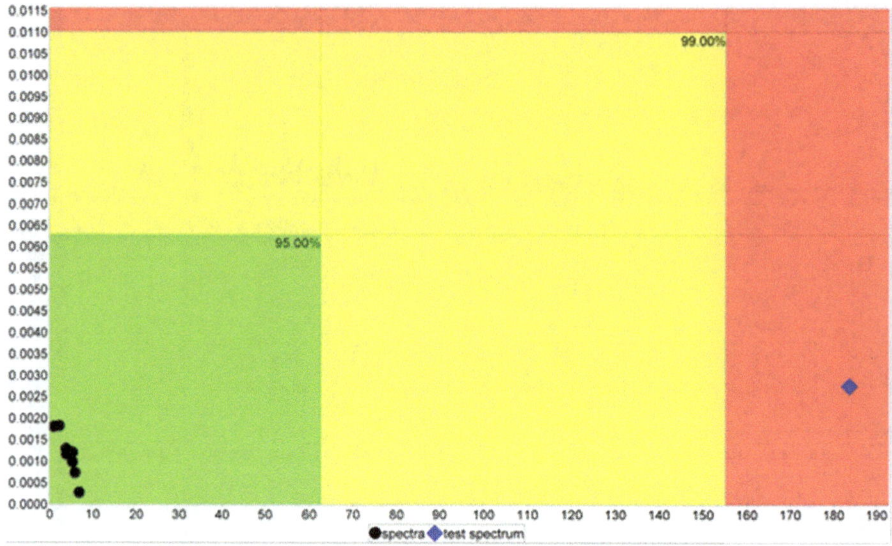

Figure 12.18 Vitamin B tablet containing vitamin C against a SIMCA model of a vitamin B tablet commercial product that does not contain vitamin C.

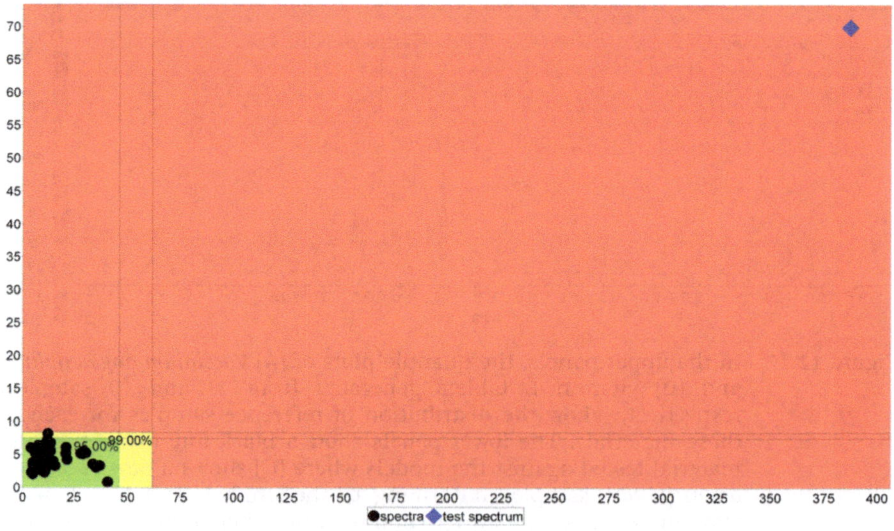

Figure 12.19 Huckleberry (*Vaccinium gaylussacia*) against a SIMCA model of *Vaccinium angustifolium*.

SIMCA classification is based on distance measures in the reduced data space. First, the SIMCA model needs to be defined by (1) selection of the reduced model space and (2) the definition of maximum allowed distances. This is usually done by defining one or more confidence intervals. The PCA and SIMCA methods assume that the model spectra follow a Gaussian distribution. A sample is tested by being projected into the reduced model space. Two distances are taken into account: distance to the model (the residual standard deviation, "off-model components") and the distance to model center (within reduced model space).[40,42,43] If the sample distance is outside the defined confidence interval, it is regarded as an outlier.

Figures 12.18 and 12.19 show the results of such a classification: The model spectra are in black and within the 95% confidence bound whereas the test spectra are outside the 99% confidence bound. The *x*-axis shows the distance to the model center and the *y*-axis the distance to the model.

12.3.2.5 PLS Regression

Partial least-squares (PLS) regression is a linear multivariate method that finds the correlation between the data tables *X* (factors/predictors) and *Y* (responses) for quantitative analysis.[49,50] The data table *X* are the NMR data of the calibrated samples and the data table *Y* are the known quantitative values such as concentrations of chemical components. The advantage of using PLS regression is the ability to focus on only the data analysis that relates the variables of *X* to *Y*. This allows the analysis of noisier, collinear, and even incomplete data sets for both *X* and *Y*.[50] In general, there are two modes that are used in practice: PLS-1 and PLS-2. PLS-1 uses just one column in the *Y* table, *e.g.* the concentration of a single component. PLS-2 uses more than one column of the *Y* table. Usually PLS-1 gives more precise results than PLS-2, but in the case of PLS-1 one needs one model for each component whereas in PLS-2 a single model will quantify several components at the same time. The precision for any PLS model increases when one provides more relevant *X*-variables. With a properly calibrated PLS model, chemical concentration can be predicted for new samples. Using the whole NMR spectrum for concentration prediction offers increased precision compared with using smaller regions. It is possible to quantify components that are not directly visible, *e.g.* overlapped or a not defined NMR pattern (=mixture of compounds or structures). For nutraceuticals, PLS regression can be used for the analysis of edible oil content (borage, safflower, walnut, hazelnut, olive oil, *etc.*). In recent years, there has been an increasing trend for the adulteration of edible oils by mixing a cheaper alternative with the original product.[51,52] Owing to the complexity of the NMR spectrum of edible oil samples, multivariate approaches such as PLS regression fit well. For example, borage oil is a widely used dietary supplement for the treatment of various degenerative diseases such as osteoporosis, diabetes, and cancer.[53] However, in the market, it has been known to be adulterated with other similar materials such as safflower oil. From a past study in our laboratory (unpublished results), PLS regression was able to determine the adulteration of safflower oil in borage oil at levels as low as 0.25%. This was done using a calibrated set of carefully measured standard borage oil samples with various concentrations of safflower oil (0.0–10%). Using PLS regression, a calibration was performed on the standard samples to the safflower oil concentrations (*Y* table). With the calibrated PLS model, various mixtures of borage oil and safflower oil (0.25 and 0.52%) were predicted after NMR acquisition and the results calculated were 0.25 and 0.53% with a root mean square error of calibration (cross-validated) of 0.03%. A cautionary note, however, about using PLS regression is that this type of analysis is considered a supervised approach. Since the model was adjusted to fit a particular direction using the *Y* table, there is a possibility of overfitting the data due to the biased nature of the test. PLS regression is different to unsupervised approaches such as PCA, where no prior information is used and differences between samples are mainly based on the intrinsic properties of the sample itself. To ensure that the PLS model of one's NMR samples does not overfit, various cross-validation procedures, validating using an independent data set, and permutation tests can be carried out, and readers are advised to check relevant references for further information.[53–56]

12.4 Conclusion

The growth in the global trade of nutraceuticals over the past couple of decades has resulted in increased access to potentially beneficial products and also increased regulation to protect consumers. The regulation poses new challenges to suppliers and manufacturers to validate the material being sold. Determining the identity, strength, composition, and purity of a nutraceutical may be challenging using traditional analytical methodology. An emerging technology for this application is NMR spectroscopy, which offers high reproducibility, absolute quantitation, and compound specificity. Using a metabolomics approach, NMR employs both targeted and non-targeted screening approaches, which may be conducted simultaneously from a single NMR spectrum, to provide valuable insight into the identity, purity, strength, and composition of the nutraceutical product. Developmental efforts for specific analyses, such as the development of NMR SBASEs and statistical models, may be shared with other laboratories, enhancing the utility of NMR in product validation and as a new addition to the nutraceutical analysis toolbox.

References

1. K. L. Wrick, in *Regulation of Functional Foods and Nutraceuticals: A Global Perspective*, ed. C. M. Hasler, Blackwell Publishing, Ames, Iowa, 2005, p. 8.
2. Dietary Supplement Current Good Manufacturing Practices (CGMPs), U.S. Food and Drug Administration, Rule 2007, http://www.fda.gov/Food/GuidanceRegulation/CGMP/ucm110858.htm; also see Guidance for Industry: Current Good Manufacturing Practice in Manufacturing, Packaging, Labeling, or Holding Operations for Dietary Supplements; Small Entity Compliance Guide, December 2010, http://www.fda.gov/Food/

GuidanceRegulation/GuidanceDocumentsRegulatoryInformation/DietarySupplements/ucm238182.htm
3. K. L. Wrick, in *Regulation of Functional Foods and Nutraceuticals: A Global Perspective*, ed. C. M. Hasler, Blackwell Publishing, Ames, Iowa, 2005, p. 16.
4. K. Walker and W. Applequist, *Econ. Bot.*, 2012, **66**(4), 321.
5. E. Sanzini, M. Badea, A. Dos Santos, P. Restani and H. Sievers, *Food Funct.*, 2011, **2**(12), 740.
6. P. Jiao, Q. Jia, G. Randel, B. Diehl, S. Weaver and G. Milligan, *J. AOAC Int.*, 2010, **93**(3), 842.
7. J. Edwards, *Aloe Vera Leaf, Aloe Vera Leaf Juice, Aloe Vera Inner Leaf Juice: Standards of Identity, Analysis, and Quality Control*, ed. R. Upton, American Herbal Pharmacopoeia® • Aloe Vera Leaf Monograph, Scotts Valley, CA, USA, 2012, p. 33.
8. M. A. Markus, S. M. Luchsinger, J. Yuk, J. Ferrier, J. M. Hicks, K. B. Killday, C. W. Kirby, F. Berrue, R. G. Kerr, K. Knagge, T. Goedecke, B. E. Ramirez, D. C. Lankin, G. F. Pauli, I. W. Burton, T. K. Karakach, J. T. Arnason and K. L. Colson, *Planta Med.*, 2014, **80**, 732.
9. J. M. Hicks, A. Muhammad, J. Ferrier, A. Saleem, A. Currier, J. T. Arnason and K. L. Colson, *J. AOAC Int.*, 2012, **95**(5), 1406.
10. F. Gerber, M. Krummen, H. Potgeter, A. Roth, C. Siffrin and C. Spoendlin, *J. Chromatogr. A*, 2004, **1036**(2), 127.
11. H. Pham-Tuan, L. Kaskavelis, C. A. Daykin and H. G. Janssen, *J. Chromatogr. B: Anal. Technol. Biomed. Life Sci.*, 2003, **789**(2), 283.
12. A. W. Bristow, W. F. Nichols, K. S. Webb and B. Conway, *Rapid Commun. Mass Spectrom.*, 2002, **16**(24), 2374.
13. L. M. Jackman and S. Sternhell, *Applications of Nuclear Magnetic Resonance Spectroscopy in Organic Chemistry*. 2nd edn, Pergamon Press: New York, 1969.
14. G. Wider and L. Dreier, *J. Am. Chem. Soc.*, 2006, **128**(8), 2571.
15. I. W. Burton, M. A. Quilliam and J. A. Walter, *Anal. Chem.*, 2005, 77, 3123.
16. J. Ferrier, PhD Thesis, *Ethnobotany, Pharmacology, and Metabolomics of Antidiabetic Plants used by the Eeyou Istchee Cree, Lukomir Highlanders, and Q'eqchi' Maya*, University of Ottawa, Ontario, Canada, 2014.
17. E. Saude, C. Slupsky and B. D. Sykes, *Metabolomics*, 2006, **2**(3), 113.
18. B. Halford, *Bosutinib Buyer BewareMolecule Mix-up: The wrong isomer has been sold under the name of the cancer-fighting compound*, Chem. Eng News, Washington, DC, USA, May 11, 2012, Web edition.
19. S. Dentali, Co-Chair of session entitled *New Analytical Trends and Techniques for Evaluating Botanicals* as part of opening statements at the 2012 AOAC International Meeting in Las Vegas, NV, USA, Oct. 2, 2012.
20. H. M. Heyman and J. J. M. Meyer, *S. Afr. J. Bot.*, 2012, **82**, 21.
21. A. Tomassini, G. Capuani, M. Delfini and A. Michheli, NMR-Based Metabolomics in *Food Quality Control in Data Handling in Science and Technology*, 2013, **28**, 411–447.

22. F. van der Kooy, F. Maltese, Y. H. Choi, H. K. Kim and R. Verpoorte, *Planta Med.*, 2009, 75(7), 763.
23. O. Hendrawati, Q. Yao, H. K. Kim, H. J. M. Linthorst, C. Erkelens, W. M. Lefeber, Y. H. Choi and R. Verpoorte, *Plant Sci.*, 2006, **170**(6), 1118.
24. J. Kang, S. Lee, S. Kang, H. N. Kwon, J. H. Park, S. W. Kwon and S. Park, *Arch. Pharmacal Res.*, 2008, **31**(3), 330.
25. J. Schripsema, *Phytochem. Anal.*, 2010, **21**(1), 14.
26. J. Yuk, K. L. McIntyre, C. Fischer, J. Hicks, K. L. Colson, E. Lui, D. Brown and J. T. Arnason, *Anal. Bioanal. Chem.*, 2013, **405**(13), 4499.
27. K.-H. Ott and N. Aranibar, *Metabolomics*, ed. W. Weckwerth, Humana Press, Totowa, NJ, USA, 2007, p. 247.
28. D. G. Cox, J. Oh, A. Keasling, K. L. Colson and M. T. Hamann, *Biochim. Biophys. Acta*, 2014, **1840**, 3460.
29. H. K. Kim, Y. H. Choi and R. Verpoorte, *Nat. Protoc.*, 2010, **5**, 536.
30. C. F. Chen, Y. D. Li and Z. Xu, *Yaoxue Xuebao*, 2010, **45**(4), 422.
31. H. K. Kim, Y. H. Choi and R. Verpoorte, *Methods Mol. Biol.*, 2013, **1011**, 267.
32. S. van der Sar, H. K. Kim, A. Meissner, R. Verpoorte and Y. H. Choi, *The Handbook of Plant Metabolomics*, ed. W. Weckwerth and G. Kahl, Wiley-VCH Verlag GmbH & Co., Weinheim Germany, 2013, ch. 3, p. 57.
33. M. Piotto, F.-M. Moussallieh, A. Imperiale, M. A. Benahmed, J. Detour, J.-P. Bellocq, I. J. Namer and K. Elbayed, *Methodologies for metabolomics : experimental strategies and technique*, ed. N. Lutz, J. V. Sweedler, and R. Wevers, Cambridge University Press, New York, NY, 6th edn, 2013, vol. 482–483, pp. 505–507.
34. A. M. Liberty, J. W. Amoroso, C. C. Neto and P. E. Hart, *ISHS Acta Hortic.*, 2007, **841**, 61.
35. C. C. Neto, *Mol. Nutr. Food Res.*, 2007, **51**(6), 652.
36. K. D. Patel, F. J. Scarano, M. Kondo, R. A. Hurta and C. C. Neto, *J. Agric. Food Chem.*, 2011, **59**(24), 12864.
37. R. Upton, *Aloe Vera leaf, Aloe Vera Leaf Juice, Aloe Vera Inner Leaf: Standards of Identity, Analysis and Quality Control*, American Herbal Pharmacopoeia, Scotts Valley, CA, USA, 2012, 1.
38. G. F. Pauli, B. U. Jaki and D. C. Lankin, *J. Nat. Prod.*, 2005, **68**(1), 133.
39. H. K. Kim, K. S. Saifullah, E. G. Wilson, S. D. Kricun, A. Meissner, S. Goraler, A. M. Deelder, Y. H. Choi and R. Verpoorte, *Phytochemistry*, 2010, **71**(7), 773.
40. M. Otto, *Chemometrics*, John Wiley and Sons, New York, NY, USA, 1999.
41. I. T. Jollife, *Principal Component Analysis*, Springer, New York, NY, USA, 2nd edn, 2002.
42. B. G. M. Vandeginste and S. C. Rutan, *Handbook of Chemometrics and Qualimetrics: Part B*, Elsevier, Amsterdam, 1998.
43. K. Esbensen, S. Schoenkopf and T. Midtgaard. *Multivariate Analysis in Practice, CAMO*, Trondheim, Norway, 1996.

44. M. Spraul, B. Schuetz, P. Rinke, S. Koswig, E. Humpfer, M. Moertter, F. Fang, U. C. Marx and A. Minoja, *Nutrients*, 2009, **1**(2), 148.
45. M. I. Georgiev, K. Ali, K. Alipieva, R. Verpoorte and Y. H. Choi, *Phytochemistry*, 2011, 72(16), 2045.
46. I. T. Jollife, *Principal Component Analysis*, Springer, New York, NY, USA, 2nd edn, 2002.
47. B. G. M. Vandeginste and S. C. Rutan, *Handbook of Chemometrics and Qualimetrics: Part B*, Elsevier, Amsterdam, 1998.
48. D. E. Axelson, *Data Preprocessing For Chemometric and Metabonomic Analysis*, MRi Consulting, Illinois, USA, 2010.
49. L. Eriksson, E. Johansson, N. Kettaneh-Wold and S. Wold, *Introduction to Multi- and Megavariate Data Analysis using Projection Methods (PCA and PLS)*, Umetrics, Umea, Sweden, 1999, p. 69.
50. J. Trygg, E. Holmes and T. Lundstedt, *J. Proteome Res.*, 2007, **6**(2), 469.
51. F. Ge, C. Chen, D. Liu and S. Zhao, *Food Anal. Methods*, 2014, 7(1), 146.
52. Q. Zhang, A. S. M. Saleh and Q. Shen, *Food Bioprocess Technol.*, 2013, **6**(9), 2562.
53. I. Tasset-Cuevas, Z. Fernandez-Bedmar, M. D. Lozano-Baena, J. Campos-Sanchez, A. de Haro-Bailon, A. Munoz-Serrano and A. Alonso-Moraga, *PLoS One*, 2013, **8**(2), e56986.
54. D. M. Hawkins, S. C. Basak and D. Mills, *J. Chem. Inf. Comput. Sci.*, 2003, **43**(2), 579.
55. J. A. Westerhuis, H. C. J. Hoefsloot, S. Smit, D. J. Vis, A. K. Smilde, E. J. J. van Velzen, J. P. M. van Duijnhoven and F. A. van Dorsten, *Metabolomics*, 2008, **4**(1), 81–89.
56. L. Eriksson, E. Johansson, S. Kettapeh-Wold, *Introduction to multi- and megavariate data analysis using projection methods (PCA and PLS)*. Umetrics, Umea, Sweden 1999, vol. **8**(2), p. 69.

CHAPTER 13

Prospects and Challenges in Molecular Structure Identification by Atomic Force Microscopy

BRUNO SCHULER,[a] FABIAN MOHN,[a] LEO GROSS,*[a] GERHARD MEYER[a] AND MARCEL JASPARS*[b]

[a] IBM Research – Zurich, CH-8803 Rüschlikon, Switzerland; [b] Marine Biodiscovery Centre, Department of Chemistry, University of Aberdeen, Old Aberdeen AB24 3UE, UK
*Email: lgr@zurich.ibm.com; m.jaspars@abdn.ac.uk

> If you have a strange substance and you want to know what it is, you go through a long and complicated process of chemical analysis. ... It would be very easy to make an analysis of any complicated chemical substance; all one would have to do would be to look at it and see where the atoms are.
>
> *There's Plenty of Room at the Bottom*
> Richard P. Feynman
> December 1959

13.1 Structure Determination Using Spectroscopic Methods

When faced with a compound of unknown structure, a chemist has a number of different possibilities to determine its molecular configuration.

Modern NMR Approaches To The Structure Elucidation of Natural Products,
Volume 1: Instrumentation and Software
Edited by Antony J. Williams, Gary E. Martin and David Rovnyak

Published by the Royal Society of Chemistry, www.rsc.org

The first choice should be crystallization of the material followed by X-ray crystallography to determine its atomic connectivities, together with its relative stereochemistry and in many cases its absolute stereochemistry. However, the application of this technique is limited to materials that crystallize easily and are available in sufficient quantities to attempt crystallization. For complex compounds of natural origin, often only very small quantities of material are available and crystallization is impractical, if not impossible. In these cases, the combined use of spectroscopic methods in conjunction with structural and spectroscopic property databases can yield a successful outcome. If a compound has been reported previously, the process of dereplication can be effected using databases and a number of search parameters such as molecular mass, molecular formula and molecular fragments (substructures), amongst others (see Chapter 8 for examples). A large proportion of the remaining compounds are analogues of known compounds and dereplication techniques based on substructures and other features can give an indication of the molecular framework involved. For unknown compounds comprised of a novel molecular skeleton, several stages can be discerned in the process of *ab initio* structure determination.[1] The first step is the determination of the molecular framework, which for organic compounds means the connectivity of all the heavy atoms such as C, N and O. This step will also involve the correct identification and placement of all functional groups, such as alcohol groups and amide functionalities (regiochemistry). Completion of this step will give the framework of the molecule under study. Once this has been accomplished, the correct relative orientation of atoms in space must be determined (relative stereochemistry), but the correct orientation of remote elements of stereochemistry can be difficult to define. When merited, the complete three-dimensional orientation of all atoms in space should be determined (absolute stereochemistry).[2]

The steps described above can be performed using a combination of spectroscopic techniques, heavily biased towards the use of nuclear magnetic resonance (NMR) spectroscopy and closely followed by mass spectrometry (MS). Both techniques are capable of providing information far in excess of that needed for the structural solution of organic compounds and without a defined strategy, it can be difficult to know how to balance the differing and sometimes (apparently) conflicting information. There are a number of approaches that can be applied that will favour a successful outcome, if there is indeed sufficient information to derive a single, unique structure. The heuristic depends on a number of steps, which are described in detail elsewhere in this two-volume set, but will be summarized here to clarify the subsequent discussions[1,2] (see Figure 13.1).

Ideally, the first step should be the determination of the molecular formula, either purely from an accurate mass measurement or by using this in combination with ^{13}C/DEPT-135 NMR data. Obtaining a basic array of 1D- and 2D-NMR spectra, for instance ^{1}H,^{13}C/DEPT-135 HSQC, COSY and HMBC experiments can be sufficient to determine the planar structure and

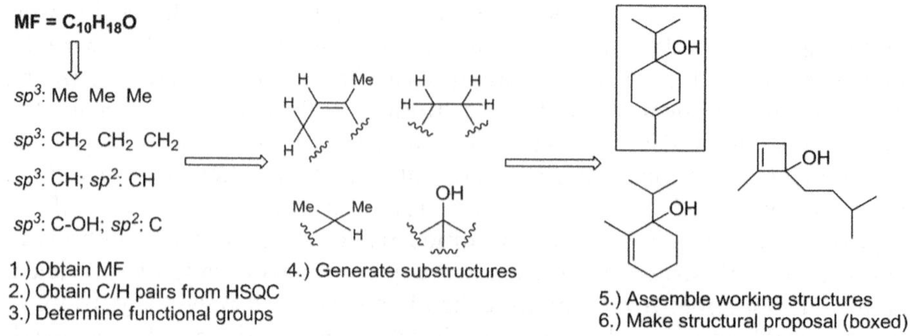

Figure 13.1 An example of the heuristic used for *ab initio* structure determination of organic compounds using spectroscopic techniques.

some elements of the relative stereochemistry. The first step is always the assignment of each proton to its directly attached carbon using an (edited) HSQC spectrum, and together with a ^{13}C spectrum this gives a list of all ^{1}H and ^{13}C shifts with carbon multiplicities. In sufficiently protonated molecules, the next step is to construct spin systems – contiguous chains of protonated carbons – using a COSY spectrum. These substructures can then be combined with the known functional groups, or alternatively the remaining quaternary carbons and additional heteroatoms, to compile a complete list of substructures adding up to the molecular formula of the compound under study. The HMBC NMR spectrum gives information regarding long-range C–H coupling, in effect providing two- or three-bond (and sometimes longer) C–H correlations (*e.g.*, C–C–H and C–C–C–H). This information can be used to combine the substructures from the full list into a number of possible working structures, each of which must be tested against, and be consistent with, all of the spectroscopic data. For example, an assessment of all the ^{1}H and ^{13}C chemical shifts would be compared with literature, database and calculated values, ensuring all the H–H and H–C correlations are consistent with the proposed structure amongst a range of methods for more specific examples. The final step is the determination of relative and absolute stereochemistry, which normally proceeds with a case-by-case approach but may involve the use of coupling constants, through-space correlations and molecular modelling approaches. One of the major problems encountered with structure determination using spectroscopic data is that there is an inherent operator bias. It may be that a particular structural type fits the data, but is not the only possible solution given the data. This bias can lead to incorrect solutions or may lengthen the process of reaching the correct solution significantly. Developing a completely bias-free methodology to solve organic structures is problematic, although several promising systems exist.[3]

For proton-poor structures, those where the heavy atom count equals or exceeds the count of protons or where this is true for a substructure of a larger structure, serious problems arise and consequently this class of compounds suffers from the greatest rate of structural corrections.[4] Problems arise from the lack of C–H correlations that can be used to compose the molecular framework. In some cases, the problem is so extreme that many carbons have no correlations at all to protons in the molecule, *i.e.*, they are further than three bonds away from the nearest proton. When this occurs, a unique solution cannot be reached based on connectivities derived from 2D-NMR spectra. Some methods exist that can define C–C correlations and thus the carbon skeleton of a molecule, but these are inherently insensitive and cannot bridge heteroatoms. In such cases, the structure determination begins with the enumeration of all possible structures consistent with the connectivities obtained from 2D-NMR spectra. The next step is to evaluate these using predicted or calculated shifts derived from expert systems[3] or computer modelling. In many cases, a unique solution cannot be reached easily or is impossible to attain. In such cases, alternative methods must be found to derive accurate structures. One alternative approach that has proven expedient in this respect is atomic force microscopy, which will be discussed next.

13.2 Atomic Resolution on Molecules with Atomic Force Microscopy

In the following, the functional principles of non-contact atomic force microscopy (AFM) are outlined. Next, the experimental prerequisites and the experimental setup for atomic resolution with AFM on molecules are described. Additionally, an estimation of the quantity of material needed for AFM analysis is included. To conclude, the origin of the atomic contrast in AFM images of molecules is discussed.

13.2.1 Experimental Setup

An atomic force microscope uses a flexible beam (called a cantilever) with a very sharp tip at its end to probe the force between the tip and a sample surface with high lateral resolution (Figure 13.2a). The cantilever is characterized by its spring constant k, its eigenfrequency f_0 and its quality factor Q. In non-contact AFM (also called dynamic AFM), the cantilever is mechanically excited to oscillate at its resonant frequency f_0 in the direction of the surface normal (z). This permits stable operation at close tip–sample distances, without the tip making contact with the sample.

For the application of molecular imaging, the frequency modulation (FM-AFM) mode is used,[6] where a feedback loop ensures that the cantilever is oscillating with constant amplitude A. In FM-AFM, the deflection signal is routed through a bandpass filter, phase shifted and fed back to the actuator, as shown in Figure 13.2b. A phase-locked loop (PLL) determines the oscillation frequency $f = f_0 + \Delta f$, and the frequency shift Δf, which is closely related to the force acting on the cantilever, is used to generate the image.

Atomic resolution on molecules has been achieved with sensors in the qPlus geometry, introduced by Giessibl:[7] instead of using silicon cantilevers,

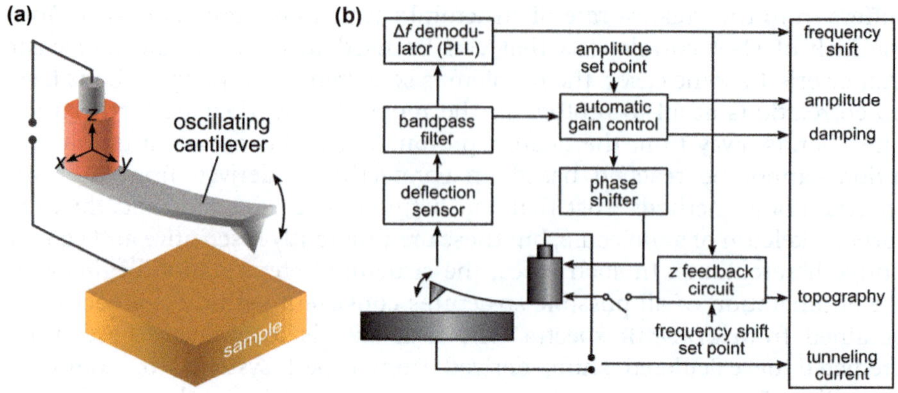

Figure 13.2 Functional principles of non-contact AFM. (a) A sharp tip mounted on a flexible cantilever is scanned over the sample surface. The cantilever is mechanically excited to oscillate at its resonant frequency and the shift of this frequency (Δf) induced by tip–sample forces is recorded as the imaging signal. (b) Schematic diagram of the frequency modulation AFM feedback loop. The physical observables are listed in the box on the right. The z feedback loop can be open (constant-height mode) or closed (constant-frequency mode).
Reproduced with permission from Mohn.[5]

a conducting tip is mounted to one prong of a quartz tuning fork (Figure 13.3d and e). This setup allows the simultaneous collection of the tunnelling current and the frequency shift, hence combined scanning tunnelling microscopy (STM) and AFM operation is possible. Owing to the high stiffness of the tuning fork ($k \approx 1800$ N m^{-1}), stable imaging with very small oscillation amplitudes down to about 10 pm can be achieved. In consequence, the force detection is predominantly sensitive to short-range forces, which ultimately results in atomic-scale contrast.[8]

The experiments are conducted in ultrahigh vacuum (UHV) and at cryogenic temperatures, in our case at $T = 5$ K. Photographs of the laboratory-built system based on an earlier design by Meyer[9] are shown in Figure 13.3. Low temperatures are required to freeze out surface diffusion, increase measurement stability and allow tip preparation by atomic manipulation techniques. Vacuum conditions are needed to obtain and maintain a clean sample preparation. On the one hand, vacuum conditions ensure that the molecule of interest is imaged and not some contaminant; on the other, a clean sample is required for the tip preparation, as described below.

13.2.2 Sample and Tip Preparation

As the substrate, a Cu(111) single crystal cleaned by standard sputtering and annealing cycles was used. This crystal was then partially covered with thermally evaporated NaCl, which grows in the form of mostly two-monolayer-thick islands [NaCl(2ML)/Cu(111)]. Because the molecules to be

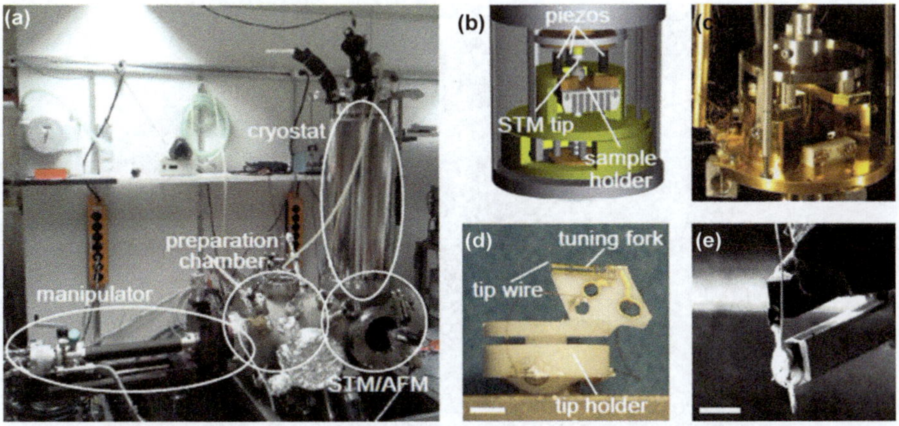

Figure 13.3 Low-temperature STM/AFM system for molecular structure determination. (a) The UHV chamber. The preparation part of the chamber, the part containing the STM/AFM, the liquid nitrogen/liquid helium bath cryostat and the manipulator used for transferring samples are indicated. (b) Schematic drawing of the scanner stage. (c) The scanner stage. (d) The tuning fork sensor. (e) Focused ion beam microscope image of the tuning fork. Scale bars: 2 mm in (d) and 200 μm in (e). Reproduced with permission from Mohn.[5]

investigated might stick with different adsorption energies and in different adsorption geometries to different surfaces, having both the pristine Cu surface and the NaCl islands on the sample increases the chance of obtaining suitable AFM imaging conditions.

To obtain atomic resolution in AFM images, the tip has to be functionalized by controlled termination with a certain atom or molecule using atomic manipulation techniques.[10] First, a clean and stable metal tip is formed by indentations into the metal substrate or by picking up individual metal atoms from the surface.[11] Once a good metal tip is obtained, the tip can be functionalized, *e.g.*, by picking up a single CO molecule: the tip is positioned above a CO molecule on the surface and the distance is decreased until the CO is transferred from the sample to the tip.[12] Note also that tip functionalizations other than CO can yield atomic resolution, *e.g.*, Cl-terminated tips.[13]

In Figure 13.4a, an STM image of a typical sample surface is shown, including partial NaCl coverage, dosed CO molecules, evaporated Au atoms and several different known molecules. The molecules and the Au atoms have been adsorbed at a sample temperature of about $T = 10$ K, thus freezing out surface diffusion.

13.2.3 Amount of Material Needed

An advantage of the AFM method is that only a very small amount of the substance to be investigated is needed to obtain a suitable sample. It is

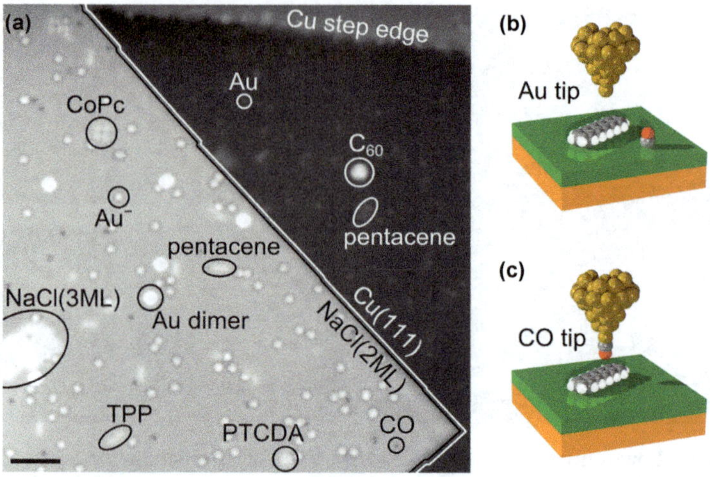

Figure 13.4 Sample and tip preparation. (a) STM overview image of a typical sample preparation. The Cu(111) substrate and a two-monolayer NaCl island (with a small patch of the third layer on top) can be identified. Different adsorbates have been deposited on the surface and can be distinguished from their appearance in the STM topography: Au monomers and dimers, CO, C_{60} fullerenes, terphenylpyridine (TPP), perylenetetracarboxylic acid dianhydride (PTCDA), cobalt phthalocyanine (CoPc) and pentacene. Scale bar: 5 nm. (b), (c) Schematic representation of the creation of a CO tip. Upon approach of the sharp metal tip to a CO molecule on the NaCl(2ML)/Cu(111) surface, the molecule is transferred to the tip apex.
Reproduced with permission from Mohn.[5]

relatively straightforward to calculate the actual quantity of material used in these studies and also to estimate the minimum quantity of material that might be used in an ideal situation. Due consideration must be given to sample handling and also the number of molecules needed to obtain a suitable surface coverage for reliable AFM measurements. The limiting factor is likely to be the handling of small amounts of sample while maintaining a high degree of purity. In a typical experiment, 0.5 mg of organic compound of molecular weight 500 was dissolved in 2 mL of solvent giving a 0.5 mM solution, of which 10 μL were drop-cast on a small piece of a silicon wafer. The wafer was then transferred into the UHV chamber and, by resistive heating of the wafer, the molecules were evaporated directly onto the cold sample at $T = 10$ K. STM imaging was used to determine the surface coverage of the deposited molecules, which amounted to ~500 molecules within an overview area of 800×800 Å under the above conditions. Assuming that one would like to have at least one molecule per overview area, the required minimum amount of material is then found by linear extrapolation to be approximately 5 ng or 10 pmol. Compared with the sample quantities needed for NMR analysis – typically milligram quantities in standard 5 mm probes, but down to tens of micrograms for low-volume probes – this figure is remarkable. Only MS operates in a comparable range with sensitivity in the picomole region, but providing only limited structure information.

The AFM system used in these studies has not been optimized with respect to material consumption and an improvement by a factor of 100–1000 could be achieved, for example, by arranging the sample parallel to the wafer or reducing the distance between the wafer and the sample. Moreover, as the AFM method can probe individual molecules and not entire ensembles like spectroscopic techniques, it is almost unaffected by impurities within the compound. The only prerequisite is that the molecule under study and the impurity can be distinguished by their appearance in the AFM images.

13.2.4 Origin of Atomic Contrast

To interpret the measured AFM images and compare them with theoretical simulations, it is important to look at the relationship between the frequency shift $\Delta f(z)$, the force acting between the tip and the sample and the corresponding interaction energy. In the limit of small oscillation amplitudes, the measurement signal, *i.e.*, the frequency shift, is given by

$$\Delta f = -\frac{f_0}{2k} \times \frac{\partial F}{\partial z}$$

where $\partial F/\partial z$ is the vertical force gradient. The effect of finite amplitudes can also be deconvolved.[14] However, in our case, with a typical oscillation amplitude of $A \approx 0.5\,\text{Å}$, the small amplitude limit is already a very good approximation. In this way, the force F can be extracted from the $\Delta f(z)$ spectrum. The interaction energy E between tip and sample can then be easily calculated using the relation $F = -\partial E/\partial z$.

For the interpretation of AFM images, it is instructive to distinguish the different forces that contribute to the interaction between tip and sample. The attractive interaction between tip and molecule is usually dominated by the van der Waals (vdW) force and the electrostatic force stemming from the contact potential difference between the tip and the substrate. Density functional theory (DFT) calculations show that these contributions result mainly in a diffuse background and do not contribute to the atomic contrast in AFM images of single molecules.[13] Rather, Pauli repulsion was found to be responsible for the short-range repulsive component of the force that gives rise to the atomic contrast. The Pauli repulsion is a consequence of the Pauli exclusion principle, which states that no two fermionic particles can occupy the same quantum mechanical state of the system. This leads to a repulsive force when the electron density of the tip and the sample overlap significantly. Therefore, the observed contrast reflects the electron density in the molecule and hence shows the positions of atoms and bonds, because of their increased electron density.[15]

13.3 AFM-aided Structure Determination

This section begins with an explanation of the interpretation of AFM images, exemplified by two known sample molecules. To illustrate the capabilities of this new technique for molecular structure elucidation, two further examples are presented subsequently. The use of atomically resolved AFM images to assist with the unambiguous structure identification in these two examples and a general procedure for AFM-assisted structure elucidation are proposed.

13.3.1 Polycyclic Aromatic Hydrocarbons

The possibility of imaging molecules on a surface with atomic resolution suggests that in principle this method is suitable for bias-free structure determination of hydrogen-poor planar compounds. For planar polycyclic aromatic hydrocarbons (PAHs), which have a contiguous carbon skeleton, the interpretation is straightforward. As the molecular backbone consists only of carbon and is planar, the effects of topography and chemical contrast can be neglected and the image corresponds to the geometry of the aromatic backbone because the highest electron density is found above the positions of the carbon atoms and the C–C bonds. As shown for pentacene in Figure 13.5a and b, the repulsive part of the tip–molecule interaction leads to bright features in the constant-height AFM image at the positions of the atoms and bonds in the molecule, whereas the long-range vdW part of the interaction leads to the dark halo surrounding the molecule. Note that the different bonds in the pentacene molecule appear different in length and brightness in the AFM image. On the one hand, this is due to a non-planar adsorption geometry of pentacene on Cu(111)[16] and a non-constant vdW background, resulting in the enhanced brightness at the molecular ends. On the other hand, variations of the brightness of different bonds in a molecule can in certain cases also be attributed to differences in the bond order as demonstrated, for example, for hexabenzo[*bc*,*ef*,*hi*,*kl*,*no*,*qr*]coronene (HBC), shown in Figure 13.5c and d.[17] Note that the bonds of the central ring of HBC, labelled *i* in Figure 13.5c, are of greater bond order and are imaged with greater brightness in Figure 13.5d compared with the bonds connecting the central ring to the outer rings, labelled *j*.

13.3.2 Cephalandole A

The interpretation of AFM images becomes more challenging for compounds containing heteroatoms and non-planar substructures. This is because deconvolving the influence of molecular geometry and chemical composition on the image contrast is not straightforward. However, if a compound can be deposited on a surface and imaged by AFM with CO tip functionalization, the resulting image can be a powerful aid to structure determination.

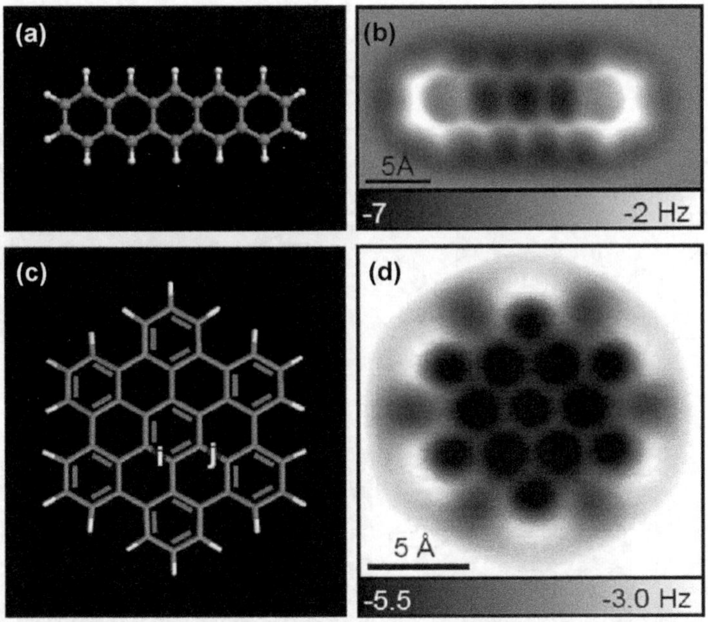

Figure 13.5 Pentacene and hexabenzocoronene (HBC) imaged with AFM. (a), (c) Ball-and-stick models of pentacene and HBC, respectively. (b), (d) Constant-height AFM images of pentacene and HBC, respectively, both on Cu(111), recorded with a CO-terminated tip. Reproduced from Gross *et al.*[13,17] with permission from AAAS.

In the only examples of the use of AFM to assist in structure determination so far, we used a heuristic similar to that used for structure determination by spectroscopic methods, that is, the assembly of substructures into all possible solutions and then checking these for fidelity with the AFM image (see Figure 13.1). The process is summarized in Figure 13.6 for an unknown natural product later identified as cephalandole A, which follows the workflow presented in Figure 13.1 fairly closely. Cephalandole A has the molecular formula $C_{16}H_{10}N_2O_2$ and hence a ratio of heavy atoms to protons of 2 : 1, predicting that structure determination based on NMR might not lead to an unambiguous solution, as confirmed by the initial publication of an incorrect structure, which was later corrected by synthesis.[3,18] This makes cephalandole A an excellent test case for the application of AFM to structure determination.

The process was started by obtaining 1D- and 2D-NMR data, which were used to compose a list of substructures, which in this case could be combined into four different working structures. The absence of long-range C–H correlations or nuclear Overhauser effect (NOE) information made it impossible to determine the correct structure unambiguously. After recording constant-height AFM images of the compound on NaCl(2ML)/Cu(111), these images were overlaid with molecular models of the working

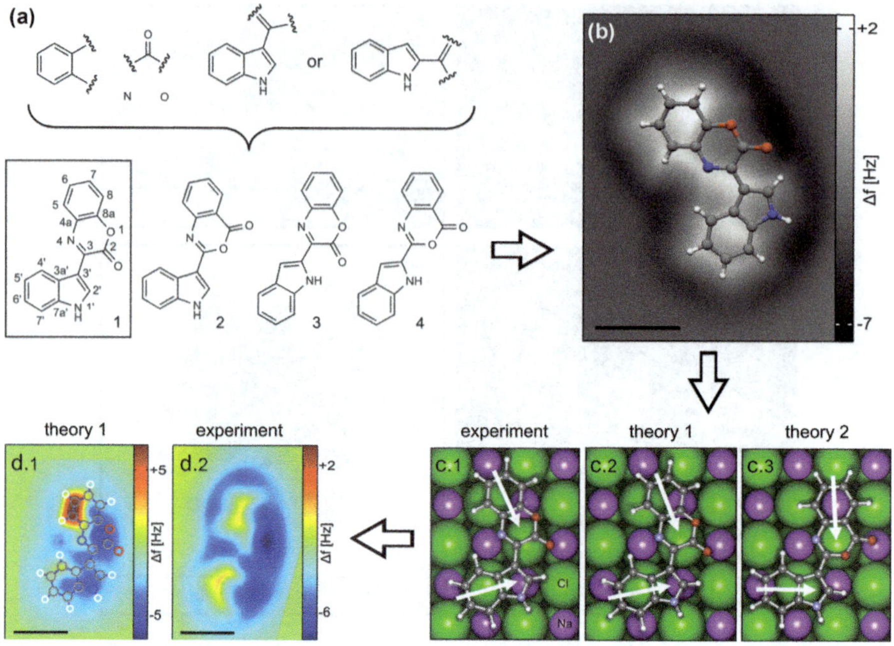

Figure 13.6 Schematic representation of the workflow [from (a) to (d)] used to determine the structure of cephalandole A (boxed) using a combination of NMR and AFM. (a) Molecule substructures identified by NMR and molecule working structures composed out of these substructures labelled **1** to **4**. Structure **1** is the final solution. (b) Constant-height AFM image with molecular model of structure **1** overlaid. (c) Determination of the adsorption position. White arrows indicating the orientation of the bicyclic systems have been added as a guide to the eye. (c.1) Experimental adsorption position deduced from AFM images with constant-current feedback (see ref. 19 for details). (c.2), (c.3) DFT-calculated adsorption geometry of structures **1** and **2**, respectively. The agreement between (c.1) and (c.2) and the disagreement between (c.1) and (c.3) lead to structural assignment as **1**. (d) Comparison between experimental and simulated AFM images. (d.1) DFT simulation for structure **1**. (d.2) Constant-height AFM image. The good agreement between DFT and experiment validates the structural assignment as **1**. Scale bars: 5 Å.
Adapted from Gross *et al.*[19] with permission from Macmillan Publishers Ltd: Nature Chemistry. Copyright 2015.

structures. Working structures **3** and **4** in Figure 13.6a could immediately be discarded owing to the 2-substituted indole moiety. The remaining two solutions could be overlaid equally well on the AFM image, hence further information was necessary to separate the two possibilities. Therefore, the adsorption orientation and position of the molecule with respect to the NaCl substrate were measured with STM and AFM. DFT calculations of the adsorption geometry of the two final candidate structures revealed a difference in their expected orientation and position with respect to the NaCl substrate. Only one of the candidates showed an adsorption geometry that matched the experimental findings, thus highlighting the preferred candidate. A final confirmation of the assigned structure was obtained by simulating the AFM frequency shift map using *ab initio* calculations[19] and comparing it with the experimentally obtained map, which showed excellent agreement. This example provided the first proof of principle that AFM with atomic resolution can be used as a powerful adjunct to other structure elucidation methods.[19]

13.3.3 Breitfussin A

In a second example, AFM imaging was important in the structure determination of the novel compound breitfussin A.[20] Spectroscopic and computational techniques were used together with AFM to demonstrate a new paradigm in organic structure analysis. Although the structure of breitfussin A was solved using a combination of techniques, we show here how AFM was used to obtain many of the connectivities between substructures needed to derive the overall topology, *i.e.*, the planar structure of the molecule (Figure 13.7).

The molecular formula of breitfussin A was established as $C_{16}H_{11}N_3O_2BrI$ by high-resolution MS. Analysis of the fragmentation pattern provided evidence for an MeO moiety. Once the halogens were accounted for, this left an aromatic skeleton with a molecular formula $C_{15}H_8N_3O_1$, the structure of which we tried to deduce using AFM data as shown in Figure 13.7. The centres of the aromatic rings give rise to the most negative Δf values in AFM images, which allowed us to propose a tetracyclic system containing five- and six-membered rings. One five-membered ring and two connecting bonds were readily distinguishable from the AFM image (bold bonds) whereas the other three rings could not be resolved unambiguously (dashed lines), leading to the proposed heavy atom topology depicted as **A** in Figure 13.7c.

Next, the direction of the linking bonds between the rings and to the side groups were used to define the framework more completely. The feature with complex contrast at the top centre of the AFM image was assigned to the Me of the MeO group, although the substitution position of the MeO on the ring was not clear. The bond angles of the bond directed from the topmost ring to the top left-hand side halogen indicated a six-membered ring at the top of the bicyclic system, therefore the remaining rings had to be five-membered. At this point, four substitution patterns remained possible and are labelled **B1**–**B4** in Figure 13.7d. Based on known contrast mechanisms in AFM, we can also propose structure **B1** as the most probable: iodine is expected to give rise to a larger Pauli repulsive force owing to its additional filled electron shell and therefore a higher Δf value compared with Br. Therefore, Br was proposed to be connected to the bicyclic system and I to the central five-membered ring. The brightest feature in the MeO substituent is proposed to indicate the Me as it protrudes from the plane. The position of the Me with respect to the bicyclic system suggested that MeO is connected

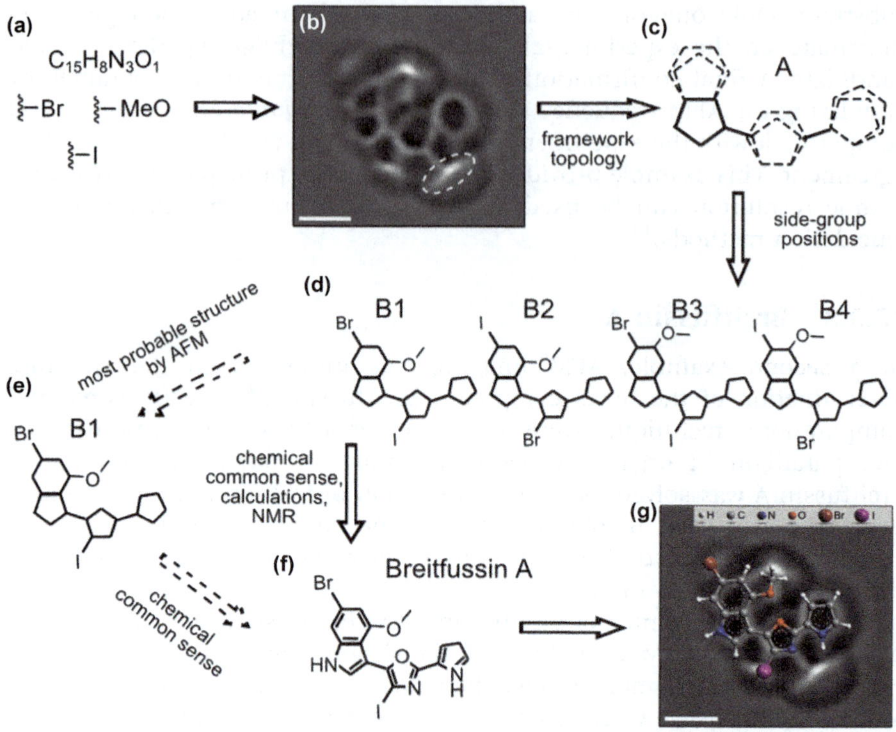

Figure 13.7 Workflow applied in the AFM-assisted structure determination of breitfussin A [from (a) to (g)].[20] (a) Molecule substructures derived by high-resolution MS. (b) Low-pass and Laplace filtered constant-height AFM image. The white encircled region marks a non-intrinsic molecular feature (see ref. 20 for details). (c) Aromatic molecule backbone proposed from (b). Bold lines mark distinct bonds, whereas dashed lines indicate the ambiguity between a five- and a six-membered ring at the respective positions. (d) Remaining structure possibilities from (c) respecting linking bond directions between the rings and to the side groups, labelled **B1–B4**. (e) Most probable (although speculative) structure according to AFM. (f) Final structure assignment, named breitfussin A. Although the final assignment was mediated by spectroscopic and computational techniques [directly from structures in (d)], in retrospect the complete structure could be proposed based on known contrast mechanisms in AFM [step (d)–(e)] and chemical common sense [step (e)–(f)]. (g) Breitfussin A model overlaid on (b). Scale bars: 5 Å.

to the indole at the 4-position, in accordance with the final structure assignment. Although AFM can be used to determine the molecular topology, identifying the positions of the heteroatoms within the aromatic network has so far been possible only *via* chemical common sense or the use of complementary spectroscopic and computational techniques.

These two examples, using the AFM images in different ways to assist the structure elucidation process, show how such a process might be made more generally applicable. The strategy can be described as model building and comparison of these models with the AFM image. The models are based on the chemical formula identified by MS and can be assembled either from substructures derived from NMR data (as in the case of cephalandole A) or from the molecular topology derived *ab initio* from the AFM images (with information from image contrast and bond angles used to determine ring sizes and substituent positions, as in the case of breitfussin A). The different proposed structures can then be compared with the AFM image by overlay and simple visual inspection or by calculating the electron density of the structural proposal on the surface followed by AFM image simulation.

13.4 Conclusion and Outlook

Because of the extremely small amount of sample needed and the valuable information that can be gained from AFM imaging, our approach is certainly a powerful addition to the chemist's toolbox for molecular structure determination, especially when the amount of sample material is very limited.

However, it is clear that additional work needs to be carried out to understand the origin of the contrast effects and shapes observed for different elements and functional groups. In particular, a method with chemical sensitivity for molecules using AFM, as was demonstrated on semiconductor surfaces,[21] would further enhance the applicability of AFM to structure determination. However, this task is extremely challenging for molecules, because atomic species can appear in different coordination states and with different bond orders, hence atoms with different hybridization will appear differently in NC-AFM. Further challenges arise from the fact that molecules are usually not planar and even small deviations from a planar geometry complicate both measurement and analysis. Nevertheless, a possible route towards chemical sensitivity for molecules is the use of different tip functionalizations and comparison of the resulting interaction forces. In addition, the information about the molecular structure gained from AFM could also be complemented by applying STM with functionalized tips,[22,23] which yields information about the structure of the molecular frontier orbitals[24,25] and about the intramolecular charge distribution from Kelvin probe force microscopy,[26] to facilitate an unambiguous structural assignment. Furthermore, for studying less planar molecules, three-dimensional data acquisition techniques can be used,[27,28] and to image the different facets of molecules, atomic manipulation techniques could be applied to turn over individual molecules[29] or parts of molecules.[30] Finally, the preparation techniques have to be improved for the study of larger molecules. In the examples described above, molecules were thermally evaporated, hence fragmentation becomes a problem if the desorption temperature becomes equal to or greater than the temperature for fragmentation. Electrospray deposition might be used to circumvent this problem.

References

1. M. Jaspars, *Nat. Prod. Rep.*, 1999, **16**, 241.
2. P. Crews, J. Rodriguez and M. Jaspars, *Organic Structure Analysis*, Oxford University Press, New York, 2nd edn, 2010.
3. M. E. Elyashberg, A. J. Williams and K. A. Blinov, *Nat. Prod. Rep.*, 2010, **27**, 1296.
4. T. Amagata, *Comprehensive Natural Products II*, 2010, vol. 2, p. 581.
5. F. Mohn, Probing electronic and structural properties of single molecules on the atomic scale, Ph.D. thesis, Universität Regensburg, 2012.
6. T. R. Albrecht, P. Grütter, D. Horne and D. Rugar, *J. Appl. Phys.*, 1991, **69**, 668.
7. F. J. Giessibl, *Appl. Phys. Lett.*, 2000, **76**, 1470.
8. F. J. Giessibl, *Rev. Mod. Phys.*, 2003, **75**, 949.
9. G. Meyer, *Rev. Sci. Instrum.*, 1996, **67**, 2960.
10. D. M. Eigler, C. P. Lutz and W. E. Rudge, *Nature*, 1991, **352**, 600.
11. J. Repp, G. Meyer, F. E. Olsson and M. Persson, *Science*, 2004, **305**, 493.
12. L. Bartels, G. Meyer and K.-H. Rieder, *Appl. Phys. Lett.*, 1997, **71**, 213.
13. L. Gross, F. Mohn, N. Moll, P. Liljeroth and G. Meyer, *Science*, 2009, **325**, 1110.
14. J. E. Sader and S. P. Jarvis, *Appl. Phys. Lett.*, 2004, **84**, 1801.
15. N. Moll, L. Gross, F. Mohn, A. Curioni and G. Meyer, *New J. Phys.*, 2010, **12**, 125020.
16. B. Schuler *et al.*, *Phys. Rev. Lett.*, 2013, **111**, 106103.
17. L. Gross *et al.*, *Science*, 2012, **337**, 1326.
18. J. Mason, J. Bergman and T. Janosik, *J. Nat. Prod.*, 2008, **71**, 1447.
19. L. Gross *et al.*, *Nat. Chem.*, 2010, **2**, 821.
20. K. O. Hanssen *et al.*, *Angew. Chem., Int. Ed.*, 2012, **51**, 12238.
21. Y. Sugimoto *et al.*, *Nature*, 2007, **446**, 64.
22. R. Temirov, S. Soubatch, O. Neucheva, A. C. Lassise and F. S. Tautz, *New J. Phys.*, 2008, **10**, 053012.
23. G. Kichin, C. Weiss, C. Wagner, F. S. Tautz and R. Temirov, *J. Am. Chem. Soc.*, 2011, **133**, 16847.
24. G. Repp, G. Meyer, S. M. Stojkovic, A. Gourdon and C. Joachim, *Phys. Rev. Lett.*, 2005, **94**, 026803.
25. L. Gross *et al.*, *Phys. Rev. Lett.*, 2011, **107**, 086101.
26. F. Mohn, L. Gross, N. Moll and G. Meyer, *Nat. Nanotechnol.*, 2012, 7, 227.
27. M. Z. Baykara, T. C. Schwendemann, E. I. Altman and U. D. Schwarz, *Adv. Mater.*, 2010, **22**, 2838.
28. F. Mohn, L. Gross and G. Meyer, *Appl. Phys. Lett.*, 2011, **99**, 053106.
29. D. L. Keeling *et al.*, *Phys. Rev. Lett.*, 2005, **94**, 146104.
30. F. Moresco *et al.*, *Phys. Rev. Lett.*, 2001, **86**, 672.

Subject Index